CAMBRIDGE LIBRARY COLLECTION

Books of enduring scholarly value

Mathematics

From its pre-historic roots in simple counting to the algorithms powering modern desktop computers, from the genius of Archimedes to the genius of Einstein, advances in mathematical understanding and numerical techniques have been directly responsible for creating the modern world as we know it. This series will provide a library of the most influential publications and writers on mathematics in its broadest sense. As such, it will show not only the deep roots from which modern science and technology have grown, but also the astonishing breadth of application of mathematical techniques in the humanities and social sciences, and in everyday life.

A Treatise on Analytical Statics

As senior wrangler in 1854, Edward John Routh (1831–1907) was the man who beat James Clerk Maxwell in the Cambridge mathematics tripos. He went on to become a highly successful coach in mathematics at Cambridge, producing a total of twenty-seven senior wranglers during his career – an unrivalled achievement. In addition to his considerable teaching commitments, Routh was also a very able and productive researcher who contributed to the foundations of control theory and to the modern treatment of mechanics. This two-volume textbook, which first appeared in 1891–2 and is reissued here in the revised edition that was published between 1896 and 1902, offers extensive coverage of statics, providing formulae and examples throughout for the benefit of students. While the growth of modern physics and mathematics may have forced out the problem-based mechanics of Routh's textbooks from the undergraduate syllabus, the utility and importance of his work is undiminished.

Cambridge University Press has long been a pioneer in the reissuing of out-of-print titles from its own backlist, producing digital reprints of books that are still sought after by scholars and students but could not be reprinted economically using traditional technology. The Cambridge Library Collection extends this activity to a wider range of books which are still of importance to researchers and professionals, either for the source material they contain, or as landmarks in the history of their academic discipline.

Drawing from the world-renowned collections in the Cambridge University Library and other partner libraries, and guided by the advice of experts in each subject area, Cambridge University Press is using state-of-the-art scanning machines in its own Printing House to capture the content of each book selected for inclusion. The files are processed to give a consistently clear, crisp image, and the books finished to the high quality standard for which the Press is recognised around the world. The latest print-on-demand technology ensures that the books will remain available indefinitely, and that orders for single or multiple copies can quickly be supplied.

The Cambridge Library Collection brings back to life books of enduring scholarly value (including out-of-copyright works originally issued by other publishers) across a wide range of disciplines in the humanities and social sciences and in science and technology.

A Treatise on Analytical Statics

With Numerous Examples

VOLUME 1

EDWARD JOHN ROUTH

CAMBRIDGE
UNIVERSITY PRESS

CAMBRIDGE
UNIVERSITY PRESS

University Printing House, Cambridge, CB2 8BS, United Kingdom

Published in the United States of America by Cambridge University Press, New York

Cambridge University Press is part of the University of Cambridge.
It furthers the University's mission by disseminating knowledge in the pursuit of
education, learning and research at the highest international levels of excellence.

www.cambridge.org
Information on this title: www.cambridge.org/9781108050289

This edition first published 1896
This digitally printed version 2013

ISBN 978-1-108-05028-9 Paperback

A TREATISE ON

ANALYTICAL STATICS.

London: C. J. CLAY AND SONS,
CAMBRIDGE UNIVERSITY PRESS WAREHOUSE,
AVE MARIA LANE.
GLASGOW: 263, ARGYLE STREET.

LEIPZIG: F. A. BROCKHAUS.
NEW YORK: MACMILLAN AND CO.
BOMBAY: GEORGE BELL AND SONS.

A TREATISE ON

ANALYTICAL STATICS

WITH NUMEROUS EXAMPLES

BY

EDWARD JOHN ROUTH, Sc.D., LL.D., M.A., F.R.S., &c.,

HON. FELLOW OF PETERHOUSE, CAMBRIDGE;
FELLOW OF THE UNIVERSITY OF LONDON.

VOLUME I.

SECOND EDITION

CAMBRIDGE
AT THE UNIVERSITY PRESS
1896

First Edition 1891
Second Edition 1896

CAMBRIDGE: PRINTED BY J. AND C. F. CLAY,
AT THE UNIVERSITY PRESS.

PREFACE.

DURING many years it has been my duty and pleasure to give courses of lectures on various Mathematical subjects to successive generations of students. The course on Statics has been made the groundwork of the present treatise. It has however been necessary to make many additions; for in a treatise all parts of the subject must be discussed in a connected form, while in a series of lectures a suitable choice has to be made.

A portion only of the science of Statics has been included in this volume. It is felt that such subjects as Attractions, Astatics, and the Bending of rods could not be adequately treated at the end of a treatise without either making the volume too bulky or requiring the other parts to be unduly curtailed. These remaining portions appear in the second volume.

In order to learn Statics it is essential to the student to work numerous examples. Besides some of my own construction, I have collected a large number from the University and College Examination papers. Some of these are so good as to deserve to rank among the theorems of the science rather than among the examples. Solutions have been given to many of the examples, sometimes at length and in other cases in the form of hints when these appeared sufficient.

I have endeavoured to refer each result to its original author. I have however found that it is a very difficult task to effect this

with any completeness. The references will show that I have
searched many of the older books and memoirs as well as some
of those of recent date to discover the first mention of a theorem.

In this edition I have made many additions and have also
omitted several things which on after consideration appeared
to be of minor importance. The explanations also have been
simplified wherever there appeared to be any obscurity. For
the convenience of reference I have retained the order of the
articles as far as that was possible.

The latter part of the chapter on forces in three dimensions
has been enlarged by the addition of several theorems and the
portions on five and six forces re-arranged. The chapter on
graphical statics also has been almost entirely rewritten.

An index has been added which it is hoped will be found
useful.

                                        EDWARD J. ROUTH.

PETERHOUSE,
    May, 1896.

CONTENTS.

CHAPTER I.

THE PARALLELOGRAM OF FORCES.

CHAPTER II.

FORCES ACTING AT A POINT.

CHAPTER III.

PARALLEL FORCES.

CHAPTER IV.

FORCES IN TWO DIMENSIONS.

CHAPTER V.

ON FRICTION.

CHAPTER VI.

THE PRINCIPLE OF WORK.

CHAPTER VII.

FORCES IN THREE DIMENSIONS.

CHAPTER VIII.

GRAPHICAL STATICS.

CHAPTER IX.

CENTRE OF GRAVITY.

CHAPTER X.

ON STRINGS.

CHAPTER XI.

THE MACHINES.

CONTENTS OF VOLUME II.

ATTRACTIONS.

THE BENDING OF RODS.

ASTATICS.

CHAPTER I.

1. THE science of Mechanics treats of the action of forces on bodies. Under the influence of these forces the bodies may either be in motion or remain at rest. That part of mechanics which treats of the motion of bodies is called Dynamics. That part of mechanics in which the bodies are at rest is called Statics.

If the determination of the motion of bodies under given forces could be completely and easily solved, there would be no obvious advantage in this division of the subject into two parts. It is clear that statics is only that particular case of dynamics in which the motions of the bodies are equated to zero. But the particular case in which the motion is zero presents itself as a much easier problem than the general one. At the same time this particular case is one of great importance. It is important not merely for the intrinsic value of its own results but because these are found to assist in the solution of the general case by the help of a theorem due to D'Alembert. It has therefore been generally found convenient to lead up to the general problem of dynamics by considering first the particular case of statics.

2. Since statics is a particular case of dynamics we may begin by discussing the first principles of the more general science. We should consider how the mass of a body is measured, how the velocity and acceleration of any particle are affected by the action of forces. The general principles having been obtained we may then descend to the particular case by putting these velocities equal to zero. In this way the relationship of the two great branches of mechanics is clearly seen and their results are founded on a common basis.

3. There is another way of studying statics which has its own advantages. We might begin by assuming some simple axioms relating to the action of forces on bodies without introducing any properties of motion. In this method we introduce no terms or principles but those which are continually used in statics, leaving to dynamics the study of those terms which are peculiar to it.

Whether this is an advantageous method of studying statics or not depends on the choice of the fundamental axioms. In the first place they must be simple in character. In the second place they must be easily verified by experiment. For example we might take as an axiom the proposition usually called the parallelogram of forces or we might, after Lagrange, start from the principle of work. But neither of these principles satisfies the conditions just mentioned, for they do not seem sufficiently obvious on first acquaintance to command assent.

If we found the two parts of mechanics on a common basis, that basis must be broader than that which is necessary to support merely the principles of statics. We have to assume at once all the experimental results required in mechanics instead of only those required in statics. Now there is an advantage in introducing the fundamental experiments in the order in which they are wanted. We thus more easily distinguish the special necessity for each, we see more clearly what results are deduced from each experiment. The order of proceeding would be to begin with such elementary axioms about forces as will enable us to study their composition and resolution. Presently other experimental results are introduced as they are required and finally when the general problem of dynamics is reached, the whole of the fundamental axioms are summed up and consolidated.

In a treatise on statics it is necessary to consider both these methods. We shall examine first how the elementary principles of statics are connected with the axioms required for the more general problem of dynamics, and secondly how they may be made to stand on a base of their own.

4. In mechanics we have to treat of the action of forces on bodies. The term force is defined by Newton in the following terms.

An impressed force is an action exerted on a body in order to change its state either of rest or of uniform motion in a straight line.

5. Characteristics of a Force. When a force acts on a body the action exerted has (1) a point of application, (2) a direction in space, (3) magnitude.

Two forces are said to be equal in magnitude when, if applied to the same particle in opposite directions, they balance each other. The magnitudes of forces are measured by taking some one force as a unit, then a force which will balance two unit forces is represented by two units and so on.

6. The simplest appeal to our experience will convince us that many at least of the ordinary forces of nature possess these three characteristics. If force be exerted on a body by pulling a string attached to it, the point of attachment of the string is the point of application, and the direction of the string is the direction of the force. The existence of the third element of a force is shown by the fact that we may exert different pulls on the string.

All the causes which produce or tend to produce motion in a body are not known. But as they are studied, it is found that they can be analysed into simpler causes, and these simpler causes are seen to have the three characteristics of a force. If there be any causes of motion which cannot be thus analysed, such causes are not considered as forces whose effects are to be discussed in the science of statics.

7. There are other things besides forces which possess these three characteristics. These other things may be used to help us in our arguments about forces so far as their other properties are common also to forces.

The most important of these analogies is that of a finite straight line. Let this finite straight line be AB. One extremity A will represent the point of application. The direction in space of the straight line will represent the direction of the force and the length of the line will represent the magnitude of the force.

Other things besides forces may also be represented graphically by a finite straight line. Thus in dynamics it will be seen that both the velocity and the momentum of a particle have direction and magnitude and may in the same way be represented by a finite straight line. One extremity A is placed at the particle, the direction of the straight line represents the direction of the velocity and the length represents the magnitude. Generally this analogy is useful whenever the things considered obey what we shall presently call the *parallelogram law*.

8. In order to represent completely the direction of a force by the direction of the straight line AB, it is necessary to have some convention to determine whether the force pulls A in the direction AB or pushes A in the direction BA. This convention is supplied by the use of the terms positive and negative. The positive and negative directions of straight lines being defined by some convention or rule, the forces which act in the positive directions of their lines of action are called positive and those in the opposite directions are called negative. These conventions are often indicated by the conditions of the problem under consideration, but they usually agree with the rules adopted in the differential calculus. Thus the direction of the radius vector drawn from the origin is usually taken as the positive direction, and so on for all lines.

Sometimes instead of using the term positive, the direction or sense of a force is indicated by the order of the letters, thus a force AB is a force acting in the direction A to B, a force BA is a force acting from B towards A.

9. The third element of a force is its magnitude. This is represented by the length of the representative straight line. A unit of force is represented by a unit of length on any scale we please; a force of n such units of force is then represented by a straight line of n units of length.

10. Measure of a force. A force must be measured by its effects. Since a force may produce many effects there are several methods open to us. If we wish the measure of two equal forces acting together to be twice that of a single force equal to either, the effect which is to measure the force must be properly chosen.

We may measure a force by the weight of the mass which it will support. Placing two equal masses side by side, they will be supported by equal forces. Joining these together we see that a double force will support a double mass. Thus the effect is proportional to the magnitude of the cause.

We may also measure a force by the motion it will produce in a given body in a given time. If by motion is here meant velocity

then it may be shown by the experiments usually quoted to prove
the second law of motion that a double force will produce a double
velocity.   So here also the effect chosen as the measure is proportional to the magnitude of the cause.   This measure requires
some experimental results, necessary for dynamics, but not used
afterwards in statics.

If we agree to measure a force by the weight it will support
the unit will depend on the force of gravity at the place where
the experiment is made.   Such a unit will therefore present
several inconveniences.   If also we measure a force by the velocity
generated in a unit of mass in a unit of time, it is necessary
to discuss how these other units are to be chosen.

It is not necessary for us, at this stage of our argument, to decide on the
best method of measuring a force.   It will be presently seen that our equations
are concerned for the most part with the ratios of forces rather than with the
forces themselves.   The choice of the actual unit is therefore unimportant at
present, and we can leave this choice until the proper occasion arrives.   The
comparative effects of forces will then have been discussed, and the reader will
the better understand the reasons why any particular choice is made.

When therefore we speak of several forces equal to the weight of one, two or
three pounds &c., acting on a body and determine the conditions of equilibrium,
we shall find that the same conditions are true for forces equal to the weight of
one, two or three oz. &c., and generally of all forces in the same ratio.

11.   One system of units is that based on the foot, pound, and
second as the three fundamental units of length, mass, and time.
The unit force is that force which acting on a pound of matter for
one second generates a velocity of one foot per second.   This unit
of force is called the poundal.

The foot and the pound are defined by certain standards kept
in a place of security for reference.   Thus the imperial yard is the
distance between two marks on a certain bar, preserved in the
Tower of London, when the whole bar has a temperature of
62° Fah.   The unit of time is a certain known fraction of a mean
solar day.

The units committee of the British Association recommended
the general adoption of the centimetre the gramme and the
second as the three fundamental units of space, mass and time.
These they proposed should be distinguished from absolute units,
otherwise derived, by the letters C. G. S. prefixed, these being the
initial letters of the names of the three fundamental units.   The
C. G. S. unit of force is called a dyne.   This is the force which

acting on a gramme for a second generates the velocity of a centimetre per second.

It is found by experiment that a body, say a unit of mass, falling in vacuo for one second acquires very nearly a velocity of 32·19 feet per second. This velocity is the same as 981·17 centimetres per second. It follows therefore that a poundal is about $\frac{1}{32}$th part of the weight of one pound, and a dyne is the weight of $\frac{1}{981}$th part of a gramme. These numerical relations strictly apply only to the place of observation, for the force of gravity is not the same at all places on the earth. The difference between the greatest and least values of gravity is about $\frac{1}{196}$th of its mean value.

The relations which exist between these and other units in common use are given at length in Everett's treatise on *units and Physical Constants* and in Lupton's *numerical tables*. We have nearly

one inch = 2·54 centimetres,   one pound = 453·59 grammes

It follows from what precedes that one poundal = 13825 dynes.

12.   The parallelogram of velocities. This proposition is preliminary to Newton's laws of motion.

The velocity of a particle when uniform is measured by the space described in a given time. A straight line whose length is equal to this space will represent the velocity in direction and magnitude; Art. 8. Suppose a particle to be carried uniformly in the given time from O to C, then OC represents its velocity. This change of place may be effected by moving the particle in the same time from O to A along the straight line OA, if while this 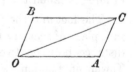 is being done we move the straight line OA (with the particle sliding on it) parallel to itself from the position OA to the position BC. The uniform motion of the particle from O to A is expressed by the statement that its velocity is represented by OA. The displacement produced by the uniform motion of the straight line is expressed by the statement that the particle has a velocity represented in direction and magnitude by either of the sides OB or AC. It is evident by the properties of similar figures that the path of the particle in space is the straight line OC.

It follows that when a particle moves with two simultaneous velocities represented in direction and magnitude by the straight lines OA, OB *its motion is the same as if it were moved with a single velocity represented in direction and magnitude by the diagonal* OC *of the parallelogram described on* OA, OB *as sides. This proposition is usually called the parallelogram of velocities.*

Let a particle move with three simultaneous velocities represented in direction and magnitude by the three straight lines OA_1, OA_2, OA_3. We may replace the two velocities OA_1, OA_2 by the single velocity represented in direction and magnitude by the diagonal OB_1 of the parallelogram described on OA_1, OA_2 as sides. The particle now moves with the two simultaneous velocities represented by OB_1 and OA_3. We may again use the same rule. We replace these two velocities by the single velocity represented in direction and magnitude by the diagonal OB_2 described on OB_1 and on OA_3 as sides. We have thus replaced the three given simultaneous velocities by a single velocity.

In the same way any number of simultaneous velocities may be replaced by a single velocity.

If the simultaneous velocities represented by OA_1, OA_2 &c. were all altered in the same ratio, it is evident from the properties of similar figures that the resulting single velocity will also be altered in the same ratio.

Let the simultaneous velocities OA_1, OA_2 &c. be such that their resulting velocity is zero. It follows that if all the velocities OA_1, OA_2 &c. are altered in any, the same, ratio the resulting velocity is still zero.

13. Newton's laws of Motion. These are given in the introduction to the Principia.

1. Every body continues in its state of rest or of uniform motion in a straight line, except in so far as it may be compelled by force to change that state.

2. Change of motion is proportional to the force applied and takes place in the direction of the straight line in which the force acts.

3. To every action there is always an equal and contrary reaction; or the mutual actions of any two bodies are always equal and oppositely directed.

The full significance of these laws cannot be understood until the student takes up the subject of dynamics. The experiments which suggest these laws, and their further verification, are best studied in connection with that branch of the science, and are to be found in books on elementary dynamics. The student who has not already read some such treatise is advised to assume the truth of these laws for the present. We shall accordingly not enter into a full discussion of them in this treatise, but we shall confine our remarks to those portions which are required in statical problems.

14. *The first law asserts the inertness of matter.* A body at rest will continue at rest unless acted on by some external force. At first sight this may appear to be a repetition of the definition of force, since any cause which tends to move a body at rest is called a force. But it is not so. Here we assert as the result of observation or experiment the inertness of each particle of matter. It has no tendency to move itself, it is moved only by the action of causes *external* to itself.

15. *In the second law of motion the independence of forces which act on a particle is asserted.* If the effect of a force is always proportional to the force impressed it is clearly meant that each force must produce its own effect in direction and magnitude as if it acted singly on the particle placed at rest.

Let us consider the meaning of this statement a little more fully. Let a given force act on a given particle placed at rest at a point O and generate in a given time a velocity which we may represent graphically by the straight line OA. Let a second force act on the same particle again placed at rest at O and generate in the same time a velocity which we may represent by OB. If both forces act simultaneously on the particle both these velocities are generated. The actual velocity of the particle is then represented by the diagonal OC of the parallelogram described on OA, OB as sides, Art. 12. In the same way, if any number of forces act simultaneously on a particle at rest, the law directs that we are to determine the velocity generated by each as if it acted alone for a given time. These separate velocities are then to be combined into a single velocity in the manner described in Art. 12. This single velocity is asserted to be the effect of the simultaneous action of the forces.

Let a system of forces be such that when they act simultaneously on a particle placed at rest the resulting velocity of the particle is zero. These forces are then in equilibrium. Let a second system of forces be also such that when they act on the particle placed at rest, the resulting velocity of the particle is again zero. Then this second system of forces is also in equilibrium. Let these two systems act simultaneously, then since the forces do not interfere with each other, the resulting velocity of the particle is still zero. We thus arrive at the following important proposition.

Let us suppose that there are two systems of forces each of which when acting alone on a particle would be in equilibrium. Then when both systems act simultaneously there will still be equilibrium.

This is sometimes called *the principle of the superposition of forces in equilibrium.* When we are trying to find the conditions of equilibrium of some system of forces, the principle enables us to simplify the problem by adding on or removing any particular forces which by themselves are in equilibrium.

Let the forces P_1, P_2 &c. acting on a given particle for a given time generate velocities v_1, v_2 &c. respectively. If the same or equal forces were made to act on a different particle the velocities generated in the same time may be different. But since the effect of each force is proportional to its magnitude the velocities generated by the several forces are to each other in the ratios of v_1 to v_2 to v_3 &c. If then a system of forces is in equilibrium when acting on any one particle, that system will also be in equilibrium when applied to any other particle (Art. 12).

16. We notice also that it is *the change of motion* which is the effect of force. A given force produces the same change of motion in a particle whether that particle is in motion or at rest.

In this way we can determine whether a moving particle is acted on by any external force or not. If the velocity is uniform and the path rectilinear there is no force acting on the particle. If either the velocity is not uniform, or the path not rectilinear, there must be some force acting to produce that change.

Let two equal forces act one on each of two particles and generate in the same time equal changes of velocity; these particles are said to have equal mass. If the force acting on one particle must be n times that on the other in order to generate equal changes of velocity in equal times, the mass of the first particle is n times that of the second. It follows that the mass of a particle is proportional to the force required to generate in it a given change of velocity in a given time. Now all bodies falling from rest in a vacuum under the attraction of the earth are found to have the same velocity at the end of the first second of time, Art. 11. We therefore infer that the masses of bodies are proportional to their weights. The units of mass and force are so chosen that the unit of force acting on the unit of mass will generate a unit of velocity in a unit of time.

The product of the mass of a particle into its velocity is called its *momentum*. It follows from what has just been said that the expression "change of motion" means change of momentum produced in a given time.

These results are peculiarly important in dynamics, but in statics, where the particles acted on are all initially at rest and remain so, they have not the same significance.

17. *In the third law the principle of the transmissibility of force is implied.* The principle is more clearly stated in the remarks which Newton added to his laws of motion. The law asserts the equality of action and reaction. If a force acting at a point *A* pull a body which has some point *B* held at rest, the reaction at *B* is asserted to be equal and opposite to the force acting at *A*. In general, when two forces act at different points of a body there will be equilibrium if the lines of action coincide, the directions of the forces are opposite, and their magnitudes equal.

From this we deduce that *when a force acts on a body, its effect is the same whatever point of its line of action is taken as the point of application, provided that point is connected with the rest of the body in some invariable manner.*

For let a force *P* act at *A* and let *B* be another point in its line of action. We have just seen that the force *P* acting at *A* may be balanced by an equal force *Q* acting at *B* in the opposite direction. But the force *Q* acting at *B* may also be balanced by an equal force *P'* acting at *B* in the same direction as *P* (Art. 15). Thus the two equal forces *P* and *P'* acting respectively at *A* and *B* in the same directions can be balanced by the same force *Q*. Thus the force *P* acting at *A* is equivalent to an equal force *P'* acting at *B*.

18. Statical Axioms. If we wish to found the science of statics on a basis independent of the ideas of motion we require some elementary axioms concerning matter and force.

In the first place we assume as before the principle of the inertness of matter.

We also require the two principles of the independence and transmissibility of force.

The first of these principles is regarded as a matter of common experience. When our attention is called to the fact, we notice that bodies at rest do not begin to move unless urged to do so by some external causes.

The other two require some elementary experiments.

Let a body be acted on by two forces, each equal to P, and having A, A' for their points of application. We may suppose these to be applied by means of strings attached to the body at A and A' and pulled by forces each of the given magnitude. Let us also suppose the body to be removed from the action of gravity and all other forces. This may be partially effected by trying the experiment on a disc placed on a smooth table or by suspending the body by a string attached at the proper point, or the experiment might be tried on some body floating in a vessel of water.

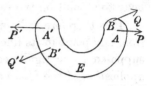

It is a matter of common experience that when the strings are pulled there cannot be equilibrium unless the lines of action of the forces acting at A and A' are on the same straight line. The body acted on will move unless this coincidence of the lines of action is exact.

This result is not to be regarded as obvious apart from experiment. In the diagram the points of application A and A' are separated by a space not occupied by the body. The forces have therefore to counterbalance each other by acting, if we may so speak, round the corner E. As the manner in which force is transmitted across a body is not discussed in this part of statics, it is necessary to have an experimental result on which to found our arguments.

Let us now suppose that two other forces each equal to Q are applied at B and B' and have their lines of action in the same straight line. These if they acted alone on the body without the forces P, P' would be in equilibrium. Then it will be seen, on trying the experiment, that equilibrium is still maintained when both the systems act. Thus it appears that the introduction of the two forces Q, Q' does not disturb the two forces P, P' so as to destroy the equilibrium.

From the results of this experiment we may deduce exactly as in Art. 17, the principle of the transmissibility of force.

19. Rigid bodies. Let two or more bodies act and react on each other and be in equilibrium under the action of any forces. The principle of the transmissibility of force asserts that any one of these forces may be applied at any point of its line of action. If the line of action of any force acting on one of the bodies be produced to cut another, it does not follow that equilibrium will be maintained if the force is transferred from a point on the first body to a point on the second.

It is therefore to be understood that when a force is transferred from any point in its line of action to another the two points are supposed to be rigidly connected together. When the points of application of the forces are connected in some invariable manner, the body acted on is said to be rigid. Such are the bodies we shall in general speak of, though for the sake of brevity we shall often refer to them simply as bodies.

20. It is sometimes convenient to form the conditions of equilibrium of the whole system (or any part of it) as if it were one body. That this may be done is evident, since the mutual actions and reactions of the several bodies are equal and opposite. But we may also reason thus ; the system being in a position of equilibrium, we may suppose the points of application of the forces to be joined in some invariable manner. This will not disturb the equilibrium. The system being now made rigid we may form the conditions of equilibrium. These are generally necessary and sufficient for the equilibrium of the system regarded as a rigid body, but though *necessary* they are *not generally sufficient* for its equilibrium when regarded as a collection of bodies.

21. When a force acts on a rigid body, the principle of the transmissibility of force asserts that the body transmits its action from one point of application to another, but does not itself alter the magnitude of the force. It appears, therefore, that so far as this principle and that of the independence of forces are concerned the conditions of equilibrium depend on the forces and not on the body.

If a system of forces be in equilibrium when acting on any body, that system will also be in equilibrium when transferred to act on any other body, provided always the points of application are connected by some kind of invariable relations.

It follows that no definition of the body acted on is necessary when the forces in equilibrium are given. The forces must have something to act on, but all we assume here about this something, is that it transmits the force so that the axioms enunciated may be taken as true. For this reason, it is sometimes said that *statics is the science which treats of the equilibrium and action of forces apart from the subject matter acted on.*

22. Resultant force. When two forces act simultaneously on a particle and are not in equilibrium, they will tend to move the particle. We infer that there is always some one force which will keep the particle at rest.

A force equal and opposite to this force is called the resultant of the two forces and is equivalent to the forces. It is obvious that the resultant of two forces acting on a particle must also act on that particle. It is also clear that its line of action is intermediate between those of the two forces.

Let $P_1, P_2, \ldots P_n$ be any number of forces acting on the same particle. The two forces P_1, P_2 have a resultant, say Q_1. We may remove P_1 and P_2 and replace them by Q_1. Again Q_1 and P_3 may be replaced by their resultant Q_2 and so on. We finally have all the forces replaced by a single force. This single force is called their resultant.

If the forces of a system do not all act at the same point, it may happen that there is no single force which could balance the system. If so, the system is not equivalent to any single resultant force.

23. *To find the resultant of any number of forces acting at a point and having their lines of action in the same straight line.*

Let O be the point of application, and first let all the forces act in the same direction Ox. Since each acts independently of the others, the resultant is clearly the sum of the separate forces and it acts in the direction Ox.

If some of the forces act in one direction Ox and others in the opposite direction say Ox', we sum the forces in these two directions separately. Let X and X' be these separate sums, and let X be the greater. Then by Art. 15 we can remove the force X' from both sets of forces. The whole system is therefore equivalent to the single force $X - X'$ acting in the direction of X.

By the rule of signs this is also equivalent to a single force represented by the negative quantity $X' - X$ acting in the opposite direction, viz. that of X'.

The necessary and sufficient condition that a system of forces acting at a point and having their lines of action in the same straight line should be in equilibrium is that the algebraic sum of the forces should be zero.

24. Parallelogram of Forces. *To find the resultant of two forces acting at a given point and inclined to each other at any angle.* Let the two forces act at the point O *and let them be represented in direction and magnitude by two straight lines* OA, OB *drawn from the point* O (Art. 7). *Let us now construct a parallelogram having* OA, OB *for two adjacent sides and let* OC *be that diagonal which passes through the point* O. *Then the resultant of the two forces will be represented in direction and magnitude by the diagonal* OC.

Several proofs of this important theorem have been given. As the "parallelogram law" is the foundation of the whole theory of the composition and resolution of forces, it will be useful to consider more than one proof, though the student at first reading should confine his attention to one of them.

25. Newton's proof of the parallelogram of forces. This proof is founded on the dynamical measure of force. Its principle has already been explained in Art. 15. It is repeated here on account of its importance. The figure is the same as that used in Art. 12 for the parallelogram of velocities.

26. Suppose two forces to act on the particle placed at O in the directions OA, OB. Let the lengths OA, OB be such that they represent the velocities these forces could separately generate in the particle by acting for a given time. Since each force acts independently of the other, it will generate the same velocity whether the other acts or does not act. When both act the particle has at the end of the given time both the velocities represented by OA and OB. These are together equivalent to the single velocity OC. But this is also the measure of the force which would generate that velocity. Thus the two forces measured by OA, OB are together equivalent to the single force measured by OC.

27. Duchayla's proof of the parallelogram of forces. This proof is founded on the principle of the transmissibility of force, Art. 17. It has been shown in Art. 18 that this principle can be made to depend only on statical axioms.

To prove the proposition we shall use the *inductive proof.* We shall assume that the theorem is true for forces of p and m units inclined at any angle, and also for forces of p and n units inclined at the same angle; we shall then prove that the theorem must be true for forces of p and $m + n$ units inclined at the same angle.

Let the forces p and m act at the point O and be represented in direction and magnitude by the straight lines OA and OB. On the same scale let BD represent the force n in direction and magnitude. Let BD be in the same straight line with OB, then the length OD will repre-

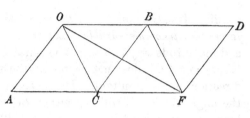

sent the force $m + n$ in direction and magnitude, Art. 23. Let the two parallelograms $OBCA$, $BDFC$ be constructed and let OC, OF, BF be the diagonals.

By hypothesis the resultant of the two forces p and m acts along OC. By Art. 18, we transfer the point of application to C. We now replace this resultant force by its two components p and m. These act at C, viz. p along BC produced and m along CF. Transfer the force p to act at B and the force m to act at F.

Since BC is equal and parallel to OA, the force p acting at B is represented by BC. The force n may be supposed also to act at B and is represented by BD. Hence by our hypothesis the resultant of these two acts along BF. Transfer the point of application to the point F.

The two forces p and $m + n$ are therefore equivalent to two forces acting at F. Their resultant must therefore pass through F, Art. 22. For the same reason the resultant passes through O, and the forces have but one resultant, Art. 22. Hence the resultant must act along OF. But this is the diagonal of the parallelogram constructed on the sides OA, OD which represent the forces p and $m + n$.

It is clear that the resultant of two equal forces makes equal angles with each of these forces. The resultant of two equal forces therefore acts along the diagonal of the parallelogram constructed on the equal forces in the manner already described. Thus the hypothesis is true for the equal forces p and p. By what has just been proved it is true for the forces p and $2p$ and therefore for p and $3p$ and so on. Thus it is true for forces p and rp where r is any integer. Again the hypothesis has just been proved true for forces rp and p; hence it is true for rp and $2p$ and so on. Thus the hypothesis is true for forces rp and sp, where r and s are any integers. Thus the proposition so far as the *direction* of the resultant is concerned is established for any commensurable forces.

28. *We have now to find the direction of the resultant when the forces are incommensurable.* Let OA, OB represent in direction and magnitude any two incommensurable forces p and q, then if the diagonal OC does not represent the resultant, let OG be the direction of the resultant. The straight line OG must lie within the angle AOB and will cut either BC between B and C or AC between A and C; Art. 22. Let it cut BC between B and C.

Divide OB into a number of equal parts each less than GC and measure off from OA beginning at O portions equal to these until we arrive at a point K where AK is less than GC. Draw GH, KL parallel to AC. Since OB and OK are commensurable the forces represented by these have a resultant which acts along the diagonal OL. Thus the forces p and q acting at O are equivalent to two forces, one of which acts

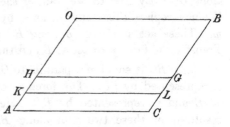

along OL and the other is the force represented by KA. The resultant of these two must act at O in a direction lying between OL and OA. But OG lies outside the angle AOL, hence the assumption that the direction of the resultant is OG is impossible. But OG represents any direction other than OC for then only is it impossible to divide OB into equal parts each less than CG. Thus the resultant force must act along the diagonal whether the forces be commensurable or incommensurable.

We have given a separate proof for incommensurable forces. But this is unnecessary. The theorem has been proved for all forces whose ratio can be expressed by a fraction. In the case of incommensurable forces we can still find a fraction which differs from their true ratio by a quantity less than any assigned difference. In the limit the theorem must be true for incommensurable forces.

29. *To prove that the diagonal represents the magnitude of the resultant as well as its direction.*

Let OA and OB represent the two forces, and let OC be the diagonal of the parallelogram $OACB$. Take OD in CO produced of such length as to represent the resultant in magnitude. Then the three forces OA, OB, OD are in equilibrium and each of them is equal and opposite to the resultant of the other two.

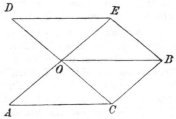

Construct on OB, OD the parallelogram $OBED$. Since OA is equal and opposite to the resultant of OB and OD, OE is in the same straight line with OA and therefore OE is parallel to CB. By construction OC is in the same straight line with OD and is therefore parallel to EB. Thus EC is a parallelogram. Hence OC is equal to EB and therefore to DO.

Thus the diagonal OC represents the resultant of the two forces OA, OB in magnitude.

30. Ex. Assuming that the diagonal represents the *magnitude* of the resultant, show that it also represents the *direction*.

As before, let OA, OB, OD represent forces in equilibrium. It is given that $OA = OE$, $OC = OD$, and it is to be proved that AOE, DOC are straight lines. Since AB and BD are parallelograms, $OA = BC$, $OD = BE$. Hence in the quadrilateral $EOCB$ the opposite sides are equal in length. The quadrilateral is therefore a parallelogram. (For the triangles OEB, BCO have their sides equal each to each.) It follows that OE is parallel to BC, and is therefore in the same straight line with OA.

31. Historical Summary. The principles on which the science of statics has been founded in former times may be reduced to three.

There is first the principle used by Archimedes, viz., that of the lever. It is assumed as self-evident or as the result of an obvious experiment, (1) that a straight horizontal lever charged at its extremities with equal weights will balance about a support placed at its middle point, (2) that the pressure on the support is the sum of the equal weights. Starting with this elementary principle, and measuring forces by the weights they would support, the conditions of equilibrium of a straight lever acted on by unequal forces were deduced. From this result by the addition of some simple axioms the other proposition of statics may be made to follow. The truth of the first elementary principle named above is perhaps evident from the symmetry of the figure, but Lagrange points out that the second is not equally evident with the first.

The second principle which has been used as the foundation of statics is that of the parallelogram of forces. In 1586, Stevinus enunciated the theorem of the triangle of forces. Till this time the science of statics had rested on the theory of the lever, but then a new departure became possible. The simplicity of the principle and the ease with which it may be applied to the problems of mechanics caused it to be generally adopted. The principle finally became the basis of modern statics. For an account of its gradual development we refer the reader to *A Short History of Mathematics*, by W. W. R. Ball.

Many writers have given or attempted to give proofs of this principle which are independent of the idea of motion. One of these, that of Duchayla, has been reproduced above, as that is the one which seems to have been best received. There is another, that of Laplace, which has attracted considerable attention. This is founded on principles similar to the proofs of Bernoulli and D'Alembert. It is assumed as evident that if two forces be increased in any, the same, ratio the magnitude of their resultant will be increased in the same ratio, but its direction will be unaltered.

In comparing these proofs with that founded on the idea of motion, we must admit the force of a remark of Lagrange. He says that, in separating the principle of the composition of forces from the composition of motions, we deprive that principle of its chief advantages. It loses its simplicity and its self-evidence, and it becomes merely a result of some constructions of geometry or analysis.

The third fundamental principle which has been used is that of virtual velocities. This principle had been used by the older writers, but Lagrange gave, or attempted to give, an elementary proof and then made it the basis of the whole science of mechanics. This proof has not been generally received as presenting the simplicity and evidence which he had admired in the principle of the composition of forces.

CHAPTER II.

The triangle of forces.

32. In the last chapter we arrived at a fundamental proposition, usually called the parallelogram of forces, which we
shall be continually using. Experience shows it is not always
convenient to draw the parallelogram, for this complicates the
figure and makes the solution cumbersome. Several artifices
have been invented to enable us to use the principle with facility
and quickness. In this chapter we shall consider these in turn.

33. If OA, OB represent two forces P and Q acting at a
point O, we know that their resultant is represented by the

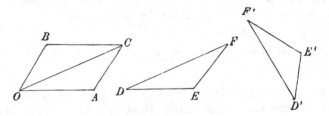

diagonal OC of the parallelogram constructed on those sides.
Now it is evident that AC will represent the force Q in direction
and magnitude as well as OB, though it will not represent the
point of application. This however is unimportant if the point of
application is otherwise indicated. Thus the triangle OAC may
be used instead of the parallelogram $OACB$.

As the points of application are supposed to be given independently it is no longer necessary to represent the forces by
straight lines passing through O. Thus we may represent the

forces P, Q, R acting at O both in direction and magnitude by the sides of a triangle DEF provided these sides are parallel to the forces and proportional to them in magnitude.

It is clear that all theorems about the parallelogram of forces may be immediately transferred to the triangle. We therefore infer the following proposition called the *triangle of forces*.

If two forces acting at some point are represented in direction and magnitude by the sides DE, EF of any triangle, the third side DF will represent their resultant.

If three forces acting at some point are represented in direction and magnitude by the three sides of any triangle taken in order viz., DE, EF, FD, the three forces are in equilibrium.

34. When three forces *in one plane* are given and we wish to determine whether they are in equilibrium or not, we see that there are two conditions to be satisfied.

1. If they are not all parallel two of them must meet in some point O. The resultant of these two will also pass through the same point. The third force must be equal and opposite to this resultant and must therefore also pass through the same point. Hence the lines of action of the three forces must meet in one point or be parallel.

2. If the forces are not all parallel, straight lines can be drawn parallel to the forces so as to form a triangle. The magnitudes of the forces must be proportional to the sides of that triangle taken in order.

The case in which the forces are all parallel will be considered in the next chapter.

35. We may evidently extend this proposition further. Suppose we turn the triangle DEF through a right angle into the position $D'E'F'$, the sides will then be perpendicular instead of parallel to the forces. Also if the forces act in the directions DE, EF, FD they now act all three outwards with regard to the triangle $D'E'F'$. If the forces were reversed they would all act inwards. We have thus a new proposition.

If three forces acting at some point be represented in magnitude by the sides of a triangle, and if the directions of the forces be perpendicular to those sides and if they act all inwards or all outwards, the three forces are in equilibrium.

Instead of turning the triangle through a right angle, we might turn it through any acute angle. We thus obtain another theorem. If three forces acting at a point be represented in magnitude by the sides of a triangle and if their directions make equal angles with the sides taken in order, the three forces are in equilibrium.

In using this theorem, it is sometimes found to be inconvenient to sketch the triangle. We then put the theorem into another form. The sides of the triangle are proportional to the sines of the opposite angles. This relation must therefore also hold for the forces. Hence we infer the following theorem.

Three forces acting on a body in one plane are in equilibrium if (1) *their lines of action all meet in one point,* (2) *their directions are all towards or all from that point,* (3) *the magnitude of each is proportional to the sine of the angle between the other two.*

36. Polygon of forces. We may further extend the triangle of forces into a polygon of forces. If several forces act at a point O we may represent these in magnitude and direction by the sides of an unclosed polygon DE, EF, FG, GH &c. taken in order. The resultant of DE, EF is represented by DF. That of DF and FG is DG and so on. *Thus the resultant is represented by the straight line closing the polygon.* It is clear that the sides of the polygon need not all be in the same plane.

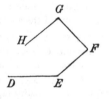

If several forces acting at one point be represented in direction and magnitude *by the sides of a closed polygon taken in order, they are in equilibrium.*

37. Ex. 1. Forces in one plane, whose magnitudes are proportional to the sides of a closed polygon, act perpendicularly to those sides at their middle points all inwards or all outwards. Prove that they are in equilibrium.

Let $ABCD$ &c. be the polygon. Join one corner A to the others C, D &c. Consider the triangle ABC thus formed. The forces across AB, BC meet in the centre of the circumscribing circle, and have therefore for resultant a force proportional to AC acting perpendicularly to it at its middle point. Taking the triangles ACD, ADE &c. in turn, the final resultant is obviously zero.

Ex. 2. Forces in one plane, whose magnitudes are proportional to the cosines of half the internal angles of a closed polygon, act inwards at the corners in directions bisecting the angles. Prove that they are in equilibrium.

Apply along each side of a polygon two equal and opposite forces, say each equal to F, and let these act at the corners. The two which act at the corner A have a resultant $2F \cos \frac{1}{2}A$ whose direction bisects that angle. These resultants must therefore be in equilibrium.

38. **Ex. 1.** Forces represented by the numbers 4, 5, 6 are in equilibrium; find the tangents of the halves of the angles between the forces.

By drawing parallels to these forces we construct a triangle of the forces. The angles of this triangle can be found by the ordinary rules of trigonometry.

Ex. 2. Forces represented by 6, 8, 10 lbs. are in equilibrium; find the angle between the two smaller forces. How must the least force be altered that the angle between the other two may be halved?

Ex. 3. If OA, OB represent two forces, show that their resultant is represented by twice OM, where M is the middle point of AB.

Ex. 4. Two constant equal forces act at the centre of an ellipse parallel to the directions SP and PH, where S and H are the foci and P is any point on the curve. Show that the extremity of the line which represents their resultant lies on a circle.   [Math. Tripos, 1883.]

Ex. 5. Forces P, Q act at a point O, and their resultant is R; if any transversal cut the directions of the forces in the points L, M, N respectively, show that

$$\frac{P}{OL} + \frac{Q}{OM} = \frac{R}{ON}.$$ [Math. Tripos, 1881.]

Clear of fractions and the equation reduces to the statement that the area LOM is the sum of the areas LON, MON.

Ex. 6. A particle O is in equilibrium under three forces, viz., a force F given in magnitude, a force F' given in direction, and a force P given in magnitude and direction. Find the lines of action of F by a geometrical construction.

If OA represent P, draw AB parallel to F', and describe a circle whose centre is O and whose radius represents F in magnitude.

Ex. 7. $ABCD$ is a tetrahedron, P is any point in BC, and Q any point in AD. Prove that a force represented in magnitude, direction, and position by PQ, can be replaced by four components in AB, BD, DC, CA in one and in only one way, and find the ratios of these components.   [St John's Coll., 1887.]

Ex. 8. Lengths BD, CE, AF are drawn from the corners along the sides BC, CA, AB of a triangle ABC; each length being proportional to the side along which it is drawn. If forces represented in magnitude and direction by AD, BE, CF acted on a point, show that they would be in equilibrium. Conversely if the forces AD, BE, CF act at a point and are in equilibrium, then BD, CE, AF are proportional to the sides.

39. **Parallelepiped of forces.** *Three forces acting at a point O are represented in direction and magnitude by three straight lines OA, OB, OC not in one plane. To show that the resultant is represented in direction and magnitude by the diagonal of the parallelepiped constructed on the three straight lines as sides.*

Consider the parallelogram constructed on OA, OB, the resultant of these two forces is represented by OD. If CE be the parallel diagonal of the opposite face, it is clear by geometry that $OCED$ will be a parallelogram. The resultant of the forces represented by OC, OD will therefore be OE, i.e. the diagonal of the parallelepiped.

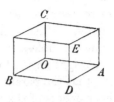

We may also deduce the theorem from Art. 36. The resultant of the three forces represented by OA, AD, DE is represented by the straight line which closes the polygon $OADE$, i.e. it is OE.

40. Three methods of oblique resolution.

(1) Any three directions (not all in one plane) being given, a force R represented by OE may be replaced by three forces X, Y, Z, acting in the given directions. The force R is then said to be resolved in these directions and the forces X, Y, Z are called its components. The magnitudes of the components are found geometrically by constructing the parallelepiped whose diagonal is R and whose sides OA, OB, OC have the given directions.

(2) *When the resultant OE is given, each component may be found by resolving perpendicularly to the plane containing the other two.* Thus suppose the component along OC of a force R acting along OE is required. Let OC, OE make angles θ, γ respectively with the plane AOB, then, since the perpendiculars from C and E on that plane are equal, $OC \sin \theta = OE \sin \gamma$. The component Z along OC is therefore given by $Z \sin \theta = R \sin \gamma$.

(3) A third method of effecting an oblique resolution is given in Arts. 51 and 53.

Ex. 1. If six forces, acting on a particle, be represented in magnitude and direction by the edges of a tetrahedron, the particle cannot be at rest. [Math. T., 1859.]

Ex. 2. Four forces acting at a point O are in equilibrium, and equal straight lines are drawn from O along their directions. Prove that each force is proportional to the volume of the tetrahedron described on the lines drawn along the other three forces.

Method of Analysis.

41. We have seen that any force may be replaced by two others, called its components, which are inclined at any angle to each other which may appear suitable. But it is found by experience that when a force has to be resolved it is generally more useful to resolve it into two components which are at right angles. When therefore the component of a force is spoken of it is meant, unless it is otherwise stated, that the other component is at right angles to it. By referring to the figure of Art. 33, we see that the parallelogram $OACB$ becomes a rectangle. The two components of the force OC are $OC \cdot \cos COA$ and $OC \cdot \sin COA$.

We may put this result into the form of a working rule. *If a force R act at O in the direction OC, its component in any direction Ox is $R \cos COx$. Its component in the opposite direction Ox' is $R \cos COx'$.* In the same way the component of R perpendicular to Ox is $R \sin COx$.

It is convenient to have some short name to distinguish the rectangular components of a force from its oblique components. The name *resolute* for the components in the first case has been suggested in Lock's *Elementary Statics*.

42. *Two forces P_1, P_2 act at a point O. To find the position and magnitude of their resultant.*

Let Ox, Oy be any two rectangular axes, and let α_1, α_2 be the angles the forces P_1, P_2 make with the axis of x. The sums of the components parallel to the axes are

$$X = P_1 \cos \alpha_1 + P_2 \cos \alpha_2,$$
$$Y = P_1 \sin \alpha_1 + P_2 \sin \alpha_2.$$

If these are the components of a force R whose line of action makes an angle $\bar{\alpha}$ with the axis of x, we have

$$X = R \cos \bar{\alpha}, \qquad Y = R \sin \bar{\alpha}.$$

We easily find by adding together the squares of X and Y that

$$R^2 = P_1^2 + P_2^2 + 2P_1 P_2 \cos \theta,$$

where $\theta = \alpha_1 - \alpha_2$, so that θ is the angle between the directions of the forces P_1, P_2. This result also follows from the parallelogram of forces. For the right-hand side is evidently the square of the diagonal of the parallelogram whose sides are P_1 and P_2.

The direction of the resultant is also easily found, for we have

$$\tan \bar{\alpha} = \frac{Y}{X} = \frac{P_1 \sin \alpha_1 + P_2 \sin \alpha_2}{P_1 \cos \alpha_1 + P_2 \cos \alpha_2}.$$

43. Ex. 1. Two forces P, Q act at an angle a and have a resultant R. If each force be increased by R, prove that the new resultant makes with R an angle whose tangent is $\dfrac{(P-Q)\sin a}{P+Q+R+(P+Q)\cos a}$.   [St John's Coll., 1880.]

Take the line of action of the resultant R for the axis of x.

Ex. 2. Forces each equal to F act at a point parallel to the sides of a triangle ABC. If R be their resultant, prove that $R^2 = F^2(3 - 2\cos A - 2\cos B - 2\cos C)$.

Ex. 3. The resultant of P and Q is R, if Q be doubled R is doubled, if Q be reversed, R is also doubled; show that $P:Q:R::\sqrt{2}:\sqrt{3}:\sqrt{2}$.   [St John's Coll.]

44. *Any number of forces act at a point O in any directions. It is required to find their resultant.*

Take any rectangular axes Ox, Oy, Oz. Let P_1, P_2, P_3 &c. be the forces, $(\alpha_1\beta_1\gamma_1)$, $(\alpha_2\beta_2\gamma_2)$ &c. their direction angles. The sums of the components of these parallel to the axes are

$$X = P_1\cos\alpha_1 + P_2\cos\alpha_2 + \ldots = \Sigma P\cos\alpha,$$
$$Y = P_1\cos\beta_1 + P_2\cos\beta_2 + \ldots = \Sigma P\cos\beta,$$
$$Z = P_1\cos\gamma_1 + P_2\cos\gamma_2 + \ldots = \Sigma P\cos\gamma.$$

If these are the components of a force R whose direction angles are $(\bar\alpha\bar\beta\bar\gamma)$ we have

$$R\cos\bar\alpha = X, \quad R\cos\bar\beta = Y, \quad R\cos\bar\gamma = Z.$$

By a known theorem in solid geometry

$$\cos^2\bar\alpha + \cos^2\bar\beta + \cos^2\bar\gamma = 1.$$

Hence $R^2 = X^2 + Y^2 + Z^2$,

$$\frac{\cos\bar\alpha}{X} = \frac{\cos\bar\beta}{Y} = \frac{\cos\gamma}{Z} = \frac{1}{(X^2+Y^2+Z^2)^{\frac{1}{2}}}.$$

Thus both R and its direction cosines have been found.

If the conditions of equilibrium are required it is sufficient and necessary that $R = 0$. This gives the three conditions

$$X = \Sigma P\cos\alpha = 0, \quad Y = \Sigma P\cos\beta = 0, \quad Z = \Sigma P\cos\gamma = 0.$$

45. *If the resolved parts of the forces P_1, P_2 &c. along any three directions OA, OB, OC not all in one plane are zero, they are in equilibrium.*

Let the axis Oz coincide with OC and let the plane xOz contain OA. Since the resolved part along Oz is zero, we have $Z=0$. Since the resolved part along OA is zero, we have $X\cos xOA = 0$. Since xOA cannot be a right angle without making OA, OC coincide, we have $X=0$. Lastly since the resolved part along OB is zero we find $Y\cos yOB = 0$. This gives $y=0$.

46. *The magnitude and direction of R may also be expressed in a form independent of coordinates in the following manner.*

By a known theorem in solid geometry if θ_{12} be the angle between the straight lines whose direction angles are $(\alpha_1\beta_1\gamma_1)$ $(\alpha_2\beta_2\gamma_2)$ with the usual convention as to direction, then

$$\cos\theta_{12} = \cos\alpha_1\cos\alpha_2 + \cos\beta_1\cos\beta_2 + \cos\gamma_1\cos\gamma_2.$$

Adding together the squares of the expressions for X, Y, Z we have $R^2 = P_1^2(\cos^2\alpha_1 + \cos^2\beta_1 + \cos^2\gamma_1) + \&c.$

$$+ 2P_1P_2(\cos\alpha_1\cos\alpha_2 + \cos\beta_1\cos\beta_2 + \cos\gamma_1\cos\gamma_2) + \&c.$$

$$= P_1^2 + P_2^2 + \&c. + 2P_1P_2\cos\theta_{12} + \&c.$$

This gives the *magnitude* of R.

To determine the line of action of R, we shall find the angles ϕ_1, ϕ_2 &c. which its direction makes with the directions of the forces P_1, P_2 &c. The axes of coordinates being perfectly arbitrary; let us take the axis of x to be coincident with the line of action of the force P_1. Then $\bar{a} = \phi_1$, $\alpha_1 = 0$, $\alpha_2 = \theta_{12}$ &c., the equations

$$R\cos\bar{a} = X = \Sigma P\cos\alpha$$

become $R\cos\phi_1 = P_1 + P_2\cos\theta_{12} + P_3\cos\theta_{13} + \&c.$

In the same way by taking the axis of x along the force P_2 we find $R\cos\phi_2 = P_1\cos\theta_{12} + P_2 + P_3\cos\theta_{23} + \ldots$

and so on. Thus the direction of R has been found.

47. Polyhedron of forces. The equations of Art. 44 have a geometrical meaning which is often useful. Let any closed polyhedron be constructed, let A_1, A_2 &c. be the areas of its faces. Let normals be drawn to these faces, each from a point in the face all outwards or all inwards, and let θ_1, θ_2 &c. be the angles these normals make with any straight line which we may call the axis of z. Let us now project orthogonally all these areas on the plane of xy. The several projections are $A_1\cos\theta_1$, $A_2\cos\theta_2$ &c. Since the polyhedron is closed the total projected area which is positive is equal to the total negative projected area. We therefore have $A_1\cos\theta_1 + A_2\cos\theta_2 + \ldots = 0.$

Similar results hold for the projection on the other coordinate planes. Thus we obtain three equations which are the same as the equations of equilibrium already found, except that we have A_1, A_2 &c. written for P_1, P_2 &c. We therefore have the following theorem. *If forces acting at a point be represented in magnitude by the areas of the faces of a closed polyhedron and if the directions of the forces be perpendicular to those faces respectively, acting all inwards or all outwards, then these forces are in equilibrium.*

48. *By using the theory of determinants we may put the results of Art. 46 into a more convenient form.* Let it be required to find the resultant of any three forces acting at a point. To obtain a symmetrical result we shall reverse the resultant and speak of *four forces in equilibrium.*

Let P_1, P_2, P_3, P_4 be four forces in equilibrium. Putting $R = 0$, we have found in Art. 46 four linear equations connecting these. Eliminating the forces, we have the determinantal equation

$$\begin{vmatrix} 1 & \cos\theta_{12} & \cos\theta_{13} & \cos\theta_{14} \\ \cos\theta_{21} & 1 & \cos\theta_{23} & \cos\theta_{24} \\ \cos\theta_{31} & \cos\theta_{32} & 1 & \cos\theta_{34} \\ \cos\theta_{41} & \cos\theta_{42} & \cos\theta_{43} & 1 \end{vmatrix} = 0.$$

This is the relation connecting the mutual inclinations of any four straight lines in space*. If all these angles except one (say θ_{12}) are given, we have a quadratic to find the two possible values which $\cos\theta_{12}$ could then have. If three of the angles say θ_{12}, θ_{23}, θ_{31} are right angles this determinant reduces to the well-known form $\cos^2\theta_{14} + \cos^2\theta_{24} + \cos^2\theta_{34} = 1.$

If the angles between the four directions in which the forces act are given, the ratios of the forces are found from any three of the four linear equations above mentioned. It follows that the forces are in the ratio of the minors of the constituents in *any row* of the determinant.

49. Ex. Show that the *squares* of the forces are in the ratio of the minors of the constituents in the leading diagonal.

For let I_{rs} be the minor of the rth row and sth column, then by a theorem in Salmon's *Higher Algebra* $I_{11}I_{22} = I_{12}^2$. But we have shown above that

$$P_1 : P_2 = I_{11} : I_{12}$$

hence we deduce at once $P_1^2 : P_2^2 = I_{11} : I_{22}.$

For the sake of reference we state at length the minor of the leading constituent.

It is $I_{11} = 1 - \cos^2\theta_{23} - \cos^2\theta_{34} - \cos^2\theta_{42} + 2\cos\theta_{23}\cos\theta_{34}\cos\theta_{42}.$

This expression is easily recognized as one which occurs in many formulæ in spherical trigonometry. For example, if unit lengths are drawn from any point O parallel to the directions of any three of the forces (say P_2, P_3, P_4) the volume of the tetrahedron so formed is one sixth of the square root of the corresponding minor (viz. I_{11}).

50. Sometimes it is necessary to refer the forces to oblique axes. In this case we replace the direction cosines of each force by its direction ratios. Let the direction ratios of P_1, P_2 &c. be $(a_1 b_1 c_1)$, $(a_2 b_2 c_2)$ &c. Then by the same reasoning as before, the sums of the components of the forces parallel to the axes are

$$X = \Sigma Pa, \quad Y = \Sigma Pb, \quad Z = \Sigma Pc.$$

If these are the components of a force R with direction ratios (l, m, n) we have

$$Rl = X, \quad Rm = Y, \quad Rn = Z.$$

The relations between the direction ratios of a straight line and the angles that straight line makes with the axes are given in treatises on solid geometry or on spherical trigonometry. They are not nearly so simple as when the axes of reference are rectangular. For this reason oblique axes are seldom used.

The mean centre.

51. There is another method of expressing the magnitude and direction of the resultant of any number of forces acting at a

---

* Another proof is given in Salmon's *Solid Geometry*, Ed. IV., Art. 54.

point which will be found very useful both in geometrical and analytical reasoning.

Let us represent the forces P_1, P_2 &c. in direction by the straight lines OA_1, OA_2 &c. To represent their magnitudes we shall take lengths measured along these straight lines, thus the force along OA_1 is represented by $p_1 . OA_1$, that along OA_2 by $p_2 . OA_2$, and so on. The advantage of introducing the numerical multipliers p_1, p_2 &c. is that the extremities A_1, A_2 &c. of the straight lines may be chosen so as to suit the figure of the problem under consideration. It is evident that this is equivalent to representing the forces by straight lines on different scales, viz. the scales p_1, p_2 &c. of force to each unit of length.

Taking O for origin, let $(x_1 y_1 z_1)$, $(x_2 y_2 z_2)$ &c. be the coordinates of the points A_1, A_2 &c. We have already proved that the components of the resultant are

$$\left. \begin{aligned} X &= \Sigma P \cos \alpha = \Sigma p . OA_1 \cos \alpha = \Sigma px \\ Y &= \Sigma P \cos \beta \qquad\qquad = \Sigma py \\ Z &= \Sigma P \cos \gamma \qquad\qquad = \Sigma pz \end{aligned} \right\} \dots\dots\dots(1).$$

Let us take a point G whose coordinates $(\bar{x}\bar{y}\bar{z})$ are given by the equations $\qquad \bar{x} = \dfrac{\Sigma px}{\Sigma p}, \qquad \bar{y} = \dfrac{\Sigma py}{\Sigma p}, \qquad \bar{z} = \dfrac{\Sigma pz}{\Sigma p} \dots\dots\dots\dots(2).$

It follows at once that

$$X = \bar{x} \Sigma p, \quad Y = \bar{y} \Sigma p, \quad Z = \bar{z} \Sigma p.$$

These equations imply that the resultant of the forces is represented in direction and magnitude by $OG . \Sigma p$.

This point G is known by a variety of names. It is called the *centre of gravity*, or *centroid* or *mean centre* of a system of particles placed at A_1, A_2 ...... whose masses or weights are proportional to p_1, p_2 &c.

The result is, *if forces acting at a point O be represented in direction by the straight lines OA_1, OA_2 &c. and in magnitude by $p_1 . OA_1$, $p_2 . OA_2$ &c., then their resultant is represented in direction by OG and in magnitude by $\Sigma p . OG$, where G is the centroid of masses proportional to p_1, p_2 &c. placed at A_1, A_2 &c.* This theorem is commonly ascribed to Leibnitz.

We notice that *forces represented in magnitude and direction by $p_1 . OA_1$, $p_2 . OA_2$ &c., are in equilibrium when O is the centroid of masses proportional to p_1, p_2 &c., placed at A_1, A_2 &c.*

Conversely, *a force R, acting along OG, may be resolved into three forces P_1, P_2, P_3, which act along three given straight lines passing through O, by making G to be the mean centre of masses placed at convenient points A_1, A_2, A_3, on those straight lines.* If p_1, p_2, p_3 are those masses, the components P_1, P_2, P_3 are given by

$$\frac{P_1}{p_1 . OA_1} = \frac{P_2}{p_2 . OA_2} = \frac{P_3}{p_3 . OA_3} = \frac{R}{(p_1 + p_2 + p_3) OG}.$$

In using this theorem we may draw some or all of the straight lines OA_1, OA_2 &c. in the opposite directions to the forces. If this be done we simply regard the p's of those straight lines as negative.

When some of the p's are negative, it may happen that $\Sigma p = 0$. In this case the centroid is at infinity and this representation of the resultant though correct is not convenient. The components along the axes are still given by the expressions $X = \Sigma px$, $Y = \Sigma py$, $Z = \Sigma pz$ which do not contain any infinite quantities.

52. The utility of this proposition depends on the ease with which the point G can be found when A_1, A_2, &c., are given. *The working rule is that the distance of G from any plane of reference, taken as the plane of xy, is given by the formula* $\bar{z} = \dfrac{\Sigma pz}{\Sigma p}$. The properties of this point and its positions in various cases are discussed in the chapter on the centre of gravity.

53. Ex. 1. The centroid G of two particles p_1, p_2 placed at two given points A_1, A_2, lies in the straight line $A_1 A_2$ and divides it so that $p_1 . A_1 G = p_2 . A_2 G$.

Take $A_1 A_2$ as the axis of x, A_1 as origin and let $A_1 A_2 = a$. Then $x_1 = 0$, $x_2 = a$, $y_1 = 0$, $y_2 = 0$. Using the working rule we have

$$\bar{x} = \frac{p_1 x_1 + p_2 x_2}{p_1 + p_2} = \frac{p_2 a}{p_1 + p_2}, \qquad \bar{y} = \frac{p_1 y_1 + p_2 y_2}{p_1 + p_2} = 0.$$

Hence G lies in $A_1 A_2$ and since $\bar{x} = A_1 G$ we find $p_1 . A_1 G = p_2 (A_1 A_2 - A_1 G) = p_2 . A_2 G$.

This theorem enables us to resolve a force P which acts along a given straight line OG into two directions OA_1, OA_2, which are not necessarily at right angles. The components P_1, P_2 are given by

$$\frac{P_1}{p_1 . OA_1} = \frac{P_2}{p_2 . OA_2} = \frac{P}{(p_1 + p_2) OG}$$

where p_1, p_2 are the distances of G from A_2, A_1 *taken positively when measured inwards.*

Ex. 2. Prove that the centroid of three masses p_1, p_2, p_3, placed at the corners of a triangle is the point whose areal coordinates are proportional to p_1, p_2, p_3. When the masses are equal this point is briefly called the centroid of the triangle.

If a, β, γ are the distances of a point G from the sides BC, CA, AB of a triangle taken positively when measured inwards, and p, q, r are the perpendiculars from the corners on the same sides, the ratios $x = a/p$, $y = \beta/q$, $z = \gamma/r$ are called the areal coordinates of G. It is evident that x, y, z are also proportional to the areas of the triangles BGC, CGA, AGB respectively. Also $x + y + z = 1$.

Taking any side AB as the axis of reference we deduce from the working rule (Art. 52) that the distance of the centroid from it is $\gamma = p_3 r/s$ where $s = p_1 + p_2 + p_3$. Similarly $a = p_1 p/s$, $\beta = p_2 q/s$. It follows that x, y, z are proportional to p_1, p_2, p_3.

Ex. 3. A force P acting at the corner D of a tetrahedron intersects the opposite face ABC in a point G whose areal coordinates referred to the triangle ABC are (xyz). If the components of P along the edges DA, DB, DC are P_1, P_2, P_3 prove
$$\frac{P_1}{x \cdot DA} = \frac{P_2}{y \cdot DB} = \frac{P_3}{z \cdot DC} = \frac{P}{DG}.$$

Ex. 4. Any number of forces are represented in magnitude and direction by straight lines $A_1 A_1'$, $A_2 A_2'$, ...$A_n A_n'$ and G, G' are the centroids of the points A_1, A_2,...A_n and A_1', A_2',...A_n'. Show that these forces transferred parallel to themselves to act at a point have a resultant which is represented in magnitude and direction by $n \cdot GG'$. [Coll. Ex., 1889.]

The group of forces AA' is equivalent to the three groups AG, GG', $G'A'$, Art. 36. The first and last are separately in equilibrium, Art. 51.

Ex. 5. Three forces in one plane, acting at A, B, C, are represented by AD, BE, CF where D, E, F are their intersections with the sides of the triangle ABC. Show that these are equivalent to three forces acting along the sides AB, BC, CA of the triangle represented by $\left(\dfrac{CD}{a} - \dfrac{CE}{b}\right) c$, $\left(\dfrac{AE}{b} - \dfrac{AF}{c}\right) a$ and $\left(\dfrac{BF}{c} - \dfrac{BD}{a}\right) b$.

Thence show that if $BD/a = CE/b = AF/c = \kappa$, these three forces are statically equivalent to the three forces $(1 - 2\kappa) c$, $(1 - 2\kappa) a$, $(1 - 2\kappa) b$ acting along the sides of the triangle.

Prove that the centroid of equal particles placed at D, E, F, coincides with that of the triangle. Thence show that the forces represented by OD, OE, OF, (where O is any point) have a resultant whose magnitude and line of action are independent of the value of κ.

Ex. 6. A particle in the plane of a triangle is acted on by forces directed to the mid-points of the sides whose magnitudes are proportional directly to the distances from those points and inversely to the radii of the circles escribed to those sides. Find the position of equilibrium. [Math. Tripos, 1890.]

The point is the centre of the inscribed circle.

Ex. 7. A, B, C, D are four small holes in a vertical lamina, and four elastic strings of natural lengths OA, OB, OC, OD are attached to a point O in the lamina, their other ends being passed through A, B, C, D respectively and attached to a small heavy ring P. Assuming that the tension of an elastic string is a given multiple of its extension, prove that when the lamina is turned in its own plane about O the locus of P in the lamina will be a circle. [Coll. Ex., 1888.]

Ex. 8. A quadrilateral $ABCD$ is inscribed in a circle whose centre is O, forces proportional to $\triangle BCD \pm 2 \triangle OBD$, $\triangle ACD \pm 2 \triangle OAC$, $\triangle ABD \pm 2 \triangle OBD$, $\triangle ABC \pm 2 \triangle OAC$, act along OA, OB, OC, OD respectively, the signs being determined according to a certain convention, show that the forces are in equilibrium. [Math. Tripos, 1889.]

Ex. 9. Three forces P, Q, R act along three straight lines DA, DB, DC not in one plane; if their resultant is parallel to the plane ABC, prove that
$$P/DA + Q/DB + R/DC = 0. \qquad \text{[St John's Coll., 1882.]}$$

Ex. 10. Assuming that the force of the wind on a sail is proportional to some power of the difference of the velocities of the wind and boat resolved normally to the sail, determine if the boat, by properly adjusting the sail, could be made to travel quicker than the wind in a direction making a given angle with the wind, and find the limits of the angle.

Ex. 11. *ABCDEF* is a regular hexagon, and at *A* forces act represented in magnitude and direction by *AB*, 2*AC*, 3*AD*, 4*AE*, 5*AF*. Show that the length of the line representing their resultant is $\sqrt{351}AB$.　　　　[Math. Tripos, 1880.]

Equilibrium of a particle under constraint.

54. *Distinction between smooth and rough bodies.* Let a particle under the influence of any forces be constrained to slide along an infinitely thin fixed wire. There is an action between the particle and the curve. Let this force be resolved into two components, one acting along a normal to the curve and the other along the tangent. The latter of these is called friction. By common experience it is found to depend on the nature of the materials of which the wire and particle are made. When this component is zero or so small that it can be neglected the bodies are said to be *smooth*. When it cannot be neglected the conditions of equilibrium are more complicated and will be found in another chapter. For the present we shall confine our attention to smooth bodies. Similar remarks apply when a particle is constrained to remain on a surface. In all such cases *the constraining curve or surface is called smooth when the action between it and the particle is along the normal to that curve or surface.*

55. If the particle be a bead slung on the curve, the bead can only move in the direction of a tangent drawn to the curve at the point occupied by the bead. *The necessary and sufficient condition of equilibrium is that the component of the forces along the tangent to the curve at the point occupied by the particle is zero.*

If the particle rest on one side of the curve the action of the curve on the particle will only prevent motion in one direction along the normal. It is therefore also *necessary for equilibrium that the external forces should press the particle against the curve.*

If a particle rest on a smooth surface at any point, the component of the forces along every tangent to the surface at that point must be zero. In other words, the *resultant force at a position of equilibrium must act normally to the surface in such a direction as to press the particle against the surface.*

56. *The form of a curve being given by its equations; to find the positions on it at which a particle would rest in equilibrium under the action of any given forces.*

Suppose the curve to be given by its Cartesian equations, and let the axes of reference be rectangular. Let x, y, z be the coordinates of the particle when in a position of equilibrium. Let X, Y, Z be the components of the forces parallel to these axes. Let s be the arc measured from some fixed point on the curve up to the point occupied by the particle. Then resolving the forces X, Y, Z along the tangent, we have by Art. 41,

$$X\frac{dx}{ds} + Y\frac{dy}{ds} + Z\frac{dz}{ds} = 0.$$

If the equations of the curve are given in the form

$$\phi(x, y, z) = 0, \quad \psi(x, y, z) = 0,$$

we have with the usual notation for partial differential coefficients

$$\phi_x dx + \phi_y dy + \phi_z dz = 0, \quad \psi_x dx + \psi_y dy + \psi_z dz = 0.$$

Eliminating the ratios $dx : dy : dz$, we have the determinant

$$J = \begin{vmatrix} X, & Y, & Z \\ \phi_x, & \phi_y, & \phi_z \\ \psi_x, & \psi_y, & \psi_z \end{vmatrix} = 0.$$

This determinantal equation, joined to the two equations of the curve, suffice in general to find the values of x, y, z. There may be several sets of values of these coordinates, and these give all the positions of equilibrium.

57. *The form of a surface being given by its equation; to find the point or points on it at which a particle would rest in equilibrium under the action of given forces.*

Let the surface be given by its Cartesian equation $f(x, y, z) = 0$ when referred to rectangular axes. By Art. 55 the direction cosines of the resultant force must be proportional to those of the normal to the surface. We therefore have

$$X/f_x = Y/f_y = Z/f_z.$$

Joining these two equations to the given equation of the surface, we have three equations to find (x, y, z).

58. *Pressure on the curve or surface.* It follows from Art. 54 that in the position of equilibrium the resultant force acts normally and is equal to the pressure. If then R be the pressure on the curve or surface, its magnitude is given by $R^2 = X^2 + Y^2 + Z^2$ and its direction is determined by the direction cosines X/R, Y/R, Z/R.

59. In these propositions the components X, Y, Z are supposed to be given functions of the coordinates x, y, z. In many cases these components are respectively partial differential coefficients with regard to x, y, z of some function V called the potential of the forces. Thus $X = \dfrac{dV}{dx}$, $Y = \dfrac{dV}{dy}$, $Z = \dfrac{dV}{dz}$ .... (1). The condition of equilibrium of a particle resting on a smooth curve defined by its Cartesian equations $\phi = 0$, $\psi = 0$ has been found above and is equivalent to the assertion that the Jacobian of (V, ϕ, ψ) vanishes at the points of equilibrium.

If we equate the potential V to an arbitrary constant c we obtain a system of surfaces. Each of these is called a level surface. By equations (1) X, Y, Z are proportional to the direction-cosines of the normal to a level surface. The resultant force at any point, therefore, acts along the normal to the level surface which passes through that point. *If then a particle is constrained to rest on any smooth curve or surface, the positions of equilibrium are those points at which the curve or surface touches a level surface.*

A curve or surface may be such that every point of it is a position of equilibrium. In this case the resultant force is everywhere normal to the curve or surface. If then the particle be constrained by a curve, the curve must lie on one of the level surfaces, if by a surface, that surface must be a level surface.

60. Another interpretation may be found for the condition of equilibrium
$$X\,dx + Y\,dy + Z\,dz = 0.$$
Substituting for X, Y, Z from (1), this is equivalent to $dV = 0$, i.e. at a position of equilibrium the potential of the forces is a maximum or minimum.

61. Ex. 1. A heavy particle is constrained to slide on a smooth circle whose plane is vertical. A string, attached to the particle, passes through a small ring placed at the highest point of the circle and supports an equal weight at its other end. Prove that the system is in equilibrium when the string between the ring and the particle makes an angle $60°$ with the vertical.

Ex. 2. The ends of a string are attached to two heavy rings of masses m, m', and the string carries a third ring of mass M which can slide on it; the rings m, m' are free to slide on two smooth fixed rigid bars inclined at angles α and β to the horizontal. Prove that if ϕ be the angle which either part of the string makes with the vertical, then $\cot \phi : \cot \beta : \cot \alpha = M : M + 2m' : M + 2m$.   [St John's, 1890.]

Ex. 3. A weight P, attached by a cord to a fixed point O, rests against a smooth curve in the same vertical plane with O; show that, (1) if the pressure on the curve is to be independent of the position of the weight on it, the curve must be a circle; (2) if the tension in the cord is to be independent of the position of the weight, the curve must be a conic section with O as focus.   [Math. Tripos, 1886.]

The vertical OA drawn through O, the normal PA to the curve and the string PO form a triangle whose sides are proportional to the forces which act along them. In case (1) the ratio of OA to AP is constant; it follows that P lies on a circle or on a straight line passing through O. In case (2) the ratio of OA to OP is constant; it follows that P lies on a conic or on a horizontal straight line through O.

Ex. 4. Two small rings without weight slide on the arc of a smooth vertical circle ; a string passes through both rings and has three equal weights attached to it, one at each end and one between the pegs. Show that in equilibrium the rings must be 30° distant from the highest point of the circle.   [Math. Tripos, 1853.]

Ex. 5. A smooth elliptic wire is placed with its major axis vertical, and a bead of given weight W is capable of sliding on the wire but is maintained in equilibrium by two strings passing over smooth pegs at the foci and sustaining given weights, of which the higher exceeds the lower by W/e, where e is the eccentricity. Prove that the pressure on the curve will be a maximum or minimum when the bead is at the extremities of the major axis or when the focal distances have between them the same ratio as the two sustained weights.   [Christ's Coll., 1865.]

Ex. 6. If four equal particles, attracting each other with forces which vary as the distance, slide along the arc of a smooth ellipse, they cannot be in equilibrium unless placed at the extremities of the axes; but if a fifth equal particle be fixed at any point and attract the other four according to the same law, there will be equilibrium if the distances of the four particles from the semi-axis major be the roots of the equation

$$(y^2 - b^2)\left(y + \frac{b^2 q}{5a^2 - 3b^2}\right)^2 = -\frac{a^2 b^2 p^2}{(3a^2 - 5b^2)^2} y^2$$

where p and q are the distances of the fifth particle from the axis minor and axis major respectively.

Ex. 7. A surface is such that the product of the distances of any point on it from two fixed points A and B is equal to the sum of those distances multiplied by a constant. A particle constrained to remain on the surface is acted on by two equal centres of repulsive force situated at A and B. If each force varies as the inverse square of the distance, show that the particle is in equilibrium in all positions.

Ex. 8. A heavy smooth tetrahedron rests with three of its faces against three fixed pegs and the fourth face horizontal : prove that the pressures on the pegs are proportional to the areas of the corresponding faces.   [Math. Tripos, 1869.]

Work.

62. Let a force P act at a point A of a body in the direction AB and let us suppose the point A to move into any other position A' very near A. Let ϕ be the angle the direction AB of the

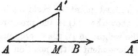

force makes with the direction AA' of the displacement of the point of application, then *the product $P . AA' . \cos \phi$ is called the work done by the force*. If for ϕ we write the angle the direction AB of the force makes with the direction $A'A$ opposite to the displacement, the product is called the work done against the force.

Let us drop a perpendicular $A'M$ on AB; *the work done by the force is also equal to the product $P \cdot AM$, where AM is to be esti-mated positive when in the direction of the force.* Let P' be the resolved part of P in the direction of the displacement; *the work is also equal to $P' \cdot AA'$.* These expressions for the work of a force are clearly equivalent, and all three are in continual use.

63. The forces which act on a particle generally depend on the position of that particle. Thus if the particle be moved from A to any point A' at a *finite distance* from A, the force P will not generally remain the same either in direction or magnitude. For this reason it is necessary to suppose the displacement AA' to be so small a quantity that we may regard the force as fixed in direction and magnitude. Taking the phraseology of the dif-ferential calculus this is expressed by saying that the displacement AA' is of the first order of small quantities.

We may suppose any finite displacement of the point A to be made along a curve beginning at A and ending at some point C. Let ds be any element of this curve, and when the particle has reached this element let P' be the resolved part of the force along ds in the direction in which s is measured. Then by the above definition $\int P'ds$ is the sum of the separate works done by the force P as the particle travels along each element in turn. *This sum is defined to be the whole work in any finite displacement.* If s be measured from any point O on the curve, the limits of this integral will evidently be $s = OA$ and $s = OC$.

64. The resolved displacement $AA' \cos\phi$ is sometimes called the *virtual velocity of the point of application.* The product $P \cdot AA' \cdot \cos\phi$ is called the *virtual moment* or *virtual work* of the force. But these terms are restricted to infinitely small displace-ments. When the displacement is finite, the integral of the virtual works is called the work.

65. It is often convenient to construct a proposed displace-ment by several steps. Thus a displacement AA' may be con-structed by moving A first to D and then from D to A' (see figure in Art. 62). Supposing AD and DA' to be infinitely small so that the direction and magnitude of the force P continue constant throughout, it is easy to see *that the work due to the whole displace-ment AA' is the sum of the works due to the displacements AD and*

DA'. For if we drop the perpendiculars DN and $A'M$ on the direction of the force, the separate works *with their proper signs* will be $P.AN$ and $P.NM$. The sum of these is $P.AM$, which is the work due to the whole displacement AA'.

If the displacement AA' is finite, and the force P remains unaltered in direction and magnitude, the work due to the resultant displacement is equal to the sum of the works due to the partial displacements AD, DA'.

66. Suppose next that several forces act at the point A ; then as A moves to A' each of these will do work. The sum of the works done by each separately is defined to be the work done by all the forces collectively.

If any number of forces act at a point A, the sum of the works due to any small displacement AA' is equal to the work done by their resultant.

The work done by any one force P is equal, by definition, to the product of AA' into the resolved part of P in the direction of AA'. The work done by all the forces is therefore the product AA' into the sum of their resolved parts. By Art. 44 this is equal to AA' into the resolved part of the resultant, i.e. is equal to the work done by the resultant.

67. This theorem leads to another method of stating the conditions of equilibrium of any number of forces P_1, P_2 &c. acting at the same point A.

Case 1. If the particle at A is free to move in all directions it is necessary for equilibrium that the resultant force should vanish. The virtual work of the forces P_1, P_2 &c. must therefore be zero in whatever direction the particle is displaced.

Conversely, if the virtual work for any displacement AA' is zero it immediately follows that the resolved part of the resultant in that direction is also zero. If then the virtual work of P_1, P_2 &c. is zero for any three different displacements not all in one plane, the three resolved parts of the resultant in those directions are zero. The particle is therefore in equilibrium.

68. Case 2. If the particle is constrained to move on some curve or surface, then besides the forces P_1, P_2 &c. the particle is acted on by a pressure R which is normal to the curve or surface. The forces which maintain equilibrium are therefore P_1, P_2 &c.

and R. Then by Case 1 their virtual work is zero for all small displacements.

If the displacement given to A is along a tangent to the curve or is situated in the tangent plane to the surface, the angle ϕ between the reaction R and the displacement is a right angle. The virtual work of that force is therefore zero. It immediately follows that for all such displacements the virtual work of P_1, P_2 &c. is zero.

Conversely, suppose the particle constrained to move on a *curve*; then if the virtual work for a displacement along the tangent is zero the resolved part of the resultant force in that direction is also zero. The particle is therefore in equilibrium.

Next, suppose the particle constrained to move on a *surface*; then if the virtual works for any two displacements, not in the same straight line, are each zero, the resolved parts of the resultant force in those directions are each zero. The particle is therefore in equilibrium.

69. Ex. 1. Deduce from the principle of virtual velocities the conditions of equilibrium obtained in Art. 56, for a particle constrained to rest on a curve.

The forces on the particle are X, Y, Z; the displacement is ds, the projections of ds on the forces are dx, dy, dz. Multiplying each force by the corresponding projection, we see at once that the condition of equilibrium is $Xdx + Ydy + Zdz = 0$.

Ex. 2. Two small smooth rings of equal weight slide on a fixed elliptical wire, of which the axis major is vertical, and are connected by a string passing over a smooth peg at the upper focus; prove that the rings will rest in whatever position they may be placed. [Math. Tripos, 1858.]

Let P, Q be the two rings, W the weight of either. Let T be the tension of the string, l its length. Let S be the peg, let x, x' be the abscissæ of P, Q measured vertically downwards from S; let $r = SP$, $r' = SQ$, then $r + r' = l$. Since the ring P is in equilibrium, we have by the principle of virtual work $Wdx - Tdr = 0$. The positive sign is given to the first term because x is measured in the direction in which W acts; the negative sign is given to the second term because T acts in the opposite direction to that in which r is measured. In the same way we find for the other ring $Wdx' - Tdr' = 0$. Since $dr = -dr'$ this gives as the condition of equilibrium $Wdx + Wdx' = 0$. As yet we have not introduced the condition that the wire has the form of an ellipse. If $2c$ be its latus rectum and e its eccentricity, we have $r = c + ex$, $r' = c + ex'$. It easily follows that $dx + dx' = 0$, so that the condition of equilibrium is satisfied in whatever position the rings are placed.

Ex. 3. A small ring movable along an elliptic wire is attracted towards a given centre of force which varies as the distance: prove that the positions of equilibrium of the ring lie in a hyperbola, the asymptotes of which are parallel to the axes of the ellipse. [Math. Tripos, 1865.]

Ex. 4. Two small rings of the same weight attracting one another with a force varying as the distance, slide on a smooth parabolic shaped wire, whose axis is vertical and vertex upwards: show that if they are in equilibrium in *any* symmetrical position, they are so in *every* one.   [Coll. Ex., 1887.]

Ex. 5. Two mutually attracting or repelling particles are placed in a parabolic groove, and connected by a thread which passes through a small ring at the focus; prove that if the particles be at rest, the line joining the vertex to the focus will be a mean proportional between the abscissæ measured from the vertex. [Math. T. 1852.]

Ex. 6. A weight W is drawn up a rough conical hill of height h and slope α and the path cuts all the lines of greatest slope at an angle β. If the friction be μ times the normal pressure prove that the work done in attaining the summit will be $Wh\,(1 + \mu \cot \alpha \sec \beta)$.   [St John's Coll. 1887.]

Astatic Equilibrium.

70. Suppose that three forces P, Q, R acting *at a point* are in equilibrium. We may clearly turn the forces round that point through any angle without disturbing the equilibrium if only the magnitudes of the forces and the angles between them are unaltered. Since a force may be supposed to act at any point of its line of action these three forces may act at any points A, B, C in their respective initial lines of action. If now we turn the forces supposed to act at A, B, C, each round its own point of application, through the same angle it is clear the equilibrium will be disturbed unless these points are so chosen that the lines of action of the forces continue to intersect in some point (Art. 34).

It is evident that instead of turning the forces round their points of application we may turn the body round any point through any angle. In this case each force preserves its magnitude unaltered, continues to act parallel to its original direction supposed fixed in space, while the point of application remains fixed in the body and moves with it. *When equilibrium is undisturbed by this rotation*, it is called *Astatic*.

71. Let A and B be the points of application of the forces P and Q. Let their lines of action intersect in O. Then as the forces turn round A and B, *in the plane AOB*, the angle between them is to remain unaltered. Hence O will trace out a circle passing through A and B. The resultant of these two forces passes through O and makes constant angles with both OA and OB. It therefore will cut the circle in a fixed point C. This resultant is equal and opposite to the force called R.

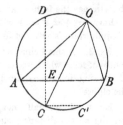

*If therefore three forces P, Q, R, acting at three points A, B, C,
intersect on the circle circumscribing ABC, and be in equilibrium,
the equilibrium will not be disturbed by turning the forces round
their points of application through any angle in the plane of the
forces.* This proof is given in Moigno's *Statics*, p. 228.

If the forces P and Q are parallel, the circle of construction becomes the straight
line AB. The point C lies on AB, and the sines of the angles AOC, BOC are
ultimately proportional to AC and CB. Hence AC is to CB inversely in the ratio
of the forces tending to A and B. If the forces P, Q, besides being parallel, are
equal and opposite, the force R acts at a point on the straight line at infinity.

72. When two forces P_1, P_2 act at given points A, B the
point at which the resultant acts, however the forces are turned
round, is called the *centre of the forces*. If a third force P_3 act at
a third given point C, we may combine the resultant of the first
two with this force and thus obtain a resultant acting at another
fixed point in the body. This is the centre of the three forces.
Thus we may proceed through any number of forces. We see that
we can obtain a single force acting at a fixed point of the body which
is the resultant of any number of given forces acting at any given
fixed points in one plane. This single force will continue to be
the resultant and to act at the same point when all the forces are
turned round their points of application through any angle. This
force is called their *astatic resultant*.

73. Astatic triangle of forces. This proposition leads us to another method
of using the triangle of forces. Referring to the figure of Art. 71, we see that the
angles ABC, AOC and BAC, BOC being angles in the same segment are equal each
to each. If therefore P, Q, R are in equilibrium, they are proportional to the sines
of the angles of the triangle ABC. It follows that P, Q, R are also proportional
to the sides of the triangle ABC. Thus

$$P : BC = Q : CA = R : AB.$$

The points A, B, C divide the circle into three segments AB, BC, CA. If O be
taken on any one of the segments, say AB, then the forces whose lines of action
pass through A and B must act both to or both from A and B. The third force
acts from or to C according as the first two act towards or from A and B. We
deduce the following proposition.

*Let three forces act at the corners of a triangle ABC; they will be in equilibrium
if* (1) *their magnitudes are proportional to the opposite sides,* (2) *their lines of action
meet in any point O on the circumscribing circle,* (3) *their directions obey the rule
given above. Also the equilibrium will not be disturbed by turning all the forces
round their points of application through any, the same angle, but without altering
their magnitudes. The forces are supposed to act in the plane of the triangle.*

74. Ex. 1. Any number of forces P, Q, R, S &c. in one plane are in equilibrium,
and their lines of action meet in one point O. Through O describe any circle

cutting the lines of action of the forces in A, B, C, D &c. If these points are regarded as the points of application of the forces, prove that the equilibrium is astatic.

Ex. 2. If CC' is drawn parallel to the opposite side AB to cut the circle in C', prove that the forces P, Q, R make equal angles with the sides BC', $C'A$, AB of the triangle $BC'A$. Thence deduce from Art. 35 the conditions of equilibrium.

Ex. 3. If a, β are the angles the forces P and Q make with their resultant R, prove that the position of the centres of the forces is given by

$$CE = \frac{AE}{\cot \beta} = \frac{BE}{\cot a} = \frac{AB}{\cot a + \cot \beta},$$

where CED is drawn from C perpendicular to AB.

Ex. 4. Let the forces act from a point O towards A and B where O is on the left or negative side of AB as we look from A towards B. If p, q are the coordinates of A, p', q' of B referred to any rectangular axes, prove that the coordinates of the central point of A and B are given by

$$(\cot a + \cot \beta) x = p \cot a + p' \cot \beta + (q' - q) \Big\}$$
$$(\cot a + \cot \beta) y = q \cot a + q' \cot \beta - (p' - p) \Big\}.$$

If the forces P and Q are at right angles, prove also that

$$(P^2 + Q^2) x = pP^2 + p'Q^2 + (q' - q) PQ \Big\}$$
$$(P^2 + Q^2) y = qP^2 + q'Q^2 - (p' - p) PQ \Big\}.$$

These are obtained by projecting AE, EC on the coordinate axes.

Stable and Unstable Equilibrium.

75. Let us suppose a body to be in equilibrium in any position, which we may call A, under the action of any forces. If the body be now moved into some neighbouring position B and placed there at rest, it may either remain in equilibrium in its new position (as in Art. 71) or the body may begin to move under the action of the forces. In the first case the position A is called one of *neutral equilibrium*. In the second case the equilibrium in the position A is called *unstable* or *stable* according as the body during its subsequent motion does or does not deviate from the position A beyond certain limits. The magnitude of these limits will depend on the circumstances of the case. Sometimes they are very restricted, so that the deviation permitted must be infinitesimal; in other cases greater latitude may be admissible.

The determination of the stability of a state of equilibrium is a dynamical problem. We must according to this definition examine the whole of the subsequent motion to determine the extent of the deviations of the body from the position of equilibrium. But sometimes we may settle this question from statical considerations. If the conditions of the problem are such that for all displacements of the body from the position A within certain limits, the forces tend to bring the body back to that position, then the position may be regarded as stable for displacements within those limits. If on the other hand the forces tend to remove the body further from the position A, that position may be regarded as unstable. This cannot however be strictly proved to be a sufficient condition until we have some dynamical equations at our disposal. Properly we should, for the present, distinguish this as the criterion of statical stability or statical instability. But for the sake of brevity we shall omit this distinction, except when we wish to draw special attention to it.

76. Two equal given forces P, Q act on a body at two given points A, B, and are in equilibrium. They therefore act along the straight line AB. Let the body be now turned round through any angle less than two right angles and let the forces continue to act at these points in directions fixed in space. It is required to find the condition of stability.

Referring to the figure, it is evident that the forces tend to restore the body to its former position if *each force acts from the point of application of the other force*, while they tend to move the body further from that position if *each force acts towards the point of application of the other*. In the first case the equilibrium is stable, in the second unstable.

If the body be turned round through two right angles, the forces will again be in equilibrium. The position of stable equilibrium will then be changed into one of unstable equilibrium and conversely.

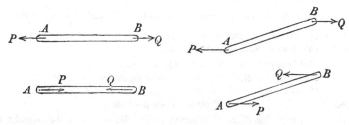

77. Ex. 1. A smooth circular ring is fixed in a horizontal position, and a small ring sliding upon it is in equilibrium when acted on by two forces in the directions of the chords PA, PB. Prove that, if PC be a diameter of the circle, the forces are in the ratio of BC to AC. If A and B be fixed points and the magnitude of the forces remain the same, show that the equilibrium is unstable. [Math. Tripos, 1854.]

Ex. 2. Three given forces P, Q, R act on a body in one plane at three given points A, B, C and are in equilibrium. When the body is disturbed, the forces continue to act at these points parallel to directions fixed in space and their magnitudes are unaltered. Find the condition of stability. See also Art. 221.

In the given position of equilibrium the lines of action of the forces must meet in some point O. If this point lie on the circle circumscribing ABC we know by Art. 71 that the equilibrium is neutral.

Next let the point O lie *within* the segment of the circumscribing circle contained by the angle ACB. Let P and Q act towards A, B while R acts from C towards O.

Describe a circle about OAB cutting OC in C'. Then since O is within the circumscribing circle, C' is without that circle. By Art. 71, the forces P and Q are astatically equivalent to a force equal and opposite to R but acting at C'. Thus the whole system is equivalent to two equal forces acting at C and C' and each tending away from the point of application of the other. The equilibrium is therefore stable for all rotatory displacements less than two right angles. In the same way if the forces P, Q act respectively from A and B towards O the equilibrium is unstable.

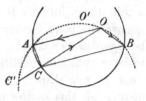

If the point O lie *outside* the circumscribing circle, but within the angle ACB, the point C' is within that circle. The conditions are then reversed, and therefore if the forces P, Q tend from O towards A, B the equilibrium is unstable.

If the point O lie within the triangle ABC, all the three forces must act from O or all the three towards O. By the same reasoning as before we may show that in the former case the equilibrium is stable, in the latter unstable.

Summing up, we have the following result. If two at least of the forces in equilibrium act from the common point of intersection O towards their points of application A, B, C; then the equilibrium is stable if O lie within the circle circumscribing ABC and unstable if O lie outside that circle. If two at least of the forces act from their points of application towards O, these conditions are reversed.

Ex. 3. A particle is in equilibrium at a point O on a smooth surface under the action of forces which have a potential, and Oz is the common normal to the surface of constraint and that level surface which passes through O. The particle being displaced through a small arc $OP = ds$, prove that the resolute F of the force of restitution in the direction of the tangent at P to OP is $F = \left(\dfrac{1}{\rho'} - \dfrac{1}{\rho}\right) Z ds$, where Z is the equilibrium pressure and ρ, ρ' are the radii of curvature of the normal sections of the two surfaces made by the plane zOP.

Let $z = PN$ be a perpendicular on the plane of xy; X', Y', Z' the resolved forces at P, and ϕ the angle xON. Since ds/ρ is the angle the tangent at P to the normal section zOP makes with ON, we have when the squares of small quantities are neglected $\qquad F = -X' \cos \phi - Y' \sin \phi - Z' ds/\rho$, where we may write for Z' its equilibrium value. Since z is of the second order X', Y', at P have the same values as at N; hence the two first terms have the same values for all surfaces which touch the plane at O. But $F = 0$ when the surface is a level surface, hence these terms $= Z ds/\rho'$.

It follows that when the level surface intersects the surface of constraint the equilibrium is stable for some displacements and unstable for others, the separating line being the intersection. If the level surface lies wholly on one side of the surface of constraint, the equilibrium is stable for all displacements or unstable for all.

We suppose that the particle is constrained, either to return to its position of equilibrium by the way it came, or to recede further on that course. The constraining force F' acts perpendicularly to the section zOP, and by considering the angle of torsion at P, we find that its magnitude is $F' = Z ds \sin \phi \cos \phi \left(\dfrac{1}{\rho_1} - \dfrac{1}{\rho_2} - \dfrac{1}{\rho_1'} + \dfrac{1}{\rho_2'}\right)$ where ρ_1, ρ_2; ρ_1', ρ_2' are the principal radii of curvature of the two surfaces.

CHAPTER III.

PARALLEL FORCES.

78. *To find the resultant of two parallel forces.*

Let the two parallel forces be P, Q and let them act at A, B, which of course are any points in their lines of action. In order to obtain a point of intersection of the forces at a finite distance let us impress at A, B in opposite directions two equal forces of any magnitude, each of which we may represent by F, Art. 15. The resultants of P, F and Q, F act respectively along some straight lines AO, BO which intersect in O.

Thus we have replaced the two given forces by two others, each of which may be supposed to act at O. Draw OC parallel to AP, BQ to cut AB in C. Consider the force acting at O along OA. We may resolve this force (as in Duchayla's proof of the parallelogram of forces) into two forces, one equal to P acting along OC and the other equal to F acting parallel to CA. In the same way the other force acting at O along OB is equivalent to Q acting along OC and F acting at O parallel to CB.

The two forces each equal to F balance each other and may be removed. The whole system is therefore reduced to the single force $P + Q$ acting along OC.

The sides of the triangle OCA are parallel to P, F and their resultant. Hence $\dfrac{OC}{CA} = \dfrac{P}{F}$. In the same way $\dfrac{OC}{CB} = \dfrac{Q}{F}$. We therefore have $$\frac{AC}{Q} = \frac{BC}{P} = \frac{AB}{P+Q}.$$

The resultant of the parallel forces P, Q is $P + Q$, and its line of action divides every straight line AB which intersects the forces in the inverse ratio of the forces.

If the forces P, Q act in opposite directions the proof is the same, but the figure is somewhat different. If Q be greater than P, BO will make a smaller angle with the force Q than OA makes

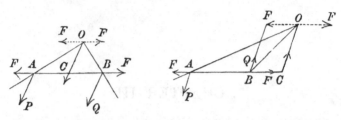

with the force P. Hence O will lie within the angle QBC. In this case the magnitude of the resultant is $Q - P$ and its line of action divides AB externally in the inverse ratio of P to Q.

We also notice that, A, B being any two points in the lines of action of the parallel forces P, Q, the point C through which the resultant acts is the centroid of two particles placed at A and B whose masses are proportional to the forces which act at those points (Art. 53).

79. Conversely *any given force R acting at a given point C may be replaced by two parallel forces acting at two arbitrary points A and B, where A, B, C are in one straight line.* Let us represent these forces by P and Q.

Let $CA = a$, $CB = b$, and let these be regarded as positive when measured from C in the same direction. We then find

$$P + Q = R, \quad P = \frac{b}{b - a} R, \quad Q = \frac{a}{a - b} R.$$

If A and B lie on the same side of C, a and b are positive; in this case the force nearer R acts in the same direction as R, the other force acts in the opposite direction and is therefore negative. If C lie between A and B, one of the two distances a, b is negative; in this case both forces act in the same direction as R.

80. *To find the resultant of any number of parallel forces P_1, P_2 &c. acting at any points A_1, A_2 &c. when referred to any axes.*

Let $(x_1 y_1 z_1)$, $(x_2 y_2 z_2)$ &c. be the Cartesian coordinates of the points A_1, A_2 &c. The forces P_1, P_2 acting at A_1, A_2 are equivalent to a single force $P_1 + P_2$ acting at a point C_1 situated in $A_1 A_2$ such that $P_1 . A_1 C_1 = P_2 . A_2 C_2$ (Art. 78). Let $(\xi_1 \eta_1 \zeta_1)$ be the coordinates of C_1. Since $A_1 C_1$, $A_2 C_1$ are in the ratio of their projections on the axes of coordinates we have

$$P_1 (\xi_1 - x_1) = P_2 (x_2 - \xi_1)$$
$$\therefore (P_1 + P_2) \xi_1 = P_1 x_1 + P_2 x_2.$$

Similar results apply for the other coordinates of C_1.

The force $P_1 + P_2$ acting at C_1 and a third force P_3 acting at A are in the same way equivalent to $P_1 + P_2 + P_3$ acting at a point C_2 whose coordinates $(\xi_2 \eta_2 \zeta_2)$ are given by

$$(P_1 + P_2 + P_3)\,\xi_2 = (P_1 + P_2)\,\xi_1 + P_3 x_3$$
$$= P_1 x_1 + P_2 x_2 + P_3 x_3$$

with similar expressions for η_2 and ζ_2.

Proceeding in this way we see that the resultant of all the forces is $P_1 + P_2 + \ldots$ and if $(\xi\eta\zeta)$ be the coordinates of its point of application, we have

$$(P_1 + P_2 + \&\text{c.})\,\xi = P_1 x_1 + P_2 x_2 + \&\text{c.}$$
$$(P_1 + P_2 + \&\text{c.})\,\eta = P_1 y_1 + P_2 y_2 + \&\text{c.}$$
$$(P_1 + P_2 + \&\text{c.})\,\zeta = P_1 z_1 + P_2 z_2 + \&\text{c.}$$

These equations are usually written

$$\xi = \frac{\Sigma Px}{\Sigma P}, \qquad \eta = \frac{\Sigma Py}{\Sigma P}, \qquad \zeta = \frac{\Sigma Pz}{\Sigma P}.$$

81. It might be supposed that this proof would either fail or require some modification if any one of the partial resultants $P_1 + P_2$, $P_1 + P_2 + P_3$ &c. were zero, for then some of the quantities ξ_1, ξ_2 &c. would be infinite. The final result also might be thought to fail if $\Sigma P = 0$. But any proposition proved true for general values of the forces must be true for these limiting cases, though its interpretation may not be understood until we come to the theory of couples.

We may avoid this apparent difficulty by a slight modification of the proof. Let us separate the forces which act in one direction from those which act in the opposite direction, thus forming two groups. Let us suppose the sums of the forces in the two groups are unequal. If we compound together first all the forces in that group in which the sum is greatest and then join to these one by one the forces of the other group, it is clear that we shall never have any of the partial resultants equal to zero and no point of application of any such partial resultant will be at infinity. If the sums of the forces in the two groups are equal, the centre of parallel forces is infinitely distant.

82. The expressions for the coordinates $(\xi\eta\zeta)$ are the same as those given in Art. 51 for the coordinates of the centroid; we therefore deduce the following rule.

To find the resultant of the parallel forces P_1, P_2 &c. we select convenient points A_1, A_2 &c. on their respective lines of action and place at these points particles whose masses are proportional to the forces P_1, P_2 &c. The line of action of the resultant passes through the centroid of these particles, its direction is parallel to that of the forces, and its magnitude is ΣP.

Conversely, any given force can be replaced by parallel forces acting at arbitrary points A_1, A_2 &c. provided the forces are such that the centroid lies on the given force.

This proposition is really the limiting case of Leibnitz's theorem. If concurrent forces act along OA_1, OA_2 &c. their resultant may be found by any of the methods considered in the last chapter. By regarding O as a point very distant from A_1, A_2 &c., the forces acting along OA_1, OA_2 &c. become parallel and the corresponding theorem follows at once. Thus in Art. 51 it is shown that the resultant of forces proportional to $P_1 . OA_1$, $P_2 . OA_2$ &c. is a force proportional to $\Sigma P . OC$ acting along OC where C is the centroid of particles P_1, P_2 &c. placed at A_1, A_2 &c. In the limit OA, OB, OC are all equal; hence the resultant of parallel forces proportional to P_1, P_2 &c. is proportional to ΣP and acts at C.

83. The point $(\xi \eta \zeta)$ determined by the equations of Art. 80 has one important property. Its position is the same whatever be the magnitudes of the angles made by the forces with the coordinate axes. If then *the points of application of the given parallel forces viz. A_1, A_2 &c. are regarded as fixed in the body, the point of application of their resultant is also fixed in the body however the forces are turned round their points of application provided they remain parallel and unaltered in magnitude.*

This point of application of the resultant is called the " *centre of parallel forces.*"

84. Ex. 1. Parallel forces, each equal to P, act at the corners A, B, C, D of a re-entrant plane quadrilateral and a fifth force equal to $-P$ acts at the intersection H of the diagonals HCA, BHD. If the centre of the five parallel forces coincide with a corner C of the quadrilateral, prove that $HC = CA$.

Ex. 2. ABC is a triangle; APD, BPE, CPF, the perpendiculars from A, B, C on the opposite sides. Prove that the resultant of six equal parallel forces, acting at the middle points of the sides of the triangle and of the lines PA, PB, PC, passes through the centre of the circle which goes through all of these middle points.

[Math. Tripos, 1877.]

Ex. 3. $ABCD$ is a quadrilateral whose diagonals intersect in O. Parallel forces act at the middle points of AB, BC, CD, DA respectively proportional to the areas AOB, BOC, COD, DOA. Prove that the centre of parallel forces is at the fourth angular point, viz. G, of the parallelogram described on OE, OF as adjacent sides where E, F are the middle points of the diagonals AC, BD of the quadrilateral. [Coll. Ex., 1885.]

Taking BD as the axis of x we find $\eta = \frac{1}{2}(p - p')$ where p, p' are the perpendiculars from A and C on BD. It follows that the centre of parallel forces lies on EG. Similarly it lies on FG.

85. *To find the conditions of equilibrium of a system of parallel forces.*

Let the forces be $P_1, \ldots P_n$; then by Art. 80 they will have a resultant unless $\Sigma P = 0$. This, though a necessary condition of equilibrium, is not sufficient.

We can find the resultant of $n-1$ of the forces by Art. 80 without introducing any forces whose lines of action are at infinity, because the sum of these $n-1$ forces is equal to $-P_n$ and therefore is not zero. It is sufficient for equilibrium that the point of application of this resultant should be situated on the line of action of P_n.

Let $(\xi\eta\zeta)$ be the coordinates of that point of application of this resultant which is found in Art. 80, then

$$\xi = \frac{P_1 x_1 + \dots + P_{n-1} x_{n-1}}{P_1 + \dots + P_{n-1}}$$

with similar expressions for η and ζ. Let $(\alpha\beta\gamma)$ be the direction angles of the forces.

Since $\xi - x_n$, $\eta - y_n$, $\zeta - z_n$ are the projections on the axes of the straight line joining the point $(\xi\eta\zeta)$ to the point of application of the force P_n, viz. $(x_n y_n z_n)$, we have

$$\frac{\xi - x_n}{\cos\alpha} = \frac{\eta - y_n}{\cos\beta} = \frac{\zeta - z_n}{\cos\gamma}$$

Substituting for $(\xi\eta\zeta)$ and remembering that the denominator of ξ is equal to $-P_n$, this reduces to

$$\frac{\Sigma Px}{\cos\alpha} = \frac{\Sigma Py}{\cos\beta} = \frac{\Sigma Pz}{\cos\gamma} \quad\dots\dots\dots\dots\dots(1).$$

Joining these two equations to the condition $\Sigma P = 0$, we have *the three necessary and sufficient conditions of equilibrium.*

If the equilibrium is to exist however the forces are turned round their points of application, the point of application of the resultant of the first $n-1$ forces as found by Art. 80 must coincide with the given point of application of the force P_n. We have therefore

$$\xi = x_n, \qquad \eta = y_n, \qquad \zeta = z_n.$$

These give $\qquad \Sigma Px = 0, \qquad \Sigma Py = 0, \qquad \Sigma Pz = 0 \dots\dots\dots\dots(2).$

Joining these three equations to $\Sigma P = 0$ we have *the four necessary and sufficient conditions that a system of parallel forces should be astatically in equilibrium.*

86. Ex. 1. Prove that any system of parallel forces can be replaced by three parallel forces acting at the corners of an arbitrary triangle ABC.

Let P be any one of the forces, intersecting the plane of the triangle in a point whose areal coordinates are x, y, z, Art. 53, Ex. 2. We may replace P by the parallel forces Px, Py, Pz, acting at the corners, Art. 82. All the forces are therefore equivalent to ΣPx, ΣPy, ΣPz acting at A, B, C, respectively.

Ex. 2. If four parallel forces balance each other, let their lines of action be intersected by a plane, and let the four points of intersection be joined by six straight lines so as to form four triangles; each force will be proportional to the area of the triangle whose corners are in the lines of action of the other three.

[Rankine's Applied Mathematics, Art. 143.]

87. *A heavy body is suspended from a fixed point without any other constraint. It is required to find the position of equilibrium.*

The body is in equilibrium under the action of the weights of all its elements and the reaction at the point of support. The weights of the elements form a system of parallel forces and are equivalent to the whole weight of the body acting vertically downwards at the centre of gravity. It easily follows that *in equilibrium, the centre of gravity must be vertically under the point of support.* It is also clear that the pressure on the point of support is equal to the weight of the body.

In applying this principle to examples, the positions of the centres of gravity of the elementary bodies are assumed to be known. The positions of these points will be stated as they are required. If the reader is not already acquainted with them, he may either assume the results given or refer to the chapter on the centre of gravity where their proofs may be found.

Ex. 1. A uniform triangular area ABC is suspended from a fixed point O by three strings attached to its corners. Prove that the tensions of the strings are proportional to their lengths.

To find the centre of gravity G of the triangle ABC, we draw the median line AM bisecting BC in M. Then G lies in AM, so that $AG = \frac{2}{3}AM$.

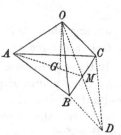

The three tensions acting along AO, BO, CO and the weight acting along OG are in equilibrium. The resultant of the tension AO and the weight is therefore equal and opposite to that of the tensions BO, CO. Since each resultant acts in the plane of the forces of which it is the resultant, their common line of action is OM.

Draw through B and C parallels to OC and OB, and let D be their point of intersection. Then, since OM bisects BC, OM passes through D. Hence the sides of the triangle OCD are parallel to the tensions CO, BO and their resultant. The tensions are therefore proportional to OC, CD, i.e. to OC, OB.

Another proof may be deduced from Art. 51. The centre of gravity of the triangular area coincides with the centre of gravity of three equal weights placed one at each corner. The components along OA, OB, OC of the force represented by $3 \cdot OG$ are therefore represented by the lengths of those lines.

Ex. 2. A heavy triangle ABC is hung up by the angle A, and the opposite side is inclined at an angle a to the horizon. Show that $2\tan a = \cot B \sim \cot C$.

[Math. Tripos, 1865.]

Ex. 3. Two uniform heavy rods AB, BC are rigidly united at B, the rods are then hung up by the end A: show that BC will be horizontal if $\sin C = \sqrt{2} \sin \frac{1}{2}B$, B and C being angles of the triangle ABC.

[Coll. Ex., 1883.]

Ex. 4. A heavy equilateral triangle, hung up on a smooth peg by a string, the ends of which are attached to two of its angular points, rests with one of its sides vertical; show that the length of the string is double the altitude of the triangle.
[Math. Tripos, 1857.]

Ex. 5. A piece of uniform wire is bent into three sides of a square $ABCD$, of which the side AD is wanting; prove that if it be hung up by the two points A and B successively, the angle between the two positions of BC is $\tan^{-1} 18$.

The distance of the centre of gravity G from BC can be shown to be equal to one third of AB. When hung up from A and B, AG and BG respectively are vertical. The angle required is therefore equal to AGB.   [Math. Tripos, 1854.]

Ex. 6. A triangle ABC is successively suspended from A and B, and the two positions of any side are at right angles to each other; prove that $5c^2 = a^2 + b^2$.
[Coll. Ex.]

Ex. 7. A uniform circular disc of weight nW has a heavy particle of weight W attached to a point on its rim. If the disc be suspended from a point A on its rim, B is the lowest point; and if suspended from B, A is the lowest point. Show that the angle subtended by AB at the centre is $2 \sec^{-1} 2 (n+1)$.   [Math. Tripos, 1883.]

Ex. 8. The altitude of a right cone is h and the radius of its base is r; a string is fastened to the vertex and to a point on the circumference of the circular base and is then put over a smooth peg: prove that if the cone rests with its axis horizontal the length of the string is $\sqrt{(h^2 + 4r^2)}$.   [Math. Tripos, 1865.]

If V be the vertex and C the centre of gravity of the base of a cone (either right or oblique), the centre of gravity of the solid cone lies in VC, so that $VG = \frac{3}{4}VC$.

Ex. 9. A string nine feet long has one end attached to the extremity of a smooth uniform heavy rod two feet in length, and at the other end carries a light ring which slides upon the rod. The rod is suspended by means of the string from a smooth peg; prove that if θ be the angle which the rod makes with the horizon, then $\tan \theta = 3^{-\frac{1}{3}} - 3^{-\frac{2}{3}}$.   [Math. Tripos, 1852.]

Ex. 10. A heavy uniform rod of length $2a$ turns freely on a pivot at a point in it, and suspended by a string of length l fastened to the ends of the rod hangs a bead of equal weight which slides on the string. Prove that the rod cannot rest in an inclined position unless the distance of the pivot from the middle point of the rod be less than a^2/l.   [Math. Tripos, 1882.]

Ex. 11. Two equal rods AB, BC of length $2a$ are connected by a free hinge at B; the ends A and C are connected by an inextensible string of length l: the system is suspended from A: prove that, in order that the angle AB makes with the vertical may be the greatest possible, l must be equal to $4a/\sqrt{3}$.   [St John's Coll., 1883.]

As l is varied the centre of gravity G of the system moves along the circle described on BE as diameter, where E is the middle point of AB. Hence the angle GAB is greatest when AG is a tangent to this circle.

Ex. 12. At the angular points A, B, C of a light rigid frame-work, three heavy particles of weights W_A, W_B, W_C are fixed and the whole is suspended from a point O by three strings OA, OB, OC; if the tensions in equilibrium be T_A, T_B, T_C respectively, prove that $\dfrac{T_A}{OA . W_A} = \dfrac{T_B}{OB . W_B} = \dfrac{T_C}{OC . W_C}$, and hence determine T_A, T_B, T_C.   [St John's Coll., 1886.]

Ex. 13. A heavy triangular lamina is suspended from a fixed point by means of three elastic strings attached to its angular points: the strings when unstretched are equal in length, but the moduli of their elasticities are different. Assuming that the tension of each is equal to the modulus multiplied by the ratio of the extension to the unstretched length, prove that the strings will be equal, if a weight be placed at a certain point on the lamina, provided the weight be not less than a certain weight: prove also that the locus of its position for different magnitudes of the weight, is a straight line. [Coll. Ex., 1887.]

Ex. 14. A uniform circular disc, whose weight is w and radius a, is suspended by three vertical strings attached to three points on the circumference of the disc separated by equal intervals. A weight W may be put down anywhere within a concentric circle of radius ma; prove that the strings will not break if they can support a tension equal to $\frac{1}{3}(2mW + W + w)$. [Trin. Coll., 1886.]

Ex. 15. A right circular cone rests with its elliptic base on a smooth horizontal table. A string fastened to the vertex and the other end of the longest generator passes round a smooth pulley above the cone, so that all parts of the string except those in contact with the pulley are vertical. If the string become gradually contracted by dampness or other causes and tend to lift the cone, show that the end of the shortest generator will remain in contact with the table provided that the diameter of the pulley be less than three times the semi-major axis of the elliptic base. [Math. Tripos, 1878.]

88. A heavy body is placed on either a smooth horizontal plane or a rough inclined plane, and its base is any polygonal area. Determine whether it will tumble over one side or remain in equilibrium.

The weights of the particles of the body constitute a system of parallel forces. These have a resultant whose position and magnitude may be found by the theorem of Art. 80 when the weights of the particles are known. This resultant acts vertically downwards through a point of the body called its centre of gravity. If equilibrium exists, this must be balanced by the pressures of the plane on the body. These pressures however distributed over the polygonal area must have a resultant which acts at some point within the polygonal area. It follows that equilibrium cannot exist unless the vertical through the centre of gravity of the body intersects the plane within the area of the base.

Ex. 1. The distance between the heels of a man's feet is $2b$, and the length of each foot is a. As the body sways, the vertical through the centre of gravity should always pass through the area contained by the feet. The toes should therefore be turned out at such an angle that the area contained by the feet is a maximum. Show (1) that a circle can be described about the feet with its centre on the straight line joining the toes, (2) that its diameter is $b + (b^2 + 2a^2)^{\frac{1}{2}}$.

Ex. 2. A heavy right cone whose height is h and semi-angle a is placed with its base on a perfectly rough plane; prove that the cone will tumble over the rim of its base if the angle θ at which the plane is inclined to the horizon is greater than that given by $\tan \theta = 4 \tan a$.

Ex. 3. A hemispherical cup of weight W is loaded by two weights w, w' attached to its rim and is then placed on a smooth horizontal plane; show that the angle θ which the principal radius of the cup makes with the vertical when the cup is in equilibrium is given by the equation

$$W \tan \theta = 2 \left\{ (w - w')^2 + 4ww' \cos^2 \beta \right\}^{\frac{1}{2}},$$

where 2β is the angle between the radii through the weights w, w', and it is assumed that the centre of gravity of the cup is at the middle point of its principal radius.

[King's Coll., 1889.]

Ex. 4. Two equal heavy particles are at the extremities of the latus rectum of a parabolic arc without weight, which is placed with its vertex in contact with that of an equal parabola, whose axis is vertical and concavity downwards. Prove that the parabolic arc may be turned through any angle without disturbing the equilibrium, provided no sliding be possible between the curves.

[Watson's Problem, Math. Tripos, 1860.]

Theory of Couples.

89. There is one case in which the theorem of Art. 80 leads to a remarkable result. Let us suppose that the parallel forces P, Q are equal and act in opposite directions. According to the theorem the magnitude of the resultant is zero, and the point of application is infinitely distant.

Two equal and opposite forces acting at two points A and B cannot balance each other unless these points are in the same straight line with the forces. Yet we have just seen that these two forces are not equivalent to any one single force at a finite distance. They therefore supply a new method of analysing forces. When a number of forces act on a body we simplify the system by reducing the forces to as few as we can. Sometimes we can reduce them to a single force acting at some point of the body. In other cases (as in the case considered in this article) the point of application is at infinity and the reduction to a single force is no longer convenient. By using a couple of equal forces, as a new elementary term, we obtain a simple method of expressing this infinitely distant force. We now have two elementary quantities, viz. a force and a couple. It may be possible to reduce a given system of forces to either or both of these constituents. With the help of both these, we may analyse a system of forces with greater completeness than with one alone.

If we regard a couple as a new element in analysis, it becomes necessary to consider the properties of such an element apart from all other combinations of forces. Since a couple can itself be

4—2

analysed into two forces we can deduce the properties of a couple
from those which belong to a combination of forces. No new axiom
is necessary in addition to those already given in the beginning of
this treatise. We proceed in the following articles to investigate
the elementary properties of a couple.

The theory of couples is due to Poinsot. In his *Elements of Statics* published
in 1803 he discusses the composition of parallel forces and deduces his new theory
of couples. On this theory he founds the general laws of equilibrium.

90. *Definitions.* A system of two equal and parallel forces
acting in opposite directions is called a *couple.*

The perpendicular distance between these two forces is called
its *arm.* It should be noticed that the arm of a couple has length,
but has no definite position in space. From any point A in the
line of action of one force, a perpendicular AB can be drawn on the
other force. Then AB is the arm. If in any case it is convenient
to regard the forces as acting at A and B, then we might regard
AB, if perpendicular to the forces, as representing the arm in
position as well as in length.

The product of the magnitude of either force into the length
of the arm is called the *moment of the couple.*

91. *The effect of a couple is not altered if it be moved parallel
to itself to any other position in its own plane or in a parallel
plane, the arm remaining parallel to itself.*

Let P, Q be the equal forces of the given couple, AB its arm.
Let $A'B'$ be equal
and parallel to AB,
we shall prove that
the couple may be
moved so that the
same forces act at
A', B'.

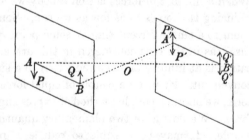

At each of the
points A', B' apply
two equal and opposite forces, each force being equal in magnitude
to P. These are represented in the figure by P', P'', Q', Q''.
Then because AB is equal and parallel to $A'B'$, $AA'BB'$ is a
parallelogram and therefore the diagonals AB', $A'B$ bisect each
other in some point O. The resultant of the forces P and Q'' is
$2P$ acting at O, the resultant of P'' and Q is $2P$ also acting at O,

but in the opposite direction. These two resultants neutralise each other. Removing them, the whole system of forces is equivalent to the couple of forces, which act at A' and B'.

92. *The effect of a couple is not altered by turning the whole couple through any angle in its own plane about the middle point of any arm.*

Let the arm AB be turned round its middle point C and let it take any position $A'B'$. At each of the points A', B' apply as before equal and opposite forces P', P'', Q', Q'', each force being equal to P. The equal forces P and P'' acting at A and A' have a resultant which acts along CE and bisects the angle ACA'. The forces Q and Q'' have an equal resultant which acts along CF and bisects the angle BCB'. These neutralise each other and may be removed. The forces remaining are the equal forces P', Q' acting

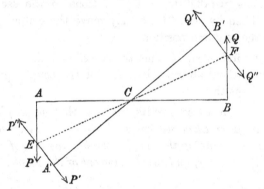

at A', B'. These together constitute a couple, which is the same as the original couple except that it has been turned round C through the angle ACA'.

93. *The effect of a couple is not altered if we replace it by another couple having the same moment, the plane remaining the same, the arms being in the same straight line and their middle points coincident.*

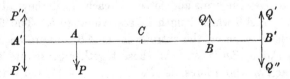

Let P, Q be the equal forces, AB the arm of the given couple. Let $A'B'$ be the new arm, P', Q' the new forces. Apply at each of the points A', B' equal and opposite forces, each equal to P'. Then by the conditions of the proposition, $P.AB = P'.A'B'$. Hence if C be the middle point of both AB and $A'B'$, we have $P.AC = P'.A'C$.

The forces P and P'' have a resultant $P - P''$ which by Art. 78 acts at C. In the same way Q and Q'' have an equal resultant, also acting at C in the opposite direction. Removing these two, it follows that the given couple is equivalent to the couple of forces $\pm P'$ acting at A', B'.

94. It follows from Arts. 91 and 92 that a couple may be transferred without altering its effect from one given position to any other given position in a parallel plane. Thus by Art. 92 we may turn a couple round the middle point of its arm until the forces become parallel to their directions in the second given position. Then by Art. 91 we may move the couple parallel to itself into the required position.

It follows from Art. 93 that the forces and the arm may also be changed without altering the effect of the couple, provided its moment is kept the same.

Summing up these results, we see that *a couple is to be regarded as given when we know,* (1) *the position of some plane parallel to the plane of the couple,* (2) *the direction of rotation of the couple in its plane, and* (3) *the moment of the couple.*

95. *To find the resultant of any number of couples acting in parallel planes.*

Let P_1, P_2 &c. be the magnitudes of the forces, a_1, a_2 &c. the arms of the couples. Let us first suppose the couples all tend to produce rotation in the same direction.

By Art. 94 we may move these couples into one plane and turn them about until their arms are in the same straight line. We may then alter the arms and forces of each until they all have a common arm AB whose length is, say, equal to b. The forces of the couples now act at the extremities of AB, and are respectively equal to P_1a_1/b, P_2a_2/b &c. All these together constitute a single couple each of whose forces is $(P_1a_1 + P_2a_2 + \&c.)/b$ and whose arm is b. This single couple is equivalent to any other couple in the same plane with the same direction of rotation whose moment is

$P_1a_1 + P_2a_2 +$ &c., i.e. whose moment is the sum of the moments of the separate couples.

If some of the couples tend to produce rotation in the opposite direction to the others, we may represent this by regarding the forces of these couples as negative. The same result follows as before.

We thus obtain the following theorem; *the resultant of any number of couples whose planes are parallel is a couple whose moment is the algebraic sum of the moments of the separate couples and whose plane is parallel to those of the given couples.*

96. Measure of a couple. We may use the proposition just established to show that the magnitude of a couple regarded as a single element is properly measured by its moment. To prove this we assume as a unit the couple whose force is the unit of force and whose arm is the unit of length. The moment of this couple is unity. By this proposition a couple whose moment is n times as great is equivalent to n such couples and its magnitude is therefore properly represented by the symbol n.

97. Axis of a couple. A couple may tend to produce rotation in one direction or the opposite according to the circumstances of the couple. One of these is usually called the positive direction and the other the negative. Just as in choosing axes of coordinates sometimes one direction is taken as the positive one and sometimes the other, so in couples the choice of the positive direction is not always the same. In trigonometry the direction of rotation opposite to the hands of a watch is taken as the positive direction. In most treatises on conics the same choice is made. In solid geometry the opposite direction is generally chosen. Having however chosen one of these two directions as the positive one it is usual to indicate the direction of rotation of a given couple in the following manner.

From any point C in the plane of the couple draw a straight line CD at right angles to the plane and on one side of it. The straight line is to be so drawn that if an observer stand with his feet at C on the plane and his back along CD, the couple will appear to him to produce rotation in what has been chosen as the positive direction. The straight line CD is called the *positive direction of the axis of the couple.*

To indicate the direction of rotation of a couple it is sufficient to give the direction in space of CD as distinguished from DC. This is effected by the convention usually employed in solid geometry. A finite straight line having one extremity at the origin of coordinates is drawn parallel to CD. The position of this straight line is defined by the angles it makes with the *positive directions* of the axes of coordinates.

The position of the straight line CD, when given, indicates at once the plane of the couple and the direction of rotation. We may also use a length measured along CD to represent the magnitude of the moment of the couple, in just the same way as a straight line was used in Art. 7 to represent the magnitude of a force.

We therefore infer that all the circumstances of a couple may be properly represented by a finite straight line measured from a fixed point in a direction perpendicular to its plane. This finite straight line is called the *axis of the couple*.

98. *To find the resultant of two couples whose planes are inclined to each other.*

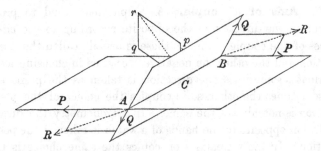

Let the two couples be moved, each in its own plane, until they have a common arm AB, which of course must lie in the intersection of the two planes. In effecting this change of arm it may have been necessary to alter the forces of the couples, but the moments of the couples must remain unaltered. Let the forces thus altered be P and Q.

At the point A we have two forces P and Q; these are equivalent to some resultant R found by the parallelogram of forces. At the point B there are two forces equal and opposite to those at A; their resultant is equal, parallel and opposite to R. Thus the two couples are equivalent to a single couple, each of whose forces is equal to R, and whose arm is AB.   Let the length of AB be b.

From any point C (which we may conveniently take in AB) draw Cp, Cq in the directions of the axes of the given couples, and measure lengths along them proportional to their moments, viz. to Pb and Qb.   These axes are perpendicular to the planes of the couples, and their lengths are also proportional to P and Q.   If we compound these two by the parallelogram law we evidently obtain an axis perpendicular to the plane of the forces $\pm R$, whose length is proportional to R.   It is evident that the parallelogram $Cpqr$ is similar to that contained by the forces PQR, but the sides of one parallelogram are perpendicular to the sides of the other.

We therefore infer the following construction for the resultant of any two couples.   *Draw two finite straight lines from any point C to represent the axes of the couples in direction and magnitude. The resultant of these two obtained by the parallelogram law represents in direction and magnitude the axis of the resultant couple.*

The rule to compound couples is therefore the same as that already given for compounding forces.   It follows that all the theorems for compounding forces deduced from the parallelogram law also apply to couples.   The working rule is that *if we represent the couples by their axes, we may compound and resolve these as if they were forces acting at a point.*

99. Ex. 1. A system of couples is represented in position and magnitude by the areas of the faces of a polyhedron, and their axes are turned all inwards or all outwards. Show that they are in equilibrium. Art. 47.   *Möbius.*

Ex. 2. Four straight lines are given in space, prove that four couples can be found, having these for the directions of their axes, which are in equilibrium. Find also their moments and discuss the case in which three of the given straight lines are parallel to a plane, Arts. 40, 48.

Ex. 3. Three couples are represented in position and magnitude by the areas of three faces OBC, OCA, OAB of the tetrahedron $OABC$, the axes of the first two being turned inwards and that of the third outwards. Prove that the resultant couple acts in the plane ODE bisecting the sides BC, CA and is represented by four times the area of the triangle ODE.

Replace each couple by another one of whose forces passes through O and the other acts along a side of ABC. The forces represented by BC, CA and BA have evidently a resultant $4DE$.

100. *A force P acting at any point A may be transferred parallel to itself, to act at any other point B, by introducing a couple*

*whose moment is Pp, where p is the perpendicular distance of B
from the line of action AF of P.   This couple acts to turn the body
in the direction AFB.*

Apply at B two equal and opposite forces P', P'', each equal to
P. One of these, viz. P',
is the force P transferred
to act at B. The two
forces P'' and P then constitute the couple whose moment is Pp.

101. Summing up the various propositions just proved on
forces and couples, we find that they fall into three classes. These
may be briefly stated thus :

1. Forces may be combined together according to the parallelogram law.

2. Couples may be combined together according to the parallelogram law.

3. A force is equivalent to a parallel force together with a
couple.

The theorems in the subsequent chapters are obtained by
continual applications of these three classes of propositions. It
is therefore evident that theorems thus obtained will apply also to
any other vectors for which these three classes of propositions are
true. Thus in dynamics we find that the elementary relations of
linear and angular velocities are governed by these three sets of
propositions. We therefore apply to these, without further proof,
all the theorems found to be true for couples and forces.

102. Initial motion of the body. If a single couple act on a body at rest, it
is clear that the body will not remain in equilibrium. It is proved in treatises on
dynamics that the body will begin to turn about a certain axis. Since a couple can
be moved about in its own plane without altering its effect, this axis cannot depend
on the position of the couple in its plane. The dynamical results are (1) the initial
axis of rotation passes through the centre of gravity of the body, (2) the axis of
rotation is not necessarily perpendicular to the plane of the couple, though this
may sometimes be the case. The construction to find the axis is somewhat
complicated, and its discussion would be out of place in a treatise on statics.

We may show by an elementary experiment that the axis of rotation is
independent of the position of the couple in its plane. Let a disc of wood be
made to float on the surface of water contained in a box. At any two points
A, B attach to the disc two fine threads and hang these over two small pullies,
fixed in the sides of the vessel at C and D, with equal weights suspended at

the other extremities. Let the strings AC, BD be parallel so that their tensions form a couple. Under the influence of this couple the body will begin to turn round. However eccentrically the points A, B are situated the body *begins* to turn round its centre of gravity. The body may not *continue* to turn round this axis for, as the body moves, the strings cease to be parallel. For this and other reasons the motion of rotation is altered.

103. Ex. 1. Forces P, $2P$, $4P$, $2P$ act along the sides of a square taken in order; find the magnitude and position of their resultant.   [St John's, 1880.]

Ex. 2. A triangular lamina ABC is moveable in its own plane about a point in itself: forces act on it along and proportional to BC, CA, BA. Prove that if these do not move the lamina, the point must lie in the straight line which bisects BC and CA.   [Math. Tripos, 1874.]

Ex. 3. Forces are represented in magnitude, direction, and position by the sides of a triangle taken in order; prove that they are equivalent to a couple whose moment is twice the area of the triangle.

If the sides taken in order represent the axes of three couples, prove that these couples are in equilibrium.

Ex. 4. If six forces acting on a body be completely represented three by the sides of a triangle taken in order and three by the sides of the triangle formed by joining the middle points of the sides of the original triangle, prove that they will be in equilibrium if the parallel forces act in the same direction and the scale on which the first three forces are represented be four times as large as that on which the last three are represented.   [Math. Tripos.]

Ex. 5. Four forces $a.AB$, $\beta.BC$, $\gamma.CD$, $\delta.DA$ act along the sides AB, BC, CD, DA of a skew quadrilateral $ABCD$; show that (1) they cannot be in equilibrium, (2) if $a=\beta=\gamma=\delta$ they form a single couple whose plane is parallel to the diagonals AC, BD, (3) if $a\gamma=\beta\delta$ they reduce to a single resultant whose line of action intersects the diagonals. Find also the magnitudes of the couple and resultant·   [Coll. Ex., 1892.]

The forces at the corners B and D have respectively resultants acting along some lines BE, DF cutting AC in E and F. Since the planes ABC, ADC do not coincide, these two partial resultants cannot act in the same straight line, and therefore cannot be in equilibrium.

If the forces are equivalent to a couple, the sum of their resolved parts along the perpendicular from B on the plane ADC is zero. This requires BE to be parallel to AC and gives $a=\beta$; similarly $\beta=\gamma$ and $\gamma=\delta$. The partial resultants at B and D are $\pm a.AC$, and act parallel to AC and CA. The plane of the couple is therefore parallel to AC, similarly it is parallel to BD. The moment of the couple is $4a$ times the area of the parallelogram whose vertices are the middle points of the sides.

If the forces are equivalent to a single resultant the points E and F on AC must coincide; but E is the mean centre of $-a$ and β at A and C, while F is the mean centre of δ and $-\gamma$ at the same points, Art. 51, hence $a\gamma=\beta\delta$. The partial resultants now intersect in the point E on the diagonal AC and are represented by $(a-\beta)EB$ and $(\gamma-\delta)ED$. The single resultant therefore passes through E and a point H on the other diagonal BD and its magnitude is $(a-\beta+\gamma-\delta).EH$.

If the quadrilateral is plane the four forces are equivalent to a single resultant except when a, β, γ, δ are equal. The forces are in equilibrium when the partial resultants are equal and opposite, i.e. when

$$a\gamma = \beta\delta, \quad a \cdot AO + \beta \cdot OC = 0, \quad \beta \cdot BO + \gamma \cdot OD = 0,$$

where O is the intersection of the diagonals.

Ex. 6. Forces are represented in magnitude, direction, and position by the sides of a skew polygon taken in order; show that they are equivalent to a couple.

If the corners of the skew polygon are projected on any plane, prove that the resolved part of the resultant couple in that plane is represented by twice the area of the projected polygon.

Ex. 7. AC, BD are two non-intersecting straight lines of constant length; prove that the effect of forces represented in every respect by AB, BC, CD, DA is the same, so long as AC, BD remain parallel to the same plane, and the angle between their projections on that plane is constant. [Coll. Ex., 1881.]

Ex. 8. If two equal lengths Aa, Bb, are marked off in the same direction along a given straight line, and two equal lengths Cc, Dd along another given line, prove that forces represented in every respect by AC, ca, CB, bc, BD, db, DA, ad are in equilibrium. [Trin. Coll.]

Ex. 9. Forces proportional to the sides a_1, a_2 ... of a closed polygon act at points dividing the sides taken in order in the ratios $m_1 : n_1$, $m_2 : n_2$, ... and each makes the same angle θ in the same sense with the corresponding side; prove that there will be equilibrium if $\Sigma \left(\dfrac{m-n}{m+n} a^2 \right) = 4\Delta \cot\theta$, where Δ is the area of the polygon. [Math. Tripos, 1869.]

Resolve each force along and perpendicular to the corresponding side and transfer the latter component to act at the middle point by introducing a couple, Art. 100. The couples balance the components along the sides, Ex. 3. The other components are in equilibrium, Art. 37.

CHAPTER IV.

104. *To find the resultant of any number of forces which act on a body in one plane, i.e. to reduce these forces to a force and a couple.*

Let the forces P_1, P_2 &c. act at the points A_1, A_2 &c. of the body. Let O be any point arbitrarily chosen in the plane of the forces, it is proposed to reduce all these forces to a single force acting at O and a couple.

Let the point O be taken as the origin of coordinates. Let the coordinates of A_1, A_2 &c. be $(x_1 y_1)$, $(x_2 y_2)$ &c. Let the directions of the forces make angles α_1, α_2 &c. with the positive side of the axis of x.

Referring to Art. 100 of the chapter on parallel forces, we see that any one of these forces as P may be transferred parallel to itself, to act at the point O, by introducing into the system a couple whose moment is Pp, where p is the length of the perpendicular ON drawn from O on the line of action of the force P. In this way all the given forces P_1, P_2 &c.

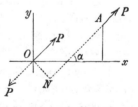

may be transferred to act at O parallel to their original directions, provided we introduce into the system the proper couples.

These forces, by Art. 44, may be compounded together so as to make a single resultant force. The couples also may be added together with their proper signs so as to make a single couple whose moment is ΣPp.

This method of compounding forces is due to Poinsot (*Éléments de Statique,* 1803).

105. It should be noticed that the argument in Art. 104 is in no way restricted to forces in two dimensions. If we refer the system to three rectangular axes Ox, Oy, Oz, having an arbitrary origin O, we may transfer the forces P_1, P_2 &c. to the point O by introducing the proper couples. The forces acting at O may be compounded into a single force, which we may call R. The couples also may be compounded, by help of the parallelogram of couples, into a single couple which we may call G. Thus the forces P_1, P_2 &c. can always be reduced to a single force R acting at an arbitrary point, together with the appropriate couple G.

106. To find the magnitude and the line of action of the resultant force we follow the rules given in Art. 44. The resolved parts of the resultant force parallel to the axes are

$$X = \Sigma P \cos \alpha, \qquad Y = \Sigma P \sin \alpha.$$

Let R be the resultant force, and let θ be the angle which its line of action makes with the axis of x, then

$$R^2 = (\Sigma P \cos \alpha)^2 + (\Sigma P \sin \alpha)^2, \qquad \tan \theta = \frac{\Sigma (P \sin \alpha)}{\Sigma (P \cos \alpha)}.$$

107. To find the moment of the resultant couple, we must find the value of Pp. By projecting the coordinates (xy) of A on ON we have $\qquad p = x \cos NOx - y \sin NOx$

$$= x \sin \alpha - y \cos \alpha.$$

Let G be the resultant couple, estimated positive when it tends to turn the body from the positive end of Ox to the positive end of Oy. Then $\quad G = \Sigma Pp = \Sigma (xP \sin \alpha - yP \cos \alpha)$

$$= \Sigma (xP_y - yP_x),$$

where P_x and P_y are the axial components of P.

108. The arbitrary point O to which the forces have been transferred may be called the *base of reference*, or more briefly the *base*. It need not necessarily be the origin, though usually it is convenient to take that point as origin.

Let some point O', whose coordinates are $(\xi\eta)$, be the base. The resultant force and the resultant couple for this new base may be deduced from those for the origin O by writing $x - \xi$ and $y - \eta$ for x and y.

The expressions in Art. 106, for the resultant force do not contain x or y. Hence *the resultant force is the same in magnitude and direction whatever base is chosen.*

The expression for the resultant couple is

$$G' = \Sigma P \{(x - \xi) \sin \alpha - (y - \eta) \cos \alpha\}$$
$$= G - \xi Y + \eta X.$$

Thus the magnitude of the couple is, in general, different at different bases.

109. *To find the conditions of equilibrium of a rigid body.*

Let the system of forces be reduced to a force R and a couple G at any arbitrary base O. Since by Art. 78 the resultant force of the couple G is a force zero acting along the line at infinity, a finite force R cannot balance a finite couple G. If it could, we should have two forces in equilibrium, though they are not equal and opposite. It is therefore necessary for equilibrium that the resultant force R and the couple G should separately vanish.

110. Since $R = 0$ in equilibrium, we have as in Art. 44,

$$\Sigma P \cos \alpha = 0, \qquad \Sigma P \sin \alpha = 0.$$

These equations are necessary and sufficient to make R vanish. But we may put this result into a more convenient form.

In order to make the resultant force R zero, it is necessary and sufficient that the sum of the resolved parts or resolutes of the forces along each of any two non-parallel straight lines should be zero.

It is obvious that these conditions are necessary, for each straight line in turn may be taken as the axis of x. To prove that the conditions are sufficient, let one of these straight lines be the axis of x, and let the other be Ox'. Let the angle $xOx' = \beta$. Equating to zero the resolved parts of the forces along these straight lines we have

$$\Sigma P \cos \alpha = 0, \qquad \Sigma P \cos (\alpha - \beta) = 0.$$

These give $\qquad X = 0, \qquad X' = X \cos \beta + Y \sin \beta = 0.$

Unless β is zero or a multiple of π, these equations give $X = 0$ and $Y = 0$, and therefore $R = 0$.

The two equations of equilibrium obtained by resolving in any two different directions are commonly called the *equations of resolution*.

111. Again, it is necessary for equilibrium that $G = 0$; this gives $\Sigma Pp = 0$. The product Pp is called *the moment of the force P about O. In order then to make $G = 0$, it is necessary and sufficient that the sum of the moments of all the forces (taken with their proper signs) about some arbitrary point should be zero.* The equation of equilibrium thus obtained is usually called briefly the *equation of moments*.

112. Thus for forces in one plane the conditions of equilibrium supply three equations, viz. two equations of resolution and one of moments. This will be better understood when we consider the different ways in which a body can move. It may be proved that every displacement of a body may be constructed by a combination of the following motions. *Firstly*, the body may be moved, without rotation, a distance h parallel to the axis of x. *Secondly*, the body may be moved, also without rotation, a distance k parallel to the axis of y. In this way some arbitrary point O of the body may be brought to another point O' whose coordinates referred to O are any given quantities h and k. *Thirdly*, the body may be turned round this point through any given angle. The two equations of resolution express the fact that the forces urging the body in the two directions of the axes are zero, and the equation of moments expresses the fact that the forces do not tend to turn the body round the origin.

113. As great use is made of moments of forces, it is important that the meaning of this term should be distinctly understood. Suppose a force P to act at any point A along any straight line AB, and let O be the point about which we wish to take the moment of P. To find this moment we multiply the force P by the length p of the perpendicular from O on its line of action, viz. AB. The product has already been defined to be the moment.

As we are now discussing the theory of forces in one plane, the line AB and the point O are all in the plane of reference. But when we speak of forces in three dimensions it will be seen that what has just been defined is the moment of the force *about a straight line* through O perpendicular to the plane OAB.

When several forces act on the body, and the sum of their moments is required, attention must be paid to their proper signs. Exactly as in elementary trigonometry we select either direction of rotation round O as the standard direction. This we call the positive direction. Thus in Art. 104 the direction opposite to that of the hands of a watch has been chosen as the positive direction. The moment of each force is to be taken positive or negative according as it tends to turn the body round O in the positive or negative direction.

114. The three equations of equilibrium may be expressed in other forms besides the three given above, viz. $X = 0$, $Y = 0$, $G = 0$.

Thus *there will be equilibrium if the sum of the moments about each of any two different points (say O and C) is zero, and the sum of the resolved parts of the forces in some one direction, not perpendicular to OC, is zero.* To prove this, take O for origin, let Ox be parallel to the direction of resolution and let (ξ, η) be the coordinates of C. The given conditions are therefore

$$G = 0, \quad G' = G - \xi Y + \eta X = 0, \quad X = 0.$$

These lead to $G = 0$, $X = 0$, and $Y = 0$, provided ξ is not zero.

In the same way it may be proved that *there will be equilibrium if the sum of the moments about three different points O, C, C', not all in the same straight line, are each zero.*

115. We may also notice that we cannot obtain more than three independent equations of equilibrium by resolving in several other directions or taking moments about several other points. All the equations thus obtained may be deduced from some three equations of equilibrium. Thus if X, Y and G are zero it follows from Arts. 108 and 110 that G' and X' are also zero.

116. Varignon's Theorem. If a system of forces be transformed by the rules of statics into any other equivalent system, then (1) the sum of the resolved parts of the forces in any given direction, and (2) the sum of the moments of the forces about any given point are equal, each to each, in the two systems.

This theorem follows easily from the results of Art. 110. Let the two systems be P_1, P_2 &c. and P_1', P_2' &c. Let O be the point about which moments have to be taken, and Ox the direction in which the resolution is to be made. Then we have to prove (1) $\Sigma P \cos a = \Sigma P' \cos a'$ and (2) $G = G'$. Since the two systems are equivalent, there will be equilibrium if all the forces of either system are reversed, and both systems, after this change, act simultaneously on the same body. Hence, resolving in the given direction and taking moments about the given point, we have, by Arts. 110 and 111

$$\Sigma (P \cos a - P' \cos a') = 0, \qquad G - G' = 0.$$

The result follows at once.

117. We may also give an elementary proof of this theorem, derived from first principles.

According to the rules of statics one system of forces is transformed into another by the use of three processes. (1) We may transfer a force from one point of its line of action to another; (2) we may remove or add equal and opposite forces, as in Art. 78; (3) we may combine or resolve forces by the parallelogram of forces.

It is evident that neither the sum of the resolved parts in any direction nor the sum of the moments of the forces about any point is altered by the first two processes. We shall now prove in an elementary manner that they are not altered by the third.

Let the forces P, Q, acting at C, be represented in direction and magnitude by CA, CB respectively, and let their resultant

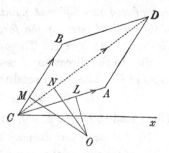

R be represented by CD. (1) Because the sum of the projections of CA, AD on any straight line (say Cx) is equal to that of CD (see Art. 65), it follows that the sum of the resolved parts of the forces P, Q along Cx is equal to the resolved part of their resultant R. (2) Let O be the point about which moments are to be taken. Draw OL, OM, ON perpendiculars on the forces. We have to prove

$$P \cdot OL + Q \cdot OM = R \cdot ON \ \ldots\ldots (1).$$

If O were on the other side of CA, say between CD and CA, the sign of the term $P \cdot OL$ would have to be changed, see Art. 113. But this change is provided for by the law of continuity, since the perpendicular from any point, as O, on a straight line, as OA, changes sign when O passes across the straight line. Such cases need not therefore be separately considered.

Dividing the equation (1) by CO, we see that it is equivalent to

$$P \sin ACO + Q \sin BCO = R \sin DCO \ldots\ldots \ldots\ldots\ldots\ldots (2).$$

This equation merely expresses that the sum of the resolved parts perpendicular to CO of the forces P, Q is equal to that of R. But if we take the arbitrary line Cx perpendicular to CO, this has just been proved true.

118. The single resultant. Any system of forces P_1, P_2 &c. can be reduced to a single force R acting at an arbitrary base together with a couple G. We shall now show that they can be further reduced to either a single force or a single couple.

The force R is zero when

$$X = \Sigma P \cos \alpha = 0, \quad Y = \Sigma P \sin \alpha = 0.$$

When this is the case, the given system of forces reduces to a single couple. It is evident that this single couple must be the same in all respects, whatever base of reference is chosen.

Supposing R not to be zero, we may by properly choosing the base of reference make the couple vanish, so that the whole system is equivalent to a single force R. Taking any convenient axes Ox, Oy, let O' be a base so chosen that the corresponding couple G' is zero. If $(\xi \eta)$ be the coordinates of O', we have by Art. 108,

$$G' = G - \xi Y + \eta X = 0 \ldots\ldots\ldots\ldots\ldots\ldots (1).$$

If then the base be chosen at any point of the straight line whose equation is (1), the resultant couple is zero. This straight line makes with Ox an angle whose tangent is Y/X; it is therefore parallel to the direction of the resultant force R. Since R acts at the new base O', this straight line is the line of action of R.

119. *Summing up;* if any set of forces be given by their resultant force and couple, viz. R and G, at any assumed base, we have the following results :

(1) The condition that the forces can be reduced to a single couple is $R = 0$. The condition that they can be reduced to a single force is that R should not be zero.

(2) If R be not zero, the given forces can be reduced to a single force whose magnitude is equal to R, and whose line of action is the straight line

$$G - \xi Y + \eta X = 0.$$

The direction in which the force acts along this straight line is indicated by the known signs of its components X and Y.

(3) Whatever system of coordinate axes is chosen this single resultant must be the same in magnitude and position. We therefore infer that this straight line is independent of all coordinates, *i.e.* is invariable in space.

120. Ex. 1. Prove that a given system of forces can be reduced to two forces acting one at each of two given points A and B, the force at A making a given angle (not zero) with AB.

Ex. 2. Show that a system of forces in one plane can be reduced to three forces which act along the sides of any triangle taken arbitrarily in that plane. Show also how to find these three forces.

(1) This resolution is possible. Let P be any one force of the system, and let it cut some one side, as AB, of the triangle ABC in M. Then P acting at M may be resolved into two forces, one acting along AB and the other along CM. The latter may be transferred to C and again resolved into two other forces acting along CA, CB respectively. Since every force may be treated in the same way, the whole system may be replaced by three forces F_1, F_2, F_3 acting along BC, CA, AB.

(2) To find the forces F_1, F_2, F_3. Let G_1, G_2, G_3 be the sums of the moments of the forces of the given system about the corners A, B, C respectively. Then if p_1, p_2, p_3 be the three perpendiculars from the corners on the opposite sides we have $F_1 p_1 = G_1$, $F_2 p_2 = G_2$, $F_3 p_3 = G_3$.

Ex. 3. Show that the trilinear equation to the single resultant of the forces F_1, F_2, F_3 acting along the sides of a triangle taken in order is $F_1 \alpha + F_2 \beta + F_3 \gamma = 0$. What is the meaning of this result when F_1, F_2, F_3 are proportional to the lengths of the sides along which they act?

Ex. 4. Two systems of three forces (P, Q, R), (P', Q', R') act along the sides taken in order of a triangle ABC: prove that the two resultants will be parallel if $(QR' - Q'R) \sin A + (RP' - R'P) \sin B + (PQ' - P'Q) \sin C = 0$.   [Math. Tripos, 1869.]

Ex. 5. Four forces in equilibrium act along tangents to an ellipse, the directions at adjacent points tending in opposite directions round the ellipse. Prove that the moment of each about the centre is proportional to the area of the triangle formed by joining the points of contact of the three other forces.

Ex. 6. A rigid polygon $A_1A_2...$ is moved into a new position $A_1'A_2'...$ and the mean centres of masses a_1, a_2,... placed at the corners in the two positions are G, G'. Prove that forces represented in direction and magnitude by $a_1.A_1A_1'$, $a_2.A_2A_2'$,... are equivalent to a force represented by $\Sigma a . GG'$ together with a couple $\sin\theta \Sigma(a . GA^2)$, where θ is the angle any side of the polygon $A_1A_2...$ makes with the corresponding side of $A_1'A_2'$ ....

Solution of Problems.

121. We shall now explain how the preceding theorems may be used to determine the positions of equilibrium of one or more rigid bodies in one plane. This can only be shown by examples. After some general remarks on the solution of statical problems a series of examples will be found arranged under different heads. The object is to separate the difficulties which occur in these applications and enable the reader to attack them one by one. A commentary is sometimes added to assist the reader in applying the same principles to other problems.

122. When the number of forces which act on a body is either three, or can be conveniently reduced to three, we can find the position of equilibrium by using the principle that these forces must meet in one point or be parallel. This is proved in Art. 34.

There are two advantages in this method, (1) the criterion that the three straight lines are concurrent may often be conveniently expressed by some *geometrical statement*, (2) the actual magnitudes of the forces are not brought into the process, so that if these are unknown, no further elimination is necessary. If the magnitudes of the forces are also required, they can be found afterwards from the principle that each is proportional to the sine of the angle between the other two. This is often called the geometrical method.

123. If there are more than three forces, or if we prefer to use an analytical method of solution even when there are only three forces, we use the results of Art. 109. We express the conditions of equilibrium (1) by resolving all the forces in some two convenient directions and equating the result of each resolution to zero, (2) by taking moments about some convenient point and equating the result to zero. Having thus obtained three equations, we must eliminate the unknown forces. Finally we shall obtain an equation expressing in an *algebraic manner* the position of equilibrium.

As we have to eliminate the unknown forces it will be convenient to *make one of the resolutions in the direction perpendicular to a force which we intend to eliminate, and to take moments about some point in its line of action.* This force will then appear only in the other resolution, which may therefore be omitted altogether. Thus by a proper choice of the directions of resolution and of the point about which moments are taken we may sometimes save much elimination.

124. When there are several bodies forming a system, we represent the mutual actions of these bodies by introducing forces called reactions at the points of contact. We may then regard each body as if it existed singly (all the others being removed) and were acted on by these reactions in addition to the given forces. We then form the equations for each body separately. Finally we must eliminate the reactions, if unknown, and the remaining equations will express the positions of equilibrium of the several bodies.

These *eliminations are sometimes avoided by expressing the conditions of equilibrium for two bodies taken together.* Afterwards we may form the equations for either separately in such a manner as to *avoid introducing the mutual reaction.*

When we come to the theory of virtual work we shall have a method of forming the equations of equilibrium free from these reactions.

125. Ex. 1. *A thin heavy uniform rod AB rests partly within and partly without a hemispherical smooth bowl, which is fixed in space. Find the position of equilibrium.*

Let G be the middle point of the rod, then the weight W of the rod may be collected at G. This should be evident from the theory of parallel forces, but it is strictly proved in the chapter on centre of gravity.

It follows from the remarks made in Art. 54, that, when two *smooth* surfaces touch each other, the pressure (if any exist) between the surfaces acts along the normal to the common tangent plane at the point of contact. If the rod be regarded as a very thin cylinder with its extremities rounded off, it is easy to see that the common tangent plane at A to the rod and the sphere coincides with the tangent plane to the sphere. The pressure at this point therefore acts along the normal AO to the sphere. We obtain the same result if we regard the rod as resting with a single terminal particle in contact with the sphere; it then follows immediately from Art. 54 that the pressure between the terminal particle and the sphere acts along the normal to the sphere.

Consider next the point C, at which the rod meets the rim of the bowl. The common tangent plane to the rod and the rim passes through both the rod and the tangent at C to the rim. The reaction is to be at right angles to both these, it therefore acts along a straight line CI drawn perpendicularly to the rod in the vertical plane containing the rod.

It will be found useful to put these remarks into the form of a *working rule*. Since the tangent plane at any point of a surface contains all the tangent straight lines at that point, the pressure between two *smooth bodies* which touch each other must be normal to every line on the two bodies which passes through the point of contact. *To find the direction of the reaction we select two lines which lie on the bodies and pass through the point of contact; the required direction is normal to both these lines.* Thus, at A, any tangent to the sphere passes through the point of contact, the reaction is therefore normal to the bowl. At C both the rod and the rim pass through the point of contact, the reaction is therefore normal both to the rod and to the tangent to the rim.

Let a be the radius of the bowl, l half the length of the rod. Let the position of equilibrium be determined by the angle $ACO = \theta$ which the rod makes with the horizon. It easily follows that $CAO = \theta$, $CA = 2a \cos \theta$.

Since the rod is in equilibrium under *three* forces, viz. R, R' and W, we use the geometrical method of solution. We have to express the condition that the three forces meet in some point I. To effect this we equate the projections of AG and AI on the horizontal. Since ICA is a right angle, I lies on the circumference produced, hence $AI = 2a$. Equating the projections, we have $\qquad l \cos \theta = 2a \cos 2\theta$,

$$\therefore \cos \theta = \frac{l}{8a} \pm \sqrt{\left(\frac{1}{2} + \frac{l^2}{64a^2} \right)}.$$

If the negative sign is given to the radical, $\cos \theta$ is negative and θ is greater than a right angle. This is excluded by geometrical considerations. The position of equilibrium is therefore given by the value of $\cos \theta$ with the positive sign prefixed to the radical.

There are however other geometrical limitations. Unless $2l$ is greater than $2a \cos \theta$ the rod will not be long enough to reach over the rim of the bowl, and unless l is less than $2a \cos \theta$ the point G at which the weight acts will fall outside the bowl. Unless the first condition is satisfied the rod will slip into the bowl, and if the second be not true the rod will tumble out. These conditions require that l should lie between $a \sqrt{\frac{2}{3}}$ and $2a$. If the half-length of the rod is less than $2a$, it is easy to prove that the value of $\cos \theta$ given above is never greater than unity.

For the sake of comparison, a solution of this problem by the analytical method is given here. We have to resolve in some directions, and take moments about some point. To avoid introducing the reaction R' into our equations, we shall resolve along AC and take moments about C. The resolution gives

$$R \cos \theta = W \sin \theta.$$

Since the perpendicular from C on AO is $a \sin COI$, and $CG = 2a \cos \theta - l$, the equation of moments is $\qquad Ra \sin 2\theta = W (2a \cos \theta - l) \cos \theta.$

Eliminating R, we have the same equation to find $\cos \theta$ as before.

The reader should notice that the value of $\cos \theta$ given by the equation of equilibrium depends only on the lengths a and l, and not on the weight of the rod. Thus all uniform rods of the same length, whatever their weights may be, will rest in equilibrium in a given bowl in the same position. This result might have been anticipated from the theory of dimensions, for a ratio like $\cos\theta$ could not be equal to any multiple of a weight, though it could be equal to the *ratio* of two weights. Now the only weight which could appear in the result is W. There is therefore no other force to make a ratio with W. It follows that W could not appear in the result.

Ex. 2. Show, by taking moments about the intersection I of the two reactions R, R' in example (1), that we arrive at the equation to find $\cos\theta$ without introducing any unknown force into the equation. Thence show that the equilibrium is stable.

If we slightly displace the rod by increasing its inclination θ to the horizon, the extremity A slides down the interior of the bowl and the rod moves a little outwards. The new position of I is therefore to the left of the vertical through the new position of G. When therefore the rod is left to itself, we see, by taking moments about the new position of I, that the weight acting at G will tend to bring the rod back to its position of equilibrium. Similar remarks apply, if the rod be displaced by decreasing θ. The equilibrium is therefore stable.

Ex. 3. A rod AB, placed with one extremity A inside a fixed wine glass, whose form is a right cone, with its axis vertical, rests over the rim of the glass at C: show that in the position of equilibrium $l\sin^2(\theta+\beta)\cos\theta = 2a\sin^2\beta$, where θ is the inclination of the rod to the horizontal, a is the radius of the rim of the cone, β the complement of the semi-vertical angle, and $2l$ the length of the rod.

Ex. 4. An open cylindrical jar, whose radius is a and weight nW, stands on a horizontal table. A heavy rod, whose length is $2l$ and weight W, rests over its rim with one end pressing against the vertical interior surface of the jar. Prove (1) that in the position of equilibrium the inclination θ of the rod to the horizon is given by $l\cos^3\theta = 2a$; (2) that the rod will tumble out of the jar if the inclination be less than this value of θ; (3) that the jar will

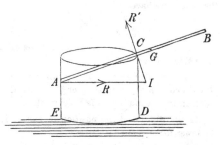

tumble over unless $l\cos\theta < (n+2)a$. Is the position of equilibrium stable or unstable?

The rod will tumble out of the jar if G lies to the right of the vertical through I in the figure. The jar will tumble over D if the moment about D of the weight of the rod acting at G is greater than that of the weight of the jar acting at its centre of gravity.

Ex. 5. Prove that the length of the longest rod which can be in equilibrium with one extremity pressing against the smooth vertical interior surface of the jar described in the last example is given by $2l^2 = a^2(n+2)^3$.

Ex. 6. A heavy rod AB, of length $2l$, rests over a fixed peg at C, while the end A presses against a smooth curve in the same vertical plane. The polar equation to the curve, referred to C as origin, is $r = f(\theta)$, θ being measured from the vertical. Show that the equilibrium value of θ satisfies the equation $(r-l)\tan\theta = dr/d\theta$.

Show, by integrating this differential equation, that the form of the curve, when the rod rests against it in equilibrium in all positions, is $(r-l)\cos\theta=a$. Thence show that the middle point of the rod always lies in a fixed horizontal straight line, and that the curve is the conchoid of Nicomedes.

If we attack this problem with the help of the principle of virtual work we arrive *first* at the result that in equilibrium the middle point must begin to move horizontally. From this geometrical fact we must then deduce the other results given above.

126. Ex. 1. *A uniform heavy rod PQ rests inside a smooth bowl formed by the revolution of an ellipse about its major axis, which is vertical. Show that in equilibrium the rod is either horizontal or passes through a focus.*

The reactions at P and Q act along the normals to the bowl. In the position of equilibrium these normals must intersect in a point I which is vertically over the middle point G of the rod.

The following geometrical property of conics is a generalization of those given in *Salmon's Conics, chap. XI, on the normal.* See also the note at the end of this volume. Let CA, CB be the semi-axes of the generating ellipse and let these be the axes of coordinates. Let $(\bar{x}\bar{y})$ be the coordinates of the middle point G of any chord PQ of a conic, and let $(\xi\eta)$ be the intersection I of the normals at P and Q. Then if p, p' be the perpendiculars from the foci on the chord and q the perpendicular from the centre, we have

$$\frac{\eta-\bar{y}}{\bar{y}}\frac{b^2}{a^2}=-\frac{pp'}{q^2}.$$

Here p and p' are supposed to have the same sign when the two foci are on the same side of the chord.

In our problem we have in equilibrium $\eta=\bar{y}$. Hence we must have either, one of the two p, p' equal to zero, or $\bar{y}=0$. In the first case the rod passes through a focus, in the second case it is horizontal.

Ex. 2. Show that the position of equilibrium in which the rod passes through the lower focus is stable.

This may be proved by finding the moment of the weight of the rod about I, tending to bring the rod back to its position of equilibrium when displaced. Another proof of this theorem, deduced from the principle of virtual work, is given in the second volume of the *Quarterly Journal* by H. G., late Bishop of Carlisle.

Ex. 3. If the bowl be formed by the revolution of an ellipse about the minor axis, which is vertical, prove that the only position of equilibrium is horizontal.

To find the positions of equilibrium we make $\xi=\bar{x}$. Since the foci on the minor axis are imaginary, we cannot immediately derive the corresponding formula for ξ from that for η by interchanging a and b. Let the chord cut the axes in L and M, then by similar triangles

$$\frac{\eta-\bar{y}}{\bar{y}}\frac{b^2}{a^2}=-\frac{CL^2-a^2+b^2}{CL^2},\qquad\therefore\ \frac{\xi-\bar{x}}{\bar{x}}\frac{a^2}{b^2}=-\frac{CM^2-b^2+a^2}{CM^2}.$$

The condition $\xi=\bar{x}$ gives $\bar{x}=0$ since the right-hand side cannot vanish.

Ex. 4. A uniform heavy rod PQ rests inside a smooth bowl formed by the revolution of an ellipse about its major axis, which is inclined at an angle a to the vertical. If the rod when in equilibrium intersect the axes CA, CB of the generating ellipse in L and M, prove that $\dfrac{CM^2+c^2}{CM} . b^2 \sin a = \dfrac{CL^2-c^2}{CL} a^2 \cos a$, where $c^2 = a^2 - b^2$.

Ex. 5. Two wires, bent into the forms of equal catenaries, are placed so as to have a common vertical directrix, and their axes in the same straight line. The extremities of a uniform rod are attached to two small rings which can freely slide on these catenaries. Show that in equilibrium the rod must be horizontal.

Ex. 6. A straight uniform rod has smooth small rings attached to its extremities, one of which slides on a fixed vertical wire and the other on a fixed wire in the form of a parabolic arc whose axis coincides with the former wire, and whose latus rectum is twice the length of the rod : prove that in the position of equilibrium the rod will make an angle of 60° with the vertical.   [Math. Tripos, 1869.]

Ex. 7. AC, BC are two equal uniform rods which are jointed at C, and have rings at the ends A and B, which slide on a smooth parabolic wire, whose axis is vertical and vertex upwards ; prove that in the position of equilibrium the distance of C from AB is one fourth of the latus rectum.   [Math. Tripos, 1871.]

Ex. 8. Two heavy uniform rods AB, BC whose weights are P and Q are connected by a smooth joint at B. The ends A and C slide by means of smooth rings on two fixed rods each inclined at an angle a to the horizon. If θ and ϕ be the inclinations of the rods to the horizon, show that $P \cot \phi = Q \cot \theta = (P+Q) \tan a$.
[Trin. Coll., 1882.]

Resolve horizontally and vertically for the two rods regarded as one system; then take moments for each singly about B.

127. Ex. 1. *Two smooth rods OM, ON, at right angles to each other are fixed in space. A uniform elliptic disc is supported in the same vertical plane by resting on these rods. If OM make an angle a with the vertical, prove that either the axes of the ellipse are parallel to the rods, or the major axis makes an angle θ with OM, given by*

$$\tan^2 \theta = \frac{a^2 \tan^2 a - b^2}{a^2 - b^2 \tan^2 a}.$$

Let P, Q be the points of contact and let the normals at P, Q meet in I. Let C be the centre, then in equilibrium either C and I must coincide, or CI is vertical.

In the former case the tangents OM, ON are parallel to the axes.

In the latter case, let D bisect PQ, then OD produced passes through C; but because the tangents are at right angles $OPIQ$ is a rectangle, therefore OD passes through I. Hence OCI is vertical.

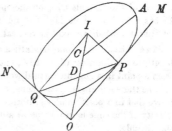

These two results follow easily from a principle to be proved in the chapter on virtual work. As the ellipse is moved round, always remaining in contact with the rods, we know by conics that C describes an arc of a circle, whose centre is O, and whose radius is $\sqrt{(a^2+b^2)}$. Hence when C is vertically over O, its altitude is a maximum. When the axes are parallel to the rods, C is at one of the extremities of its arc and its altitude is a minimum. It immediately follows from the principle of virtual work that the first of these is a position of unstable equilibrium, and that the other two are positions of stable equilibrium.

Resuming the solution, we have now to find θ when CI is vertical. The perpendicular from C on OM makes with the major axis an angle equal to the complement of θ, hence

$$a^2 \sin^2 \theta + b^2 \cos^2 \theta = OC^2 \sin^2 a = (a^2 + b^2) \sin^2 a.$$

The value of $\tan^2 \theta$ follows immediately.

Ex. 2. An elliptic disc touches two rods OM, ON, not necessarily at right angles, and is supported by them in a vertical plane. If (XY) be the coordinates of the intersection O of the rods, referred to the axes of the ellipse, prove that the major axis is inclined to the vertical at an angle θ given by $\tan \theta = - \dfrac{Y}{X} \dfrac{a^2 - X^2}{b^2 - Y^2}.$

To prove this we may use a theorem deduced from two given by Salmon in his chapter on Central Conics, Art. 180, Sixth Edition. Let (XY) be a point from which two tangents are drawn to touch a conic at P, Q. The normals at P, Q meet in a point I, whose coordinates (xy) are given by

$$\frac{x}{X} = (a^2 - b^2) \frac{b^2 - Y^2}{a^2 Y^2 + b^2 X^2}, \qquad \frac{y}{Y} = - (a^2 - b^2) \frac{a^2 - X^2}{a^2 Y^2 + b^2 X^2}.$$

The result follows, since CI must be vertical.

Ex. 3. An elliptic disc is supported in equilibrium in a vertical plane by resting on two smooth fixed points in a horizontal straight line. Prove that in equilibrium either a principal diameter is vertical, or these points are at the extremities of two conjugate diameters.

Let the principal diameters be the axes of coordinates. Let the fixed points P, Q be (xy), $(x'y')$, and let $(\xi\eta)$ be the intersection I of the normals at these points. In equilibrium IC must be perpendicular to PQ, hence $(x - x') \xi + (y - y') \eta = 0$. By writing down the equations to the normals at P, Q we find ξ, η, as is done in *Salmon's Conics*, Art. 180. This equation then becomes

$$(x - x') (y - y') \left(\frac{xx'}{a^2} + \frac{yy'}{b^2} \right) = 0.$$

One of these factors must vanish. These give the three positions of equilibrium.

That there should be equilibrium when P, Q are at the extremities of two conjugate diameters is evident; for PI, QI are perpendiculars from two of the corners of the triangle CPQ on the opposite sides, hence CI must be perpendicular to the side PQ. This is the condition of equilibrium. That there should be equilibrium when an axis is vertical is evident from symmetry.

128. **Ex. 1.** A cone has attached to the edge of its base a string equal in length to the diameter of the base, and is suspended by the extremity of this string from a point in a smooth vertical wall, the rim of the base also touching the wall. If a be the semi-angle of the cone, θ the inclination of the string to the vertical, prove that in a position of equilibrium $\tan a \tan \theta = \frac{1}{12}$. Assume that the centre of gravity of the cone is in its axis at a distance from the base equal to one quarter of the altitude.

Ex. 2. A square rests with its plane perpendicular to a smooth wall, one corner being attached to a point in the wall by a string whose length is equal to a side of the square. Prove that the distances of three of its angular points from the wall are as 1, 3 and 4. [Math. Tripos, 1853.]

By resolving vertically, and taking moments about the corner of the square which is in contact with the wall, we obtain two equations from which the inclination of any side to the wall and the tension may be found.

Ex. 3. AB is a uniform rod of length a ; a string $APBC$ is fastened to the end A of the rod and passes through a smooth ring attached to the other end B; the end C of the string is fastened to a peg C, and the portion APB is hung over a smooth peg P which is in the same horizontal plane as C at a distance $2b$ from it $(b < a)$. If AP is vertical, find the angles which the other parts of the string make with the vertical, and show that the string must have one of the lengths $\frac{8}{5}b\sqrt{3} \pm \sqrt{(a^2 - b^2)}$.   [King's Coll., 1889.]

Ex. 4. Two light elastic strings have their ends tied to a fixed point on the line joining two small smooth pegs which are in the same horizontal plane, so that when they are unstretched their ends just reach the pegs; they hang over the pegs and have their other ends fastened to the ends of a heavy uniform rod ; show that the inclination of the rod to the horizon is independent of its length, being equal to $\tan^{-1}(y_1 - y_2)/2a$, where y_1 and y_2 are the extensions of the strings when they singly support the rod, and a is the distance between the pegs. Show also that the two strings and the rod are inclined to the horizon at angles whose tangents are in arithmetical progression. It may be assumed that the tension of each string is proportional to the ratio of its extension to its unstretched length.

[Math. Tripos, 1887.]

129. Ex. 1. *A sphere rests on a string fastened at its extremities to two fixed points. Show that if the arc of contact of the sphere and plane be not less than $2 \tan^{-1}\frac{48}{55}$, the sphere may be divided into two equal portions by means of a vertical plane without disturbing the equilibrium.*   [Math. Tripos, 1840.]

It may be assumed that the centre of gravity of a solid hemisphere is on the middle radius at a distance $\frac{3}{8}$ths of that radius from the centre.

Consider the equilibrium of the hemisphere ABD and the portion AD of the string in contact with it. The mutual reactions of the string and the hemisphere may now be omitted. This compound body is acted on by (1) the tensions of the string, each equal to T, acting at A and D, (2) the weight W of the hemisphere acting at its centre of gravity G, (3) the mutual reaction R of the two hemispheres. The reaction R is the resultant of all the horizontal pressures between the elements of the plane bases and must act at some point within the area of contact. The two bases

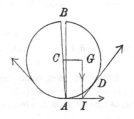

will separate unless the resultant of the remaining forces also passes inside the area of contact. The arc AD being as small as possible, this separation will take place by the hemispheres opening out at B, for the mutual pressures are then confined to the single point A at the lowest point of the sphere. The hemisphere ABD is then acted on by the three forces, T at D, $T - R$ at A, and W at G. These must intersect in a point I. Hence $CG = CA \tan \frac{1}{2}ACD$. This gives $\tan \frac{1}{2}ACD = \frac{3}{8}$ and $\tan ACD = \frac{48}{55}$.

Ex. 2. Two equal heavy solid smooth hemispheres, placed so as to look like one sphere with the diametral plane vertical, rest on two pegs which are on the same horizontal line. Prove that the least distance apart of the pegs, so that the hemispheres may not fall asunder, is to the diameter of the circle as 3 to $\sqrt{(73)}$.

[Christ's Coll.]

Ex. 3. An elliptic lamina of eccentricity e, divided into two pieces along the minor axis, is placed with its major axis horizontal in a loop of string attached to two fixed points, so that the portions of the strings not in contact with the ellipse are vertical. Show that equilibrium will not exist unless

$$(6\pi e)^2 < (9\pi - 4)(3\pi + 4).$$ [Coll. Ex., 1890.]

Each semi-ellipse is acted on by two equal tensions along the tangents at the extremities A and B of the axes. These have a resultant inclined at 45° to either axis. Let it cut the vertical through the centre of gravity G in the point H. The reaction between the semi-ellipses must pass through H. Hence the altitude of H above B must be less than the axis minor. If C be the centre, this gives at once $a - CG < 2b$. Granting that $CG = 4a/3\pi$, this leads to the result.

Ex. 4. A circular cylinder rests with its base on a smooth inclined plane; a string attached to its highest point, passing over a pulley at the top of the inclined plane, hangs vertically and supports a weight; the portion of the string between the cylinder and the pulley is horizontal: determine the conditions of equilibrium. [Math. Tripos, 1843.]

Show that the ratio of the height of the cylinder to the diameter of its base must be less than the cotangent of the inclination of the plane to the horizon.

Ex. 5. A uniform bar of length a rests suspended by two strings of lengths l and l' fastened to the ends of the bar and to two fixed points in the same horizontal line at a distance c apart. If the directions of the strings being produced meet at right angles, prove that the ratio of their tensions is $al + cl' : al' + cl$. [Math. Tripos, 1874.]

Ex. 6. A smooth vertical wall AB intersects a smooth plane BC so that the line of intersection is horizontal. Within the obtuse angle ABC a smooth sphere of weight W is placed and is kept in contact with the wall and plane by the pressure of a uniform rod of length l which is hinged at A, and rests in a vertical plane touching the sphere. Show that the weight of the rod must be greater than

$$\frac{Wh \cos a \cos \tfrac{1}{2}a}{2l \sin \tfrac{1}{2}\theta \sin \tfrac{1}{2}(a-\theta)\cos^2 \tfrac{1}{2}(a-\theta)},$$

where a and θ are the acute angles made by the plane and rod with the wall, and $h = AB$. [Math. Tripos, 1890.]

Ex. 7. A set of equal frictionless cylinders, tied together by a fine string in a bundle whose cross section is an equilateral triangle, lies on a horizontal plane. Prove that, if W be the total weight of the bundle, and n the number of cylinders in a side of the triangle, the tension of the string cannot be less than $\frac{W}{4\sqrt{3}}\left(1+\frac{1}{n}\right)^{-1}$ or $\frac{W}{4\sqrt{3}}\left(1-\frac{1}{n}\right)$, according as n is an even or an odd number, and that these values will occur when there are no pressures between the cylinders in any horizontal row above the lowest. [Math. Tripos, 1886.]

Ex. 8. A number n of equal smooth spheres, of weight W and radius r, is placed within a hollow vertical cylinder of radius a, less than $2r$, open at both ends and resting on a horizontal plane. Prove that the least value of the weight W' of the cylinder, in order that it may not be upset by the balls, is given by

$$aW' = (n-1)(a-r)W \text{ or } aW' = n(a-r)W,$$

according as n is odd or even. [Math. Tripos, 1884.]

Ex. 9. The circumference of a heavy rigid circular ring is attached to another concentric but larger ring in its own plane by n elastic strings ranged symmetrically round the centre along common radii. This second ring is attached to a third in a similar manner by $2n$ strings, and this to a fourth by $3n$ strings and so on. Supposing all the rings to have the same weight, and the strings at first to be without tension, show that, if the last ring be lifted up and held horizontal, all the other rings will be on the surface of a right cone.   [Pet. Coll., 1862.]

Ex. 10. Two spheres of densities ρ and σ, and whose radii are a and b, rest in a paraboloid of revolution whose axis is vertical and touch each other at the focus : prove that $\rho^3 a^{10} = \sigma^3 b^{10}$.   [Curtis' problem. *Educational Times*, 5460.]

130. Equilibrium of four repelling particles. Ex. 1. Four free particles situated at the corners of a quadrilateral are in equilibrium under their mutual attractions or repulsions; the forces along the sides AB, BC, CD, DA being attractive, those along the diagonals AC, BD being repulsive. If the forces are proportional to the sides along which they act, prove that the quadrilateral is a parallelogram.

In this case the forces on the particle A are represented by the sides AB, AD and the diagonal AC. The result follows at once from the parallelogram of forces.

Ex. 2. If the quadrilateral formed by joining the four particles can be inscribed in a circle, show that the attracting force along any side is proportional to the opposite side, and the repelling force along a diagonal to the other diagonal.

Ex. 3. If the quadrilateral be any whatever, prove that when the particles at the corners are in equilibrium

$$\frac{f(AB)}{AB \cdot OC \cdot OD} = \frac{f(BC)}{BC \cdot OD \cdot OA} = \&c. = \frac{f(BD)}{AC \cdot OB \cdot OD} = \frac{f(AC)}{BD \cdot OA \cdot OC},$$

where O is the intersection of the diagonals BD, AC, and the mutual force along any line, as AB, is represented by $f(AB)$.

To prove this, consider the equilibrium of the particle A.

$$\frac{f(AC)}{f(AB)} = \frac{\sin DAB}{\sin DAO} = \frac{\text{area } DAB}{\text{area } DAO} \cdot \frac{AD \cdot AO}{AD \cdot AB} = \frac{DB}{DO} \cdot \frac{AO}{AB};$$

all the results follow by symmetry.

Ex. 4. Whatever be the form of the quadrilateral, prove that (1) the moments about O of the forces which act along the sides are equal, and (2),

$$ABf(AB) + BCf(BC) + CDf(CD) + DAf(DA) = ACf(AC) + BDf(BD).$$

Reactions at Joints.

131. When two beams are connected together by a smooth hinge-joint or are fastened together by a very short string, the mutual action between them will be equivalent to a single force acting at the point of junction. In some cases the direction of this force is at once apparent, in other cases its *direction as well as its magnitude* must be deduced from the equations of equilibrium.

There are two cases in which the direction is apparent. *Firstly* let the body and the external forces be both symmetrical about some straight line through the hinge. In this case the action and

reaction between the two beams must also be symmetrically situated. Since they are equal and opposite, they must each be
perpendicular to the line of symmetry.

Secondly let the body be hinged at two points A and B, and let
it be acted on by no other forces except the reactions at A and B.
Since the body is in equilibrium under these two reactions, they
must act along the straight line joining the hinges and be equal
and opposite.

Ex. 1. Two equal beams AA', BB', without weight, are hinged together at
their common middle point C, and placed in a
vertical plane on a smooth horizontal table. The
upper ends A, B of the rods are connected by a
light string ADB, on which a small heavy ring
can slide freely. Show that in equilibrium a
horizontal line through the ring D will bisect AC
and BC.   [Coll. Ex.]

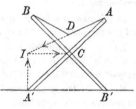

The action at C is horizontal, because the
system is symmetrical about the vertical through
C. The action at A' is vertical because, when the end of a rod rests on a surface,
the action is normal to the surface (Art. 125). The tension of the string acts along
AD. These three forces keep the rod AA' in equilibrium. They therefore meet in
some point I. By similar triangles DC is half IA'. The result follows immediately.

Ex. 2. If the weight of each rod in the last example be n times the weight of
the ring, prove that in equilibrium a horizontal line through the ring will cut CA in
a point P such that $CP = (2n + 1) PA$.

Ex. 3. Two equal heavy rods CA, CB are hinged at C, and their extremities
A, B rest on a smooth horizontal table. A third rod, attached to their middle
points E, F by smooth hinges, prevents the rods CA, CB from opening out. Find
the reactions at the hinges (1) when the rod EF has no weight, and (2) when it has
a weight W'.

The reaction R at C is horizontal by the rule of symmetry. If the weight of
the rod EF is neglected, the reactions at E
and F act along EF by the second rule of this
Article. Let this be X. The reaction R' at
A is vertical. The weight of the rod CA acts
vertically at E. These are all the forces which
act on the rod CA. By resolving horizontally
and vertically, and by taking moments about E
we easily find that R and $-X$ are each equal to
$W \tan a$, where a is half the angle ACB.

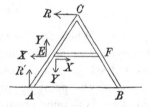

When the roof of a house is not high pitched, the angle ACB between the beams
is nearly equal to two right angles, so that $\tan a$ is large. The reactions at C and E
become therefore much greater than the weight of the beams. It is therefore
necessary to give great strength to the mode of attachment of the beams.

If the weight W' of the beam EF cannot be neglected, the reactions at E and F
will not be horizontal. Let the components of the action at E on the rod EF be

X, Y when resolved horizontally to the right and vertically downwards. It will be noticed that they have been put in directions opposite to those in which we should expect them to act. This is done to avoid confusing the figure. They should therefore appear as negative quantities in the result. The reactions on the rod AC are of course exactly opposite. The equations of equilibrium are as follows:

Resolve ver. for EF, $\qquad 2Y + W' = 0$,

Res. ver. for the system, $\qquad 2R' = W' + 2W$,

Mts. about E for AC, $\qquad Ra \cos a = R'a \sin a$,

Res. hor. for AC, $\qquad X + R = 0$,

where $2a$ is the length of either CA or CB. These four equations determine X, Y, R, R'.

Ex. 4. Two rods AB, BC, of equal weight but of unequal length, are hinged together at B, and their other extremities are attached to two fixed hinges A and C in the same vertical line. Prove that the line of action of the reaction at the hinge B bisects the straight line AC.

Ex. 5. Two uniform rods AB, AC, freely jointed at A, rest with A capable of sliding on a fixed smooth horizontal wire. B and C are connected by small smooth rings with two vertical wires in the plane ABC. If the rods are perpendicular prove that $a\sqrt{}(l + l') = l\sqrt{}l' + l'\sqrt{}l$, where l, l' are the lengths of the rods and a the distance between the vertical wires. [Coll. Ex., 1890.]

132. Ex. 1. Four rods, jointed at their extremities A, B, C, D form a parallelogram. The opposite corners are joined by strings along the two diagonals, each of which is tight. Show that their tensions are proportional to the diagonals along which they act.

Let four particles be added to the figure, one at each corner. Let the sides be jointed to the particles instead of to each other, and let the strings also be attached to the particles. By this arrangement each rod is acted on only by forces at its extremities; hence by the second rule of Art. 131 these forces act along the rod. We now proceed as in Art. 130, Ex. 1. The forces on the particle A are parallel to the sides of the triangle ABC, hence, by the parallelogram of forces, they are proportional to those sides. It follows that every side in the figure measures the force which acts along it.

Another Solution. We may also arrange the internal forces otherwise. Let the rods be jointed to each other, but let the strings be attached to the extremities of the rods AB, CD. Since AD is now acted on only by the actions at the hinges, these actions act along AD (Art. 131). In the same way the reactions at B and C act along BC. Thus the rod CD is acted on by the tensions T, T' along the diagonals DB and CA, and by the reactions along AD and BC. Resolving at right angles to the latter, we have $T \sin OBC = T' \sin OCB$, where O is the intersection of the diagonals. This gives $T \cdot OC = T' \cdot OB$, *i.e.* the tensions are as the diagonals along which they act.

It should be noticed that the mutual reactions on the rods obtained in the two solutions appear not to be the same. In the first solution, the conditions of equilibrium of the rod CD and the particles at C and D are separately considered; in the second solution, they are treated as one body and the conditions of equilibrium of this compound body are found to be sufficient to determine the ratio of the tensions of the strings. Consider the reactions at the corner D. In the first solution there are two reactions at this corner, viz. those between the particle at D and the two rods AD, CD. These are proved to act along AD and CD; let them be called R_1 and R_2 respectively. In the second solution the only reaction at the corner D which is considered is R_1, the other reaction R_2 not being required. If it had been asked, as part of the question, to find the reaction at the joint D, it would have been necessary to state in the enunciation how the rods were joined to each other and to the string. It is only when this mode of attachment is given that we can determine whether it is R_1, R_2 or some combination of both that can be properly called *the* reaction at the corner D.

Ex. 2. A parallelepiped, formed of twelve weightless rods freely jointed together at their extremities, is in equilibrium under the action of four stretched elastic strings connecting the four pairs of opposite vertices. Show that the tensions of the rods and strings are proportional to their lengths.   [Coll. Ex., 1890.]

Ex. 3. Four rods are jointed at their extremities so as to form a quadrilateral $ABCD$, and the opposite corners A, C and B, D are joined by tight strings. If the tensions are represented by $f(AC)$ and $f(BD)$, prove that

$$f(AC)\left(\frac{1}{AO}+\frac{1}{OC}\right)=f(BD)\left(\frac{1}{BO}+\frac{1}{OD}\right),$$

where O is the intersection of the diagonals.

By placing particles at the four corners as in the first solution to the last example, this problem is immediately reduced to that solved in Ex. 3, Art. 130. The result follows at once. This problem is due to Euler, who gives an equivalent result in *Acta Academiæ Scientiarum Imperialis Petropolitanæ*, 1779. From this he deduces the result given in Ex. 1 for a parallelogram.

Ex. 4. If the opposite sides AD, BC (or CD, BA) are produced to meet in X, prove that the tensions of the strings are inversely proportional to the perpendiculars drawn from X on the strings.

To prove this we follow the second method of solution adopted in Ex. 1. Let the strings be attached to the extremities of the rods AB, CD. The reactions at D and C now act along AD and BC. Considering the equilibrium of the rod CD, the result follows at once by taking moments about X.

Ex. 5. *Four rods, jointed together at their extremities, form a quadrilateral* $ABCD$. *Points E, F on the adjacent sides AB, BC are joined by one string and points G, H on the adjacent sides BC, CD are joined by another string. Compare the tensions of the strings.* This is a modification of a problem solved by Euler in 1779. *Acta Academiæ Petropolitanæ.* The following solution is founded on his.

Lemma. We may replace the string EF by a string joining any other two points E', F' taken in the same two sides AB, BC without altering any reaction except the one at B, provided the moments about B of the tensions of EF, $E'F'$ are equal. To prove this, let the strings intersect in K. The tension T, acting at F on the rod BC, may be transferred to K, and then resolved into two, viz. one U which acts along KF', and which may be transferred to F', and another V which acts along KB and may be transferred to B. In the same way the tension T acting at E on the rod AB may be resolved into U acting at E' along $E'K$, and V acting at B along BK. Thus the equal forces T, T at E and F are replaced by the equal forces U, U at E', F', i.e. by the tension U of a string $E'F'$. At the same time the mutual reactions at B are altered by the superposition of the two equal and opposite forces called V. The other forces and reactions of the system are unaffected by the change. Since T is the resultant of U and V, the moments of T and U about B must be equal.

By using this lemma we may transfer the strings EF, GH until they coincide with the diagonals AC, BD. Let T, T' be the tensions of EF, GH. Then $U = nT$ is the tension of AC, where n is the ratio of the perpendiculars from B on EF and

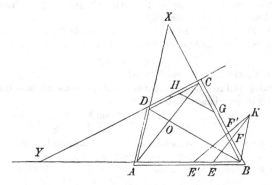

AC. So $U' = n'T'$ is the tension of BD, where n' is the ratio of the perpendiculars from C on HG and BD. The ratio of the tensions along the diagonals has been found in Ex. 3. Using that result we have

$$nT\left(\frac{1}{AO} + \frac{1}{OC}\right) = n'T'\left(\frac{1}{BO} + \frac{1}{OD}\right).$$

Ex. 6. Four rods jointed together at their extremities form a quadrilateral $ABCD$. Points E, F on the *opposite sides* AB, CD are joined by one string, and points G, H on the other two sides AD, BC are joined by a second string. If the opposite sides AD, BC meet in X, and the sides CD, BA in Y, and p, p' are the perpendiculars from X, Y on the strings EF, GH, prove that the tensions T, T' are connected by the equation

$$\frac{Tp \sin X}{AB \cdot CD} + \frac{T'p' \sin Y}{AD \cdot BC} = 0.$$

The perpendicular from X or Y on any string is to be regarded as positive when the string intersects XY at some point between X and Y.

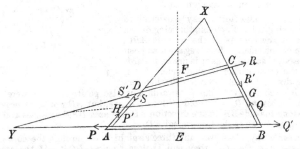

It follows that in equilibrium one string must pass between X and Y and the other outside both, contrary to what is represented in the diagram. It also follows that, if one string as GH produced passes through Y, either the tension of the other string is zero, or that string produced passes through X.

Let the reactions at each of the corners of the quadrilateral be resolved into forces acting along the adjacent sides, viz. P', P at A along DA, AB; Q', Q at B

along AB, BC; R', R at C and S', S at D. The reactions on the rods AD, BC are sketched in the figure, those acting on the rods AB, CD are equal and opposite to those drawn.

Considering the equilibrium of the rods AD and BC, we have, by taking moments about D and C respectively,

$$P . YD \sin Y = T' . DH \sin H, \qquad Q' . YC \sin Y = T' . CG \sin G.$$

Consider next the equilibrium of the rod AB, taking moments about X,

$$(P - Q') XM = Tp,$$

where XM is a perpendicular from X on AB.

Substituting, and remembering that $\sin H$, $\sin G$, and $\sin X$ have the ratio of the opposite sides in the triangle XHG, we find

$$\frac{DH . CY . XG - DY . CG . XH}{YD . YC} \cdot \frac{\sin X}{\sin Y} \cdot \frac{XM . T'}{HG} = Tp.$$

Now the numerator of the first fraction on the left-hand side is minus the sum of the products of the segments (with their proper signs) into which the sides of the triangle DCX are divided by the points G, H, Y*. The equation therefore reduces to

$$\frac{[GHY] . DC . CX . XD}{[DCX] . YD . YC} \cdot \frac{\sin X}{\sin Y} \cdot \frac{XM . T'}{HG} + Tp = 0,$$

where $[GHY]$ and $[DCX]$ represent the areas of the triangles GHY and DCX. These areas are equal to $\frac{1}{2} HG . p'$ and $\frac{1}{2} DX . CX \sin X$ respectively. Also $AB . XM$ is twice the area of the triangle AXB, and is therefore equal to $XA . XB \sin X$. Again,

$$\frac{YD}{\sin A} = \frac{AD}{\sin Y}, \qquad \frac{YC}{\sin B} = \frac{BC}{\sin Y}, \qquad \frac{XA}{\sin B} = \frac{AB}{\sin X} = \frac{XB}{\sin A}.$$

Substituting we obtain the equation connecting T, T' given in the enunciation.

---

* Let D, E, F be three arbitrary points taken on the sides of a triangle ABC. If Δ, Δ' be the areas of the triangles ABC, DEF, it may be shown that

$$\frac{\Delta'}{\Delta} = \frac{AF . BD . CE + AE . CD . BF}{abc}.$$

To form the two products $AF . BD . CE$ and $AE . CD . BF$, we start from any corner, say A, and travel round the triangle, first one way and then the other, taking on each circuit one length from each side. The sum of the two products so formed, each with its proper sign, is the expression in the numerator.

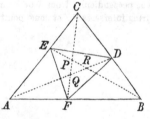

The signs of these factors may be determined by the following rule. Each length, being drawn from one of the corners of the triangle ABC, along one of the sides, is to be regarded as positive or negative according as it is drawn towards or from the other corner in that side. Thus, AF being drawn from A towards B is therefore positive, BF being drawn from B towards A is also positive. If F were taken on AB produced beyond B, AF would still be positive, but BF would be negative. If F move along the side AB, in the direction AB, the area DEF vanishes and becomes negative when F passes the transversal ED.

In the same way, if we draw any three straight lines through the corners of the triangle, say AD, BE, CF, they will enclose an area PQR. If the area of the triangle PQR is Δ'', it may be shown that

$$\frac{\Delta''}{\Delta} = \frac{(AF . BD . CE - AE . CD . BF)^2}{(ab - CE . CD) (bc - AE . AF) (ca - BF . BD)}.$$

The author has not met with these expressions for the area of two triangles which often occur. He has therefore placed them here in order that the argument in the text may be more easily understood.

133. Ex. 1. A series of rods in one plane, jointed together at their extremities, form a *closed polygon*. Each rod is acted on at its middle point in a direction perpendicular to its length by a force whose magnitude is proportional to the length of the rod. These forces act all inwards or all outwards. Show that in equilibrium (1) the polygon can be inscribed in a circle, (2) the reactions at the corners act along the tangents to the circle, (3) the reactions are all equal.

Let AB, BC, CD, &c. be the rods, L, M, N, &c. their middle points. Let $aB\beta$, $\beta C\gamma$, &c. be the lines of action of the reactions at the corners B, C &c. Since each rod is in equilibrium, the forces at the middle points of the rods must pass through a, β, γ, &c. respectively. Consider the rod BC; the triangles $BM\beta$, $CM\beta$ are equal and similar, also the reactions along $B\beta$ and $C\beta$ balance the force along $M\beta$ which

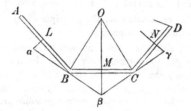

bisects the angle $B\beta C$. Hence these reactions are equal. It follows that the reactions at all the corners are equal in magnitude.

Draw BO, CO perpendicular to the directions of the reactions at B and C. These must intersect in some point O on the perpendicular through M to BC. The sides of the triangle OBC are perpendicular to the directions of the three forces which act on the rod BC, and are in equilibrium. Hence CO represents the magnitude of the reaction at C on the same scale that BC represents the force at M.

In the same way if CO', DO' be drawn perpendicular to the reactions at C and D, they will meet in some point O' on the perpendicular through N to CD. Also CO' will measure the reaction at C on the same scale that CD measures the force at its middle point. Hence by the conditions of the question $CO = CO'$, and therefore O and O' coincide. Thus a circle, centre O, can be drawn to pass through all the angular points of the polygon and to touch the lines of action of all the reactions.

Ex. 2. A series of jointed rods form an *unclosed polygon*. The two extremities of the system are constrained, by means of two small rings, to slide along a smooth rod fixed in space. If each moveable rod is acted on, as in the last problem, by a force at its middle point perpendicular and proportional to its length, prove that the polygon can be inscribed in a circle having its centre on the fixed rod.

Let A and Z be the two extremities. We can attach to A and Z a second system of rods equal and similar to the first, but situated on the opposite side of the fixed rod. We can apply forces to the middle points of these additional rods acting in the same way as in the given system. With this symmetrical arrangement the fixed rod becomes unnecessary and may be removed. The results follow at once from those obtained in the last problem.

These two problems may be derived from Hydrostatical principles. Let a vessel be formed of plane vertical sides hinged together at their vertical intersections, and let this vessel be placed on a horizontal table. Let the interior be filled with fluid which cannot escape either between the sides and the table or at the vertical joinings. The pressures of the fluid on each face will be proportional to that part of the area of each which is immersed in the fluid, and will act at a point on the median line. These pressures are represented in the two problems by the forces acting on the rods at their middle points. It will follow from a general principle, to be proved in the chapter on virtual work, that the vessel will take such a form that the altitude of the centre of gravity of the fluid above the table is the least possible. Hence the depth of the fluid is a minimum. Since the volume is given, it immediately follows that the area of the base is a maximum.

By a known theorem in the differential calculus, the area of a polygon formed of sides of given length is a maximum when it can be inscribed in either a circle or a semicircle, according as the polygon is closed or unclosed. (De Morgan's *Diff. and Int. Calculus*, 1842.) The results of the preceding problems follow at once.

We may also deduce the results from the principle of virtual work without the intervention of any hydrostatical principles.

We may notice that both these theorems will still exist if a great many consecutive sides of the polygon become very short. In the limit these may be regarded as the elementary arcs of a string acted on by normal forces proportional to their lengths. *If then a polygon be formed by rods and strings, and be in equilibrium under the action of a uniform normal pressure from within, the sides can be inscribed in a circle, and the strings will form arcs of the same circle.*

The first of these two problems was solved by N. Fuss in *Mémoires de l'Académie Impériale des Sciences de St Pétersbourg*, Tome VIII, 1822. His object was to determine the form of a polygonal jointed vessel when surrounded by fluid.

134. Ex. **Polygon of heavy rods.** *n uniform heavy rods A_0A_1, A_1A_2 &c., $A_{n-1}A_n$ are freely jointed together at A_1, A_2 &c. A_{n-1} and the two extremities A_0 and A_n are hinged to two points which are fixed in space ; it is required to find the conditions of equilibrium.*

At each of the joints A_0, A_1 &c. draw a vertical line upwards ; let θ_0, θ_1 &c. be the inclinations of the rods A_0A_1, A_1A_2 &c. to these verticals, the angles being measured round each hinge from the vertical to the rod in the same direction of rotation. Let the weights of these rods be W_0, W_1 &c.

First Method. The equilibrium will not be disturbed if we replace the weight W of any rod by two vertical forces, each equal to $\frac{1}{2}W$, acting at the extremities of the rod. In this way each rod may be regarded as separated into three parts, viz. the two terminal particles, each acted on by half the weight of the rod, and the intermediate portion thus rendered weightless. Let us first consider how these several parts act on each other. At any joint the two terminal particles of the adjacent rods are hinged together. Each particle is in equilibrium under the action of the force at the hinge, the half-weight of the rod of which it forms a part, and the reaction between itself and the intermediate portion of that rod. This last reaction is therefore a force. Since the intermediate portion of each rod has been rendered weightless, the reactions on it will act along the rod, Art. 131. Let the reactions along the intermediate portions of the rods A_0A_1, A_1A_2 &c., be T_0, T_1 &c., and let these be regarded as positive when they pull the terminal particles as if the rods were strings.

To avoid introducing the force at a hinge into our equations we shall consider the equilibrium of the two particles adjacent to that hinge as forming one system.

This compound particle is acted on by the half-weights of the adjacent rods and the reactions along the intermediate portions of those rods. The result of the argument is, that *we may regard all the rods as being without weight, and suppose them to be hinged to heavy particles placed at the joints, the weight of each particle being equal to half the sum of the weights of the adjacent rods.*

A system of weights joined, each to the next in order, by weightless rods or strings and suspended from two fixed points is usually called a *funicular polygon.*

Consider the equilibrium of any one of the compound particles, say that at the joint A_2. Resolving horizontally and vertically, we have

$$\left. \begin{array}{l} T_1 \sin \theta_1 = T_2 \sin \theta_2 \\ T_2 \cos \theta_2 - T_1 \cos \theta_1 = \tfrac{1}{2}(W_1 + W_2) \end{array} \right\} \quad \ldots\ldots\ldots\ldots\ldots\ldots (1).$$

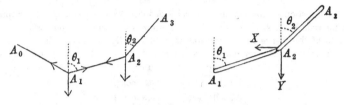

We easily find $\dfrac{\tfrac{1}{2}(W_1 + W_2)}{\cot\theta_2 - \cot\theta_1} = T_1 \sin\theta_1.$

The right-hand side of this equation is the same for all the rods, being equal to the horizontal tension at any joint, we find therefore

$$\frac{\tfrac{1}{2}(W_1 + W_2)}{\cot\theta_2 - \cot\theta_1} = \frac{\tfrac{1}{2}(W_2 + W_3)}{\cot\theta_3 - \cot\theta_2} = \&c. \quad \ldots\ldots\ldots\ldots\ldots\ldots (2).$$

If A_r, A_s be any two joints we see that each of these fractions is equal to

$$\frac{\tfrac{1}{2}W_{r-1} + W_r + \ldots\ldots + W_{s-1} + \tfrac{1}{2}W_s}{\cot\theta_s - \cot\theta_{r-1}}.$$

135. *Second Method.* In this method we consider the equilibrium of any two successive rods, say A_1A_2, A_2A_3, and take moments for each about the extremity remote from the other rod.

Let X_2, Y_2 be the resolved parts of the reaction at the joint A_2 on the rod A_2A_3. The two equations of moments give

$$\left. \begin{array}{l} -X_2 \cos\theta_2 + Y_2 \sin\theta_2 + \tfrac{1}{2}W_2 \sin\theta_2 = 0 \\ -X_2 \cos\theta_1 + Y_2 \sin\theta_1 - \tfrac{1}{2}W_1 \sin\theta_1 = 0 \end{array} \right\} \quad \ldots\ldots\ldots\ldots\ldots (3).$$

Eliminating Y_2 we find

$$X_2(\cot\theta_2 - \cot\theta_1) = \tfrac{1}{2}(W_1 + W_2) \quad \ldots\ldots\ldots\ldots\ldots\ldots (4),$$

which is equivalent to equations (2).

136. Let l_0, l_1 &c. be the lengths of the rods, h, k the horizontal and vertical coordinates of A_n referred to A_0 as origin. We then have

$$\left. \begin{array}{l} l_0 \cos\theta_0 + l_1 \cos\theta_1 + \ldots\ldots + l_{n-1} \cos\theta_{n-1} = k \\ l_0 \sin\theta_0 + l_1 \sin\theta_1 + \ldots\ldots + l_{n-1} \sin\theta_{n-1} = h \end{array} \right\} \quad \ldots\ldots\ldots\ldots (5).$$

The equations (2) supply $n-2$ relations between the angles θ_0, θ_1 &c. and the weights W_0, W_1 &c. of the rods. Joining these to (5) we have sufficient equations to find the angles when the weights are known. When the angles and the weights of two of the rods are known, the $n-2$ remaining weights may be found from (2).

137. It is evident that either of these methods may be used if the rods are not uniform or if other forces besides the weights act on them. The two equations of moments in the second method will be slightly more complicated, but they can be easily formed. In the first method the transference of the forces parallel to themselves to act at the joints is also only a little more complicated, see Art. 79.

138. *To find the reactions at the joints.* If we use the second method, these are easily found from equations (3). But if we use the first method we must transfer the weights $\frac{1}{2}W_1$ and $\frac{1}{2}W_2$ back to the extremities of the rods which meet at A_2. In the original arrangement of the rods when hinged to each other, let R_2 be the action at the joint A_2 on the rod A_2A_3. The terminal particle of the rod A_2A_3 is then acted on by the three forces R_2, $\frac{1}{2}W_2$ and T_2. We therefore have

$$R_2{}^2 = T_2{}^2 + \tfrac{1}{4}W_2{}^2 - W_2 T_2 \cos \theta_2 \dots\dots\dots\dots\dots\dots\dots\dots(6).$$

The direction of the reaction is easily deduced from equations (2). Suppose that the rods A_1A_2, A_2A_3 are joined by a short rod or string without weight. The position of this rod is clearly the line of action of R_2. Treating this rod as if it were one of the rods of the polygon, we have, if ϕ_2 be its inclination to the vertical,

$$\frac{\tfrac{1}{2}W_1}{\cot \phi - \cot \theta_1} = \frac{\tfrac{1}{2}W_2}{\cot \theta_2 - \cot \phi} \dots\dots\dots\dots\dots\dots\dots\dots(7),$$

$$\therefore\ (W_1 + W_2) \cot \phi = W_2 \cot \theta_1 + W_1 \cot \theta_2.$$

139. *The subsidiary polygon.* The lines of action of the reactions R_1, R_2 &c. at the joints will form a new polygon whose corners B_1, B_2 &c. are vertically under the centres of gravity of the rods A_1A_2, A_2A_3 &c. The weights of the rods may be supposed to act at the corners of this new polygon. Each weight will be in equilibrium with the reactions which act along the adjacent sides of the polygon.

If we suppose the corners B_1, B_2 &c. to be joined by weightless strings or rods we shall have a second funicular polygon. This funicular polygon may be treated in the same way as the former one, except that we have the weights W_1, W_2 &c. instead of $\frac{1}{2}(W_1 + W_2)$, $\frac{1}{2}(W_2 + W_3)$ &c.

140. Let $B_0B_1B_2$ &c. be any funicular polygon; W_1, W_2, &c., the weights suspended from the corners B_1, B_2 &c. From any arbitrary point O draw straight lines Ob_1, Ob_2, Ob_3 &c. parallel to the sides B_0B_1, B_1B_2, B_2B_3 &c. to meet any vertical straight line in the points b_1, b_2, b_3 &c. Since a particle at the point B_1 is in equilibrium under the action of the weight W_1 and the tensions R_1, R_2 acting

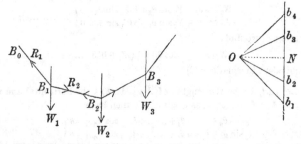

along the sides B_1B_0, B_1B_2, it follows, by the triangle of forces, that the sides of the triangle Ob_1b_2 are proportional to these forces. In the same way, the sides of the triangle Ob_2b_3 represent on the same scale the weight W_2 and the tensions acting along B_2B_1, B_2B_3. In general the straight lines Ob_1, Ob_2 &c. represent the tensions acting along the sides of the funicular polygon to which they are respectively parallel; while any part of the vertical straight line as b_2b_5 represents the sum of the weights at B_2, B_3 and B_4.

By using this figure we may find *geometrically* the relations between the tensions and the weights. If ϕ_1, ϕ_2 &c. be the inclinations of the sides B_0B_1, B_1B_2 &c. to the vertical, we have $ON (\cot \phi_1 - \cot \phi_2) = b_1b_2,$

where ON is a perpendicular drawn from O on the vertical straight line. Since ON represents the horizontal tension X at any point of the funicular polygon, this

equation gives $$\frac{W_1}{\cot \phi_1 - \cot \phi_2} = \pm X = \frac{W_2}{\cot \phi_2 - \cot \phi_3} = \&c.$$

In the same way other relations may be established.

The use of this diagram is described in Rankine's *Applied Mechanics*. Such figures are usually called *force diagrams*. We have here only considered the simple case in which the forces are parallel to each other. In the chapter on Graphics this method of solving statical problems will be again considered and extended to forces which act in any directions.

141. Ex. 1. A chain consisting of a number of equal and in every respect similar uniform heavy rods, freely jointed at their ends, is hung up from two fixed points; prove that the tangents of the angles the rods make with the horizontal are in arithmetical progression, as are also the tangents of the angles the directions of the stresses at the joints make with the same, the common difference being the same for each series.   [Coll. Ex., 1881.]

Ex. 2. OA, OB are vertical and horizontal radii of a vertical circle, A being the lowest point. A string $ACDB$ is fixed to A and B and divided into three equal parts in C and D. Weights W, W' being hung on at C and D, it is found that in the position of equilibrium C and D both lie on the circle. Prove that $W = W' \tan 15^0$.
                                                        [Trin. Coll., 1881.]

Ex. 3. Four equal heavy uniform rods AB, BC, CD, DA are jointed at their extremities so as to form a rhombus, and the corners A and C are joined by a string. If the rhombus is suspended by the corner A, show that the tension of the string is $2W$ and that the reaction at either B or D is $\frac{1}{2}W \tan \frac{1}{2}BAD$, where W is the weight of any rod.

Ex. 4. AB, BC, CD are three equal rods freely jointed at B and C. The rods AB, CD rest on two pegs in the same horizontal line so that BC is horizontal. If a be the inclination of AB, and β the inclination of the reaction at B to the horizon, prove that $3 \tan a \tan \beta = 1$.   [St John's Coll., 1881.]

Ex. 5. Three equal uniform rods are freely jointed at their extremities and rest in equilibrium over two smooth pegs, in a horizontal line at a distance apart equal to half the length of one rod. If the lowest side be horizontal, then the resultant action at the upper joint is $\frac{5}{18}\sqrt{3}W$ and at each of the lower $\frac{1}{18}\sqrt{57}W$, where W is the aggregate weight of the rods.   [Coll. Ex., 1882.]

Ex. 6. Three rods, jointed together at their extremities, are laid on a smooth horizontal table; and forces are applied at the middle points of the sides of the triangle formed by the rods, and respectively perpendicular to them. Show that, if these forces produce equilibrium, the strains at the joints will be equal to one another, and their directions will touch the circle circumscribing the triangle.
                                                        [Math. Tripos, 1858.]

Ex. 7.   Three pieces of wire, of the same kind, and of proper lengths, are bent into the form of the three squares in the diagram of Euclid I., 47, and the angles of the squares which are in contact are hinged together, so that the smaller ones are supported by the larger square in a vertical plane.   Show that in every position, into which the figure can be turned, the action, if any, between the angles of the smaller squares will be perpendicular to the hypothenuse of the right-angled triangle.

[Math. Tripos, 1867.]

Ex. 8.   Three uniform rods, whose weights are proportional to their lengths a, b, c, are jointed together so as to form a triangle, which is placed on a smooth horizontal plane on its three sides successively, its plane being vertical : prove that the stresses along the sides a, b, c when horizontal are proportional to

$$(b+c)\operatorname{cosec} 2A, \quad (c+a)\operatorname{cosec} 2B, \quad (a+b)\operatorname{cosec} 2C. \qquad \text{[Math. Tripos, 1870.]}$$

Ex. 9.   Three uniform rods AB, BC, CD of lengths $2c$, $2b$, $2c$ respectively rest symmetrically on a smooth parabolic arc, the axis being vertical and vertex upwards.   There are hinges at B and C, and all the rods touch the parabola. If W be the weight of either of the slant rods, show that its pressure against the parabola is equal to $W\,\dfrac{a^2 c}{(a^2+b^2)\,b}$, where $4a$ is the latus rectum of the parabola.

[Coll. Ex., 1883.]

Ex. 10. $ABCD$ is a quadrilateral formed by four uniform rods of equal weight loosely jointed together.   If the system be in equilibrium in a vertical plane with the rod AB supported in a horizontal position, prove that $2\tan\theta = \tan\alpha \sim \tan\beta$, where α, β are the angles at A and B, and θ is the inclination of CD to the horizon; also find the stresses at C and D, and prove that their directions are inclined to the horizon at the angles $\tan^{-1}\frac{1}{2}(\tan\beta-\tan\theta)$ and $\tan^{-1}\frac{1}{2}(\tan\alpha+\tan\theta)$ respectively.

[Math. Tripos, 1879.]

Ex. 11.   Four equal rods AB, BC, CD, DA, jointed at A, B, C, D, are placed on a horizontal smooth table to which BC is fixed, the middle points of AD, DC being connected by a string which is tight when the rods form a square.   Show that, if a couple act on AB and produce a tension T in the string, its moment must be $\frac{1}{4}T\,.\,AB\sqrt{2}$. [Coll. Ex., 1888.]

Ex. 12.   A weightless quadrilateral framework $A_1A_2A_3A_4$ rests with its plane vertical and the side A_1A_2 on a horizontal plane.   Two weights W, W' are placed at the corners A_4, A_3 respectively, while a string connecting the two corners A_1A_3 prevents the frame from closing up.   Show that the tension T of the string is given by $nT\sin\theta_2\sin\theta_4 = W\cos\theta_1\sin\theta_3 - W'\cos\theta_2\sin\theta_4$,

where θ_1, θ_2, θ_3, θ_4 are the internal angles of the quadrilateral, and n is the ratio of the side on the horizontal plane to the length of the string.

Ex. 13.   A pentagon formed of five heavy equal uniform jointed bars is suspended from one corner, and the opposite side is supported by a string attached to its middle point of such length as to make the pentagon regular. Prove that the tension of the string is equal to $4W\cos^2\frac{1}{10}\pi$, where W is the weight of any rod.   Find also the reactions at the corners.

Ex. 14.   A regular pentagon $ABCDE$, formed of five equal heavy rods jointed together, is suspended from the joint A, and the regular pentagonal form is maintained by a rod without weight joining the middle points K, L of BC and DE.   Prove that the stress at K or L is to the weight of a rod in the ratio of $2\cot 18^0$ to unity. [Math. Tripos, 1885.]

Ex. 15. The twelve edges of a regular octahedron are formed of rods hinged together at the angles, and the opposite angles are connected by elastic strings ; if the tensions of the three strings are X, Y, Z respectively, show that the pressure along any of the rods connecting the extremities of the strings whose tensions are Y and Z is $(Y+Z-X)/2\sqrt{2}$.   [Math. Tripos, 1867.]

Ex. 16. Any number of equal uniform heavy rods of length a are hinged together, and rotate with uniform angular velocity ω about a vertical axis through one extremity of the system, which is fixed; if θ, θ', θ'' be the inclinations to the vertical of the nth, $(n+1)^{\text{th}}$, $(n+2)^{\text{th}}$ rods counting from the free end, and $a\omega^2 = 3\kappa g$, prove that

$$(2n+3)\tan\theta'' - (4n+2)\tan\theta' + (2n-1)\tan\theta + \kappa\{\sin\theta'' + 4\sin\theta' + \sin\theta\} = 0.$$
     [Math. Tripos, 1877.]

Reactions at rigid connections.

142. Let AB be a horizontal rod fixed at the extremity A in a vertical wall, and let it support a weight W at its other extremity B. We may enquire what are the stresses across a section at any point C, by which the portion CB of the rod is supported.

It is evident that the reaction at C cannot consist of a single force, for then a force acting at C would balance a force W to which it could not be opposite. It is also clear that the resultant action across the section C (whatever it may be) must be equal and opposite to the force W acting at B. Let us transfer the force W from B to any point 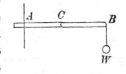 of the section C by help of Art. 100. We see that the reaction across the section is equivalent to a force equal to W, together with a couple whose moment is $W \cdot BC$.

If the portion CB of the rod is heavy, we may suppose its weight collected at the middle point of CB. Let W' be the weight of this part of the rod. Then we must transfer this weight also to the base of reference C. The whole reaction across the section of the rod will then consist of (1) a force $W + W'$ and (2) a couple whose moment is $W \cdot BC + \frac{1}{2}W' \cdot BC$.

Various names have been given to the reaction force and reaction couple at different times. The components of the force along the length of the rod and transverse to it have been called the *tension* and *shear* respectively. The former being normal to a perpendicular section of the rod is sometimes called the *normal* stress. The magnitude of the couple has been called the tendency of the forces to break the rod, or briefly, the *tendency to break*. It

is also called the *moment of flexure*, or *bending stress*. See Rankine's
Applied Mechanics. In what follows we shall restrict ourselves to
the case in which the rod is so thin that we may speak of it as a
line in discussing the geometry of the figure.

143. Generalizing this argument, we arrive at the following
result: *the action across a section at any point C of a rod is equal
and opposite to the resultant of all the forces which act on the rod
on one side of that point C.*

The action across C on CB balances the forces on CB. The
equal and opposite reaction on AC across the same section balances
those on AC. Since the forces on one side of C balance those on
the other side when there is equilibrium, it is a matter of indiffer-
ence whether we consider the forces on the one side or the other
of C provided we keep them distinct.

Thus the bending couple at C is equal to the sum of the
moments of all the forces which act on one side of C. So also the
shear at C is equal to the sum of the resolved parts of these forces
along the normal to the rod at C.

If we regard the rod as slightly elastic we may explain otherwise the origin of the force and couple. The weight W will
slightly bend the rod, and thus stretch the upper fibres and compress the lower ones. The action across the section at C will
therefore consist of an infinite number of small tensions across its
elements of area. By Art. 104 all these can be reduced to a single
force and a single couple at a base of reference at C.

144. Ex. 1. *A rod AB, of given length l, is supported in a horizontal position
by two pegs, one at each end. A heavy particle M, whose weight is W, traverses the
rod slowly from one end to the other. It is required to find the stresses at any point.*

Let $AM = \xi$, $BM = l - \xi$. Let R and R' be the pressures of the supports at A and
B on the rod. These are evidently given by

$$R'l = W . \xi, \qquad Rl = W(l - \xi).$$

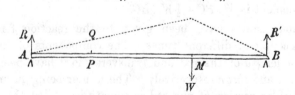

Let P be the point at which the stresses are required, and let $AP = x$. To find
these we consider the equilibrium of either the portion AP or the portion BP of
the rod. We choose the former, as the simpler of the two, because there is only

one force, viz. R, acting on it. The shear at P is therefore equal in magnitude to R, and the moment of the stress couple is equal to Rx.

If the point at which the stresses are required is on the other side of M as at P', where $AP' = x'$, it is more convenient to consider the equilibrium of BP'. The shear is here equal to R', and the bending moment to $R'(l - x')$.

As the bending couple is generally more effective in breaking a rod than either the shear or the tension, we shall at present turn our attention to the couple. If at every point P we erect an ordinate PQ proportional to the bending couple at P, the locus of Q will represent to the eye the magnitude of the bending couple at every point of the rod. In our case the locus of Q is clearly portions of two straight lines, represented in the figure by the dotted lines. The maximum ordinate is at the point M, and is represented by either $R\xi$ or $R'(l - \xi)$, according as we take moments about M for the sides AM or MB of the rod. Substituting for R or R', the bending couple at M becomes $W\xi(l - \xi)/l$. This is a maximum when M is at the middle point of AB.

This result shows in a general way that, when a man stands on a stiff plank laid across a stream, the bending couple is greatest at the point of the plank on which the man stands. Also if he walks slowly along the plank, the bending couple is greatest when he is midway between the two supports.

Ex. 2. A uniform heavy rod AB is supported at each end. If w be the weight per unit of length, prove that the bending couple at any point P will be $\tfrac{1}{2}w \cdot AP \cdot BP$.

145. When several forces act on a rod, the diagram by which the distribution of bending stress is exhibited to the eye can be constructed in a similar manner. Let forces R_1, R_2 &c. act at the points A_1, A_2 &c. of a rod in the directions indicated by the arrows. Let $A_1A_2 = a_2$, $A_1A_3 = a_3$ and so on. Then the bending moment at any point P, say between A_3 and A_4, is obtained by taking the moments of the forces which act at A_1, A_2, A_3, these being points on one side of P. Putting $A_1P = x$, the required bending moment is

$$y = R_1 x - R_2 (x - a_2) + R_3 (x - a_3).$$

Erecting an ordinate PQ to represent y, it is clear that the locus of Q between A_3 and A_4 is a straight line.

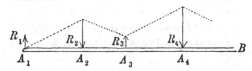

When the point P moves beyond A_4 we must add to this expression the moment of the force R_4, i.e. $- R_4 (x - a_4)$. The locus of Q is now a different straight line. It intersects the former at the point $x = a_4$, i.e. at the top of the ordinate corresponding to the point A_4, but its inclination to the rod is different.

We infer that, *when a rod is acted on only by forces at isolated points, the diagram representing the bending couple will consist of a series of finite straight lines.* This indicates an easy method of constructing the diagram. Calculate the ordinates representing the bending couples at these isolated points, and join their extremities by straight lines. In this case there can be no maximum ordinate between the isolated points A_1, A_2 &c. at which the forces act. Hence *the bending couple can be a maximum or minimum only at one of these points.*

If the rod is heavy, its weight is distributed over the whole rod. The bending couple at P will contain not merely the moments of the forces which act at A_1, A_2 &c., but also that of the weight of the portion A_1P of the rod. If w be the weight per unit of length, the bending couple at P will be

$$y = \Sigma R (x - a) - \tfrac{1}{2}wx^2,$$

for the weight of A_1P will be wx, and it may be collected at the middle point of A_1P.

This is the equation to a parabola. Hence *the diagram will consist of a series of arcs of parabolas*, each intersecting the next at the extremity of the ordinate along which an isolated force acts. All these parabolas have their axes vertical. If the different sections of the rod be of the same weight per unit of length, the latera recta of the parabolas will be equal.

This expression gives the bending moment by which the forces on the left or negative side of any point P tend to turn the portion of the rod on the positive side of P in the direction of rotation of the hands of a watch.

Suppose that any portion CD of a rod $ACDB$ has no weight, and that no point of support lies between C and D. The remaining parts of the rod on each side of CD may have any weights and any number of points of support. The bending couple at any point between C and D is always proportional to the ordinate of some straight line. But if y_1, y_2, and y are ordinates of any straight line at C, D and P, and if the distances CP and PD are l_1 and l_2, it is easy to see that

$$y (l_1 + l_2) = y_1 l_2 + y_2 l_1.$$

This equation therefore must also connect the bending couples y_1, y_2, and y at the points C, D, and any intermediate point P.

Let us next suppose that the portion CD of the rod is heavy. The bending couple at any point of this portion of the rod is now proportional to the ordinate of the parabola $y = A + Bx - \tfrac{1}{2}wx^2$, where $A = -\Sigma Ra$ and $B = \Sigma R$. If y_1, y_2 and y are the ordinates at C, D and any point P, where $CP = l_1$, $PD = l_2$, it is easy to prove that

$$y (l_1 + l_2) = y_1 l_2 + y_2 l_1 + \tfrac{1}{2}w l_1 l_2 (l_1 + l_2).$$

This equation connects the bending couples at any three points of a heavy rod provided there is no point of support within the length considered.

Ex. If y_1, y_2, y_3 be the bending couples at three consecutive points of support of a heavy horizontal rod whose distances apart are l_1, l_2, then

$$y_2 (l_1 + l_2) = y_1 l_2 + y_3 l_1 + \tfrac{1}{2}w l_1 l_2 (l_1 + l_2) - R l_1 l_2,$$

where R is the pressure at the middle point of support, and w is the weight of the rod per unit of length.

146. Since the bending couple at any point P is the *sum* of the moments of the several forces which act on one side of P, it is clear that *each force contributes its share to the bending couple as if it acted alone on the rod.* In this way it is sometimes convenient to consider the effects of the forces separately.

For example, if a heavy rod AB, supported at each end, has a weight W placed at a point M, the bending couple at any point P is the sum of the bending couples found in Art. 144 for the two cases in which (1) the rod is light and (2) there is no weight at M. The bending couple is therefore given by

$$ly = W . BM . AP + \tfrac{1}{2}wl . AP . BP.$$

147. Ex. 1. A heavy rod is supported in a horizontal position on two pegs, one at each end. A heavy particle, whose weight is n times that of the rod, is placed at a point M. If C be the middle point of the rod, show that the bending couple will be greatest either at some point between M and C or at M, according as the distance of M from C is greater or less than n times its distance from the nearer end of the rod.

Ex. 2. A semicircular wire ACB is rotated with uniform angular velocity about a tangent at one extremity A. Show that the bending couple is zero at B, is a maximum at the middle point C, vanishes at some point between C and A, and is again a maximum with the opposite sign at A. Show also that the maximum at A is greater than that at C.

It may be assumed that the effect of rotation is represented by supposing the wire to be at rest, and each element to be acted on by a force tending directly from the axis of rotation and proportional to the mass of the element and its distance from the axis.

Ex. 3. A horizontal beam AB, without weight, supported but not fixed at both ends A and B, is traversed from end to end by a moving load W distributed equally over a segment of it, of constant length PQ. Show that the bending moment at any point X of the beam, as the load passes over it, is greatest when X divides PQ in the same ratio as that in which it divides AB. Show also that this maximum bending moment is equal to $W . AX . BX (AB - \tfrac{1}{2}PQ)/AB^2$.   [Townsend.]

Let $AX=a$, $BX=b$, $AB=a+b$, $PQ=l$, $AP=x$, $BQ=\xi$. Let R be the shear at X, and y the bending moment. Since the weight of PX, viz. $w(a-x)$, may be collected at its middle point we have by taking moments about A for the portion AX of the beam $\tfrac{1}{2}w(a-x)(a+x) - y + Ra = 0$, similarly, taking moments for BX about B, $\tfrac{1}{2}w(b-\epsilon)(b+\epsilon) - y - Rb = 0$.

Eliminating R, $2l(a+b)y = W \{ab(a+b) - bx^2 - a\xi^2\}$.

Making y a maximum with the condition $x+\xi=a+b-l$, the results follow at once.

Ex. 4. A uniform horizontal beam, which is to be equally loaded at all points of its length, is supported at one end and at some other point; find where the second support should be placed in order that the greatest possible load may be placed upon the beam without breaking it, and show that it will divide the beam in the ratio 1 to $\sqrt{2} - 1$.   [Math. Tripos.]

Let ABC be the beam supported at A and B. Let wdx be the load placed on dx; wR, wR' the pressures at A, B. Let l be the length of the beam, $\xi = AB$, then $2\xi > l$. We easily find $R = l - \dfrac{l^2}{2\xi}$, $R' = \dfrac{l^2}{2\xi}$.

Let P and Q be two points in CB and BA respectively, $x=CP$, $x'=AQ$. By taking moments about P and Q respectively the bending couples y, y' at P and Q are found to be $y = -\tfrac{1}{2}wx^2$, $y' = wRx' - \tfrac{1}{2}wx'^2$.

The first parabola has its maximum ordinate at B, the second has a maximum ordinate at a point $x' = R$ which must lie between A and B. The bending couples at these points are numerically equal to $\tfrac{1}{2}w(l-\xi)^2$ and $\tfrac{1}{2}w\left(l - \dfrac{l^2}{2\xi}\right)^2$. If these are unequal, the support B can be moved so as to diminish the greater. The proper position is found by making these equal; hence $\pm(l-\xi) = l - l^2/2\xi$. Since ξ must be greater than $\tfrac{1}{2}l$, this gives $\xi\sqrt{2} = l$.

Ex. 5. Three beams AB, BC, CA are jointed at A, B, C, B being an obtuse angle, and are placed with AB vertical, and A fixed to the ground, so as to form the framework of a crane. There is a pulley at C, and the rope is fastened to AB near B and passes along BC and over the pulley. If it support a weight W, large in comparison with the weights of the framework and rope, find the couples which tend to break the crane at A and B.   [Math. Tripos.]

Ex. 6. A gipsy's tripod consists of three uniform straight sticks freely hinged together at one end. From this common end hangs the kettle. The other ends of the sticks rest on a smooth horizontal plane, and are prevented from slipping by a smooth circular hoop which encloses them and is fixed to the plane. Show that there cannot be equilibrium unless the sticks be of equal length ; and if the weights of the sticks be given (equal or unequal) the bending moment of each will be greatest at its middle point, will be independent of its length, and will not be increased on increasing the weight of the kettle.   [Math. Tripos, 1878.]

Ex. 7. A brittle rod AB, attached to smooth hinges at A and B, is attracted towards a centre of force C according to the law of nature. Supposing the absolute force to be indefinitely augmented, prove that the rod will eventually snap at a point E determined by the equation $\sin \frac{1}{2} (a + \beta) \cos \theta = \sin \frac{1}{2} (a - \beta)$, where a, β denote the angles BAC, ABC, and θ the angle AEC. *Math. Tripos*, 1854. See also the solutions for that year by the Moderators and Examiners.

Indeterminate Problems.

148. When a body is placed on a horizontal plane, the pressure exerted by its weight is distributed over the points of support. When there are more than three supports, or more than two in one vertical plane, this distribution appears to be indeterminate. Thus suppose the body to be a table with vertical legs, and let these legs intersect the plane horizontal surface of the table in the points A_1, A_2 &c. Let the projection on this plane of the centre of gravity of the body be G. The weight W of the table will then be supported by certain pressures R_1, R_2 &c. acting at A_1, A_2 &c. Let Ox, Oy be any rectangular axes of reference in this plane and let Oz be vertical. Let $(x_1 y_1), (x_2 y_2)$ &c. be the coordinates of A_1, A_2 &c. and let (xy) be those of G. Since W is supported by a system of parallel forces we have by Arts. 110 and 111

$$W = R_1 + R_2 + \dots$$
$$Wx = R_1 x_1 + R_2 x_2 + \dots$$
$$Wy = R_1 y_1 + R_2 y_2 + \dots$$

These three equations suffice to determine R_1, R_2 &c. if there are but three of them and these not all in one vertical plane, but if there are more than three, the problem appears to be indeterminate.

In this solution we have replaced the supporting power of the floor by forces R_1, R_2 &c. acting upwards along the legs. What we have really proved is that the table could be supported by such forces in a variety of different ways. Suppose there were four legs; we could choose one of these forces to be what we please, the others could then be found from these three equations. It is therefore evident that the problem of finding what forces could support the table must be indeterminate.

The actual pressures exerted by the table on the floor are not indeterminate, for in nature things are necessarily determinate. When anything appears to be indeterminate, it must be because we have omitted some of the data of the question, i.e. some property of matter on which the solution depends.

We notice that the elementary axioms relating to forces, which have been enunciated in Art. 18, make no reference to the nature of the materials of the body. We have found in the preceding Articles that the equations supplied by these axioms have in general been sufficient to determine all the unknown quantities in our statical problems. In all these problems therefore the magnitudes of the reactions and the positions of equilibrium of the bodies depended, not on the materials of the bodies, but on their geometrical forms and on the magnitudes of the impressed forces. It is evident, however, that these axioms must be insufficient to determine any unknown quantities which depend on the materials of the bodies. In such cases we must have recourse to some new experiments to discover another statical axiom. Thus, when we study the positions of equilibrium of rough bodies, another experimental result, depending on the degree of roughness of the special body considered, is found to be necessary. In the same way the mode of distribution of the pressure over the legs of the table is found to depend on the flexibility of the materials.

However slight the flexibility of the substance of the table may be, yet the weight W will produce some deformation however small. The magnitude of this will influence and be influenced by the reactions R_1, R_2 &c. The amount of yielding produced by the acting forces in any body is usually considered in that part of mechanics called *the theory of elastic solids*. No complete solution of the special problem of the table has yet been found. But when any assumed law of elasticity is given, it is easy to show by some examples, how the problem becomes determinate. *Poinsot's Éléments de Statique* and *Poisson's Traité de Mécanique*.

149. Ex. 1. *A rectangular table has the legs at the four corners alike in all respects and slightly compressible. The amount of compression in each leg is supposed to be proportional to the pressure on that leg. Supposing the floor and the top of the table to be rigid, and the table loaded in any given manner, find the pressure on the four legs. Show that when the resultant weight lies in one of four straight lines on the surface of the table, the table is supported by three legs only.* [Math. Tripos, 1860, Watson's problem, see also the Solutions for that year.]

Let the two sides AB, AD be the axes of x and y. Let the resultant weight W act at a point G whose coordinates are (xy). Let $AB=a$, $AD=b$. Since the top of the table is rigid, the surface as altered by the compression of the legs is still plane. Also, since the compression is slight, we shall neglect small quantities of the second order, and suppose the pressures at A, B, C, D to remain vertical. We have the usual statical equations

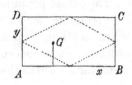

$$W=R_1+R_2+R_3+R_4,$$
$$Wx=(R_2+R_3)a, \qquad Wy=(R_3+R_4)b \Big\} \quad \dots\dots(1).$$

Because a diagonal of the table remains straight, the middle point descends a space which is the arithmetic mean of the spaces descended by its two ends. It follows that the mean of the compressions of the legs A and C is equal to the mean of the compressions of the legs B and D. But it is given that the pressures are proportional to these compressions. Hence

$$R_1+R_3=R_2+R_4 \dots\dots\dots\dots(2).$$

These four equations determine the pressures.

If we put $R_3=0$, we easily find that $2x/a+2y/b=1$, i.e. the table is supported on the three legs A, B, D when the weight W lies on the straight line joining the middle points of AB, AD. Joining the middle points of the other sides in the same way, we obtain four straight lines represented by the dotted lines. When the weight W lies within this dotted figure all the four legs are compressed; when without this figure three legs only are compressed. The equations above written are then correct, only if we suppose that some of the reactions are negative. As this cannot in general be possible, we must amend the equations (1) by putting one reaction equal to zero. The equation (2) must then be omitted.

Ex. 2. *A and C are fixed points or pegs in the same vertical line, about which the straight beams ADB and CD are freely moveable. AB is supported in a horizontal position by CD and has a weight W suspended at B. Find the pressure at C (1) when there is a hinge joint at D, and (2) when CD forms one piece with AB, the weights of the beams being in each case neglected.* [Math. Tripos, 1841.]

In the first part of the problem the action at D is a single force, in the second part it is a force and a couple, Art. 142. In both parts of the problem the action at C is a force.

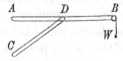

In the first part, the actions at C and D are equal and act along CD by Art. 131. Taking moments about A for the rod AD, we easily find that this action is equal to $W \cdot AB/AN$ where AN is a perpendicular on CD.

In the second part there is nothing to determine the direction of the action at C. We only know it balances an unknown force and a couple. If we write down the three equations of equilibrium for the whole body, it will be seen that we cannot find the four components of the two pressures which act at A and C. The problem is therefore indeterminate.

Ex. 3. A rigid bar without weight is suspended in a horizontal position by means of three equal vertical and slightly elastic rods to the lower ends of which are attached small rings A, B, and C through which the bar passes. A weight is then attached to the bar at any point G. Show that, on the assumption that the extension or compression of an elastic rod is proportional to the force applied to stretch or compress it, and provided the rods remain vertical, then the rod at B will be compressed if G lie in the direction of the longer of the two arms AB, BC, and be at a greater distance from B than $\dfrac{AB^2 + BC^2}{AB \sim BC}$.   [Math. Tripos, 1883.]

Ex. 4. $ABCD$ is a square; six rods AB, BC, CD, DA, AC, BD are hinged together at the angular points, and equal and opposite forces, F, are applied at B and D in the directions DB and BD respectively. The rods are elastic, but the extensions or compressions which occur may be treated as infinitesimal. e_1 is the ratio of the extension per unit length to the tension (or of the compression to the corresponding force) for the rod AB, and is a constant depending upon the material and the section of the rod. e_2, $e_3 \ldots e_6$ are similar constants for the other rods in the order written above. Prove that the tension of the rod BD is

$$\left(1 - \frac{2\sqrt{2}\,e_6}{e_1 + e_2 + e_3 + e_4 + 2\sqrt{2}\,(e_5 + e_6)}\right) F. \qquad \text{[Coll. Exam. 1886.]}$$

The rods being only slightly elastic we form the ordinary equations of equilibrium on the supposition that the figure has its undisturbed form, i.e. that $ABCD$ is a square. We then find that the thrust along every side is the same. If the thrust along any side be P and those along the diagonals BD, AC be T and T', we have also $P\sqrt{2} + T' = 0$, $P\sqrt{2} + T + F = 0$.

We next seek for a geometrical relation between the six lengths of the figure after it has been disturbed by the action of the forces F, F. If the lengths of the sides taken in the order mentioned in the question be $a\,(1+x)$, $a\,(1+y)$, $a\,(1+z)$, $a\,(1+u)$, $a\sqrt{2}\,(1+\rho')$, $a\sqrt{2}\,(1+\rho)$, we find that $2\,(\rho + \rho') = x + y + z + u$, when the squares of the small quantities are neglected. Using the law of elasticity, this geometrical condition is equivalent to $2\,(e_6 T + e_5 T') = (e_1 + e_2 + e_3 + e_4)\,P$.

We have now three equations to find P, T and T' in terms of F.

150. Stiff Framework*. Let A_1, A_2 &c. be n particles connected together by straight rods hinged to these particles. We shall suppose that all the forces which act on the system are applied to these particles, so that the reactions at the extremities of every rod are forces, both of which act along the rod. *It is proposed to ascertain whether the ordinary statical equations are or are not sufficient in number to find all these reactions*, i.e. to ascertain whether the problem of finding these pressures is determinate or indeterminate. *In the latter contingency it is further proposed to ascertain whether the equations of elasticity are sufficiently numerous to enable us to complete the solution.*

---

* The reader may consult on the subject of frameworks two papers by Maxwell in the *Phil. Mag.*, 1864 and the *Edinburgh Transactions*, 1872, also the *Statique Graphique*, by Maurice Levy, 1887.

151. *Let us first enquire what number of connecting rods could make the framework stiff.* Assuming n not to be less than 2, we start by stiffening two particles A_1 and A_2 by means of one connecting rod. The remaining $n-2$ have to be jointed to these. In order that a third particle A_3 should be rigidly connected to these two, it must be joined to both A_1 and A_2, thus requiring two more connecting rods. If a fourth A_4 is to be rigidly connected with these, it must be joined to any two out of the three particles already joined. Proceeding in this manner we see that for each particle joined to the system two additional rods are necessary. Thus *to make a system of n particles rigid, a framework of* $2(n-2)+1$, *i.e.* $2n-3$, *connecting rods is sufficient.*

When any particle, as A_3, is joined by two rods to two other particles as A_1, A_2, there must be some convention to settle on which side of the base A_1A_2 the vertex of the triangle $A_3A_1A_2$ is to be taken. If not, there may be more than one polygon having sides equal to the given lengths.

We must also notice that when the particle A_3 is joined to the fixed particles A_1, A_2 by two rods, if A_3 should happen to be in the same straight line with A_1A_2, the connection is not made perfectly rigid. The particle A_3 could make an *infinitely small displacement* perpendicular to the straight line $A_1A_2A_3$ on either side of it. This is an imaginary displacement, to be taken account of when the circumstances of the problem require that we should neglect small quantities of the second order.

If the particles are not all in the same plane, and n is not less than 3, we start with three particles requiring three rods to stiffen them. Each additional particle of the remaining $n-3$ must be attached to three of the particles already connected. Thus *to make a system of n particles rigid, a framework of* $3(n-3)+3$, *i.e.* $3n-6$, *connecting rods is sufficient.*

It is not necessary that the connections between the particles should be made in the precise way just described. All we have proved is that the system could be stiffened by $2n-3$ or $3n-6$ rods *properly placed.* These may be arranged in several different ways* so as to stiffen the system. On the other hand if the rods are not properly placed the system may not be stiff; thus one part of the system may be stiffened by more than the necessary number of rods, and another part may not have a sufficient number.

A system of particles made rigid by just the necessary number of bars is said to be *simply stiff* or *just stiff.* When there are more bars than the necessary number, the system may be called *over stiff.* When the number of bars is less than the number necessary to stiffen the system, the framework is said to be *deformable.* The shape it will assume in equilibrium is then unknown and has to be deduced, along with the reactions, from the equations of equilibrium.

152. We may infer as a corollary from this that a polygon having n corners is in general given when we know the lengths of $2n-3$ sides. If m be the number of sides and diagonals in the polygon, there must be $m-(2n-3)$ relations between their lengths. *It appears that $2n-3$ of the m lengths are arbitrary except that*

---

* The argument may be summed up as follows. Taking any fixed axes, a figure is given *in position and form* when we know the $2n$ or $3n$ coordinates of its n corners. These are the arbitrary quantities of the framework. If only *its form* is to be determinate we refer the figure to coordinate axes fixed relatively to itself, and the coordinates required to determine *the position* of a free rigid body are now no longer at our disposal. We therefore have $2n-3$ or $3n-6$ arbitrary quantities according as the body is in one plane or in space, Art. 206.

they must satisfy such conditions as will permit a figure to be formed; for instance if three of the arbitrary lengths form a triangle, any two of the lengths must together be greater than the third. The exceptional case referred to above occurs when some of these necessary conditions are only just satisfied.

If *all* the corners are joined, each to each, the number of lengths will be $\frac{1}{2}n(n-1)$. There will therefore be $\frac{1}{2}(n-2)(n-3)$ relations between the sides and diagonals of a polygon of n corners. In the same way there will be $\frac{1}{2}(n-3)(n-4)$ relations between the edges of a polyhedron.

153. *Let us next enquire how many statical equations we have.* Let us suppose the system to be acted on by any given forces whose points of application are at some or all of the particles. These we may call the external forces.

Since each particle separately is in equilibrium, we may, by resolving the forces on each parallel to the axes, obtain $2n$ or $3n$ equations of equilibrium according as the system is in one plane or in space.

However numerous the reactions along the rods may be, we can always eliminate them from these equations and obtain either three or six equations, according as the system is in one plane or in space. To prove this, we notice that, taking all the particles together as one system, the internal reactions balance each other. Resolving then the external forces in some two directions in the plane of the system and taking moments about some point, we obtain * three equations of equilibrium free from all internal reactions (Art. 112). And it is clear that no resolutions in other directions and no moments about other points will give more independent equations than three (Art. 115). In the same way, if the system is in space, it will be shown that we can obtain six equations free from internal reactions by resolving in some three directions and taking moments about some three axes. On the whole then we have either $2n-3$ or $3n-6$ equations to find the reactions. In a simply stiff framework we have just this number of independent reactions. Thus *in a framework, simply stiff, without any unknown external reactions, we have a sufficient number of equations to find all the $2n-3$ or $3n-6$ reactions.*

If the framework is subject to external constraints, for example if some points are fixed in space, the number of bars necessary to stiffen the system is altered. Whether stiff or not let there be $2n-3-k$ or $3n-6-k$ bars. It follows easily that the equations of statics will supply $k+3$ or $k+6$ equations (after elimination of the internal reactions) to find the external reactions and the position of equilibrium. If these are sufficient the problem is determinate.

154. Although the equations in statics may be sufficient in number to determine the internal reactions, yet exceptional cases may arise. *The equations thus obtained may not be independent, or they may be contradictory.*

---

* If it is not clear that these three equations must follow from the $2n$ or $3n$ equations of equilibrium of the separate particles, we may amplify the proof as follows. If any particle A_1 is acted on by a reaction R_{12} tending to A_2, then the particle A_2 is acted on by an equal and opposite reaction R_{21} tending to A_1. The resolved parts of R_{12} and R_{21} parallel to x will therefore also be equal and opposite. If then we add together all the equations obtained from all the particles separately by the resolution parallel to x, the sum will yield an equation free from all the R's. In the same way the resolution parallel to y or z will each yield another equation free from all the internal reactions.

Next since the forces on each particle balance, the sum of their moments about any straight line is zero. But by the same reasoning as before the moment of the reaction R_{12} which acts on A_1 must be equal and opposite to that of the reaction R_{21} which acts on A_2. Hence if we add all the equations obtained from all the particles by taking moments, the sum will yield an equation free from all the R's.

As an example consider the case of three rods, A_1A_3, A_3A_2, A_1A_2 jointed at A_1, A_3, A_2, and let the lengths be such that all three are in one straight line. Let the extremities A_1, A_2 be acted on by two opposite forces each equal to F. Let R_{12}, R_{23}, R_{13} be the reactions along A_1A_2, A_2A_3, A_1A_3 respectively. Here we have a simply stiff framework and we should therefore find sufficient equations to determine the reactions. The equations of equilibrium for the three corners are however

$$R_{13}+R_{12}=F, \qquad R_{13}=R_{23}, \qquad R_{12}+R_{23}=F,$$

which are evidently insufficient to determine the three reactions.

The conditions under which these exceptional cases can arise are determined algebraically by the theory of linear equations. The $2n-3$ or $3n-6$ equations to find the reactions at the corners of the framework are all linear. If a certain determinant is zero, one equation at least can be derived from the others or is contradictory to them. In the latter case some of the reactions are infinite; this of course is impossible in nature. In the former case one reaction is arbitrary, and all the others can be found in terms of it and the given external forces. In a similar manner we can find the condition that two reactions are arbitrary. These conditions can be expressed in a more definite way, but as this part of the theory follows more easily from the principle of virtual work, we shall postpone its consideration until we come to the chapter on that subject.

155. *Let us next suppose that the system of n particles has more than the number of bars necessary to stiffen it.* In this case there are not enough equations to find the reactions unless something is known about them besides what is given by the equations of statics. The rods connecting the particles are in nature elastic, and the forces acting along them are due to their extensions or compressions. Supposing the law connecting the force and the extension to be known, we have to examine whether the additional equations thus supplied are sufficient to find the reactions. The framework, being acted on by external forces, will yield, and this yielding will continue to increase until the reactions thus called into play are of sufficient magnitude to keep the frame at rest. For the sake of brevity we shall suppose that the amount of the yielding is very slight. In this case we shall assume, in accordance with Hooke's law, that the reaction along any rod is some known multiple of the ratio of the extension to the original length. This multiple depends on the nature of the material of which the rod is made.

Let the framework have m rods, where m exceeds $2n-3$ or $3n-6$ by k. Taking the case in which the framework is not acted on by any external reactions, we shall require k additional equations (Art. 153). By Art. 152 there are k relations between the lengths of these rods. Let any one of these be

$$f(l_1, l_2, \&c.) = 0 \quad\dotfill\quad (1),$$

where l_1, l_2 &c. are the lengths of the rods. Differentiating this we have

$$M_1 dl_1 + M_2 dl_2 + \&c. = 0 \quad\dotfill\quad (2),$$

where M_1, M_2 &c. are partial differential coefficients, and dl_1, dl_2 &c. are the extensions of the sides. If R_1, R_2 &c. are the reactions along the sides we may, by Hooke's law, write this equation in the form

$$M_1\lambda_1 l_1 R_1 + M_2\lambda_2 l_2 R_2 + \&c. = 0,$$

where λ_1, λ_2 &c. are the reciprocals of the known multiples.

It appears therefore that each equation such as (1) *supplies one relation between the reactions. Thus the requisite number of additional equations can be deduced from the theory of elasticity.*

In the case of the three rods mentioned in Art. 154 we notice that the relation corresponding to (1) is $l_{13} + l_{23} - l_{12} = 0$, where $l_{12} = A_1 A_2$, &c. It follows by differentiation that the three reactions are equal in magnitude if all three rods are made of the same material and are of equal sectional areas.

Astatics.

156. *Let a rigid body be acted on at given points* A_1, A_2 &c. by *forces* P_1, P_2 &c. *whose magnitudes and directions in space are given. Let this body be displaced in any manner: it is required to find how the resultant force and couple are altered.*

Choosing any base of reference O and any rectangular axes Ox, Oy fixed in the body, we may imagine the displacement made by two steps. First, we may give the body a linear displacement by moving O to its displaced position O_1, the body moving parallel to itself; secondly, we may give the body an angular displacement, by turning the body round O_1 as a fixed point until the axis Ox comes into its displaced position. Then every point of the body will be brought into its proper displaced position, for otherwise the several points of the body would not be at invariable distances from the base O and the axis Ox.

Since the forces P_1, P_2 &c. retain unaltered their magnitudes and directions in space, it is clear that the linear displacement does not in any way affect the resolved parts of the forces, or the moment about O. We may therefore disregard the linear displacement and treat O and O_1 as coincident points.

Consider next the angular displacement. It is clear that we are only concerned with the relative positions of the body and forces, for a rotation of both together will only turn the resultant force and couple through the same angle. Instead of turning the body round O through any given angle θ keeping the forces unaltered, we may turn each force round its point of application through an equal angle in the opposite direction, keeping the body unaltered. See Art. 70.

157. We are now in a position to find the changes in the resultant force and couple. Let Ox, Oy be any axes fixed in the body. Let P be any one of the forces P_1, P_2 &c. and let A be its point of application. Let α be the angle its direction makes with the axis of x. Let this force be turned round A through an angle θ in the positive direction, so that it now acts in the direction indicated in the figure by AP'.

Let X, Y, G be the resolved parts of the forces, and the moment about O before displacement; X', Y', G' the same after displacement. Then, as in Art. 106,

$$X' = \Sigma P \cos(\alpha + \theta) = X \cos\theta - Y \sin\theta,$$
$$Y' = \Sigma P \sin(\alpha + \theta) = X \sin\theta + Y \cos\theta,$$
$$G' = \Sigma P \{x \sin(\alpha + \theta) - y \cos(\alpha + \theta)\}$$
$$= G \cos\theta + V \sin\theta,$$

where $G = \Sigma(xP_y - yP_x)$, $V = \Sigma(xP_x + yP_y)$.

The symbol G represents the moment of the forces *before displacement* about the centre O of rotation. If the angle of rotation round O is a right angle, $\theta = \frac{1}{2}\pi$ and $G' = V$. Thus the symbol V *represents the moment of the forces about O after they have been rotated through a right angle**. If it is permitted to alter slightly a name given by Clausius (see *Phil. Mag.*, August 1870), V might be called the *Virial of the forces*. After a rotation through an angle θ let V' be the new value of the virial, then $V' = \Sigma P \{x \cos(\alpha + \theta) + y \sin(\alpha + \theta)\}$
$$= V \cos\theta - G \sin\theta.$$

Thus it appears that the moment G is also what the virial becomes (with the sign changed) when the forces have been rotated through a right angle.

We may find another meaning for the virial V. Let us suppose the components P_x, P_y to act at O, and let their point of application be moved to N, where $ON = x$. The work of P_x is xP_x, that of P_y is zero. Let the point of application be further moved from N to A, where $NA = y$. The additional work of P_x is zero, that of P_y is yP_y. The sum of these two for all the forces is V. Thus V *is the work of moving the forces from the base of reference O to their respective points of application*, the forces being supposed unaltered in direction or magnitude.

158. If the body is in equilibrium before displacement, we have $X = 0$, $Y = 0$, $G = 0$. Hence after a rotational displacement through an angle θ we have $X' = 0$, $Y' = 0$, $G' = V \sin\theta$. We therefore infer that the only other position in which the body can be in equilibrium is when $\theta = \pi$, i.e. when the position of the body has been reversed in space. If the body is in equilibrium in any two positions which are not reversals of each other, the body must be in equilibrium in all positions. Lastly, the analytical condition that there should be equilibrium in all positions is that $V = 0$ in some one position of equilibrium.

---

* Darboux, *Sur l'équilibre astatique*, p. 8.

159. Ex. 1. A body is placed in any position not in equilibrium, and the forces are such that the components X, Y are both zero. Find the angle through which the body must be rotated that it may come into a position of equilibrium.

Ex. 2. If a body be in a position of equilibrium under the action of forces whose magnitudes and directions in space are given, show that the equilibrium is stable or unstable according as V is positive or negative in the position of equilibrium.

160. Centre of the forces. It has been shown in Art. 118, that, provided the components of the forces (viz. X and Y) are not both zero, the whole system can be reduced to a single resultant at a finite distance from the base of reference. In any position of the forces, the equation to this single resultant is

$$G' - \xi Y' + \eta X' = 0,$$

i.e. $(G - \xi Y + \eta X) \cos \theta + (V - \xi X - \eta Y) \sin \theta = 0.$

Thus it appears that, as the forces are turned round their points of application, this single resultant always passes through a fixed point in the body, whose coordinates are given by

$$G - \xi Y + \eta X = 0,$$
$$V - \xi X - \eta Y = 0.$$

This point is called the *centre of the forces*. The first of these equations represents the line of action of the single resultant when $\theta = 0$, the second represents its line of action after a rotation through a right angle, i.e. when $\theta = \frac{1}{2}\pi$.

As every force in this theory has a point of application fixed in the body, it will be found convenient to regard the central point as the point of application of the single resultant. Thus the single resultant, like the other forces, has a fixed magnitude, a fixed direction in space, and a fixed point of application in the body. The centre of the forces may be defined in words similar to those already used in Art. 82 for parallel forces. *If the points of application of the given forces are fixed in the body, the point of application of their resultant is also fixed in the body, however the body is displaced, provided the given forces retain their magnitudes and directions in space unaltered. This fixed point is called the centre of the forces.*

Taking any one relative position of the body and forces, and any rectangular axes, the coordinates $(\xi \eta)$ of the centre of the forces are given by

$$\xi R^2 = VX + GY, \quad \eta R^2 = VY - GX,$$

where X, Y, V, G are referred to the origin as base, and R is the resultant of X and Y.

161. Ex. 1. If the forces of a system are reducible to a single resultant couple, show that the centre of the forces is at infinity.

Ex. 2. Show that, as the forces are rotated, the value of G/V at any assumed base O is always equal to the tangent of the angle which the straight line joining O to the centre C of the forces makes with the direction of the resultant force R, while the value of $G^2 + V^2$ is invariable and equal to $R^2 . CO^2$.

Since the system is equivalent to a single force R acting at C, it is evident that $G = R . ON$, where ON is a perpendicular on the line of action of R. Turning R through a right angle, we have $V = R . CN$. The results follow at once.

162. There is another method* of finding the astatic resultant of a given system which is sometimes useful. The body having been placed in any position relative to the forces which may be convenient, let two axes Ox, Oy be chosen so that the resolved parts of the forces in these directions, viz. X and Y, are neither of them zero. Consider first the resolved parts of all the forces parallel to x. By the theory of parallel forces these are equivalent to a single force, viz. $X = \Sigma P_x$, which acts at a point fixed in the body whose coordinates are $(x_1 y_1)$, where

$$x_1 X = \Sigma x P_x, \qquad y_1 X = \Sigma y P_x.$$

Consider next the resolved parts parallel to y. These also form a system of parallel forces and are equivalent to a single force $Y = \Sigma P_y$, which acts at a point fixed in the body whose coordinates are $(x_2 y_2)$, where

$$x_2 Y = \Sigma x P_y, \qquad y_2 Y = \Sigma y P_y.$$

Since the axes of coordinates are arbitrary and need not be at right angles, the forces have thus been reduced to two forces acting at two points fixed in the body in directions arbitrarily chosen but not parallel. The positions of these points depend on the directions chosen.

163. Let the fixed points thus found be called A and B. In any one relative position of the body and forces, let the two forces X and Y intersect in I, and let their resultant act along IF. Let IF intersect the circle described about the triangle ABI in C. Then, by the astatic triangle of forces, C is a point fixed in the body, and the resultant of X and Y may be supposed to act at C. The point C is therefore the centre of the forces.

Conversely, when the resultant force and the centre of the forces are known, that force may be resolved into two astatic forces by using the triangle of forces in the manner already explained in Art. 73.

---

* The method explained in this Article has been used by Darboux, *Sur l'équilibre astatique*, and by Larmor, *Messenger of Mathematics*.

CHAPTER V.

164. WHEN one body slides or rolls on another under pressure, it is found by experience that a force tending to resist motion is called into play. In order to discover the laws which govern the action of this force we begin with experiments on some simple cases of equilibrium, and then endeavour by a generalization to extend these so as to include the most complicated cases.

Let us consider the case of a box A resting on a rough table BC. A string DEH attached to the box at D passes over a small pulley E and supports a scale-pan H in which weights can be placed. By putting weights into the box A and varying the weight at H, all cases can be tried. Supposing the

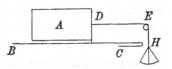

box loaded, we go on increasing the weight at H by adding sand (which can be afterwards weighed) until the box just begins to move. The result is that the box, whatever load it carries, does not move until the weight at H is a certain multiple of the weight of the box and load. Of course the experiment must be conducted with much greater attention to details than is here described. For example the friction at the pulley E must be allowed for.

165. Laws of friction. The results of this experiment suggest the following laws.

1. *The direction of the friction is opposite to the direction in which the body is urged to move.*

2. *The magnitude of the friction is just sufficient to prevent*

motion. Thus there is no friction between the box and the table until a weight applied at H begins to act on the box, and then the amount of the friction is equal to that weight.

3. *No more than a certain amount of friction can be called into play*, and when more is required to keep the body at rest, motion will ensue. This amount of friction is called *limiting friction.*

4. *The magnitude of limiting friction bears a constant ratio μ to the normal pressure* between the body and the plane on which it rests. This constant ratio μ depends on the nature of the materials in contact. It is usually called the *coefficient of friction.*

We do not here assert that the friction actually called into play is in every case equal to μ times the normal pressure, but only that this is the greatest amount which can be called into play. For smooth bodies $\mu = 0$. For a great many of the bodies we have to discuss μ lies between zero and unity.

5. *The amount of friction is independent of the area* of that part of the body which presses on the rough plane, provided that the normal pressure is unaltered.

6. *When the body is in motion, the friction called into play is found to be independent of the velocity and proportional to the normal pressure.* The ratio is not exactly the same as that found for limiting friction when the body is at rest.

It is found that the friction which must be overcome to *set the box in motion* along the table is greater than the friction between the same bodies *when in motion* under the same pressure. If the box has remained on the table for some time under pressure the friction which must be overcome is greater than if the bodies were merely placed in contact and immediately set in motion under the same pressure by the proper weight in the pan H. In some bodies this distinction between statical and dynamical friction is found to be very slight, in others the difference is considerable. The coefficient of friction μ for bodies in motion is therefore slightly less than for bodies at rest.

It should be noticed that friction is one of those forces which are usually called *resistances.* This follows from the second of the laws enunciated above. When a body is pressed against a wall, a reaction or resistance is called into play and is of just the

magnitude necessary to balance the pressing force. If there is no
pressure there is no reaction. In the same way friction acts only
to prevent sliding, not to produce it.

166. There is another method of determining the laws of
friction by which the use of the pulley and string is avoided and
which therefore presents some ad-
vantages.   Imagine the box A
placed symmetrically on an inclined
plane BC. Let the inclination of
BC to the horizon be θ. If W be
the weight of the box we easily

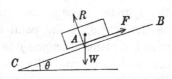

find that the normal reaction is $R = W \cos \theta$, and the friction
$F = W \sin \theta$.   Hence $\dfrac{F}{R} = \tan \theta$.   Let us now suppose the inclination θ of the plane to the horizon to be gradually increased until
the box A begins to slide. The friction F is then the limiting
friction. It is found by experiment that this inclination is the
same, whatever the weight of the box may be. It follows that
the ratio of the limiting friction to the normal pressure is independent of that pressure.

This experiment supplies us with an easy method of approxi-
mating to the value of μ for any two materials.   Place a body A
constructed of one of these materials on an inclined plane BC
constructed of the other material.   Supposing A to be at rest,
increase the inclination θ until A just begins to slide, then μ is
slightly less than the value of $\tan \theta$ thus found.   Next supposing
the inclination of the plane to be such that the body A slides, we
might decrease it until the box is just stationary, then μ is
slightly greater than the value of $\tan \theta$ thus found.   In this way
we have found two nearly equal numerical quantities between
which the coefficient of friction, viz. μ, must lie.   *The value of θ
which makes* $\tan \theta = \mu$ *is often called the angle of friction.*

Ex. Assuming that limiting friction consists of two parts, one proportional to
the pressure and the other to the surface in contact, show that if the least forces
which can support a rectangular parallelepiped whose edges are a, b, and c on
a given inclined plane be P, Q, and R when the faces bc, ca, and ab respectively
rest on the plane, then $(Q-R)\,bc + (R-P)\,ca + (P-Q)\,ab = 0$.   [Trin. Coll., 1884.]

167. The friction couple. When a wheel rolls on a rough
plane the experiment must be conducted in a different manner.

Let a cylinder be placed on a rough horizontal plane and let
its weight be W. Let two weights P and $P+p$ be suspended
by a string passing over the cylinder
and hanging down through a slit in
the horizontal plane. Let the plane
of the paper represent a section of
the cylinder through the string, let C
be the centre, A the point of contact
with the plane. Imagine p to be at first

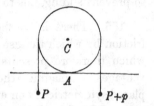

zero and to be gradually increased until the cylinder just moves.
By resolving vertically the reaction at A is seen to be equal to
$W+2P+p$. By resolving horizontally we see that there can be
no horizontal force at A. Thus the friction force is zero. Taking
moments about A we see that there must be a *friction couple* at
A whose magnitude is equal to pr.

168. The explanation of this couple is as follows. The
cylinder not being perfectly rigid yields slightly at A and is
therefore in contact with the plane over a small area. When the
cylinder begins to roll, the elements of area which are behind the
direction of motion are on the point of separating and tend to
adhere to each other, the elements in front tend to resist
compression. The resultant action across both sets of elements
may be replaced by a couple and a single force acting at some
convenient point of reference. The yielding of the cylinder at A
also slightly alters the position of the centre of gravity of the whole
mass, but this change is very insignificant and is usually neglected.
The cylinder is treated as if the section were a perfect circle
touching the plane at a geometrical point A. The whole action
is represented by a force acting at A and a couple. The resolved
parts of the force along the normal and tangent at A are often
called respectively the *reaction* and the *friction force*. In our experiment the latter is zero. The couple is called the *friction couple*.

The results of experiment show that the magnitude of p
when the cylinder just moves is proportional to the normal
pressure directly and the radius of the cylinder inversely. We
therefore state as another law of friction that the *moment of the
friction couple is independent of the curvature and proportional to
the normal pressure*. The ratio of the couple to the normal
pressure is often called the *coefficient of the friction couple*. The

magnitude of the friction couple is usually very small and its effects are only perceptible when the circumstances of the case make the friction force evanescent.

The weight p is commonly spoken of as the *friction of cohesion*, which is then said to vary inversely as the radius of the cylinder. But we have preferred the mode of statement given above.

169. It should be noticed that the laws of friction are only approximations. It is not true that the ratio of the limiting friction to the pressure is absolutely constant for all pressures and under all circumstances. The law is to be regarded as representing in a compendious way the results of a great many experiments and is to be trusted only for weights within the limits of the experiments. These limits are so extended that the truth of the law is generally assumed in mathematical calculations. If we followed the proper order of the argument, we should now enquire how nearly the laws of friction approximate to the truth, so that we may be prepared to make the proper allowance when the necessity arises. We ought also to tabulate the approximate values of μ for various substances. But these discussions would occupy too much space and lead us too far away from the theory of the subject.

170. The experimenters on friction are so numerous that only a few names can be mentioned. The earliest is perhaps Amontons in 1699. He was followed by Muschanbroek and Nollet. But the most famous are Coulomb (*Savants étrangers Acad. des Sc. de Paris* x. 1785); Ximénès (*Teoria e pratica delle resistenze de' solidi ne' loro attriti. Pisa* 1782); Vince (*Phil. Trans.* vol. 75, 1785) and Morin (*Savants étrangers Acad. des Sc. de Paris* IV. 1833). Besides these there are the experiments of Southern, Rennie, Jenkin and Ewing, Osborne Reynolds &c.

171. One of the laws of friction requires that the direction of the friction should be opposite to the direction in which the body under consideration is urged to move. When, therefore, the body can begin to move in only one way, the direction of the friction is known and only its magnitude is required. But when the body can move in any one of several ways, if properly urged, both the direction and the magnitude of the friction are unknown. It follows that problems on friction may be roughly divided into two classes. (1) We have those in which the bodies rest on one or more points of support, at all of which the lines of action of the frictions are known, but not the magnitudes. (2) There are those in which both the direction and magnitude of the friction have to be discovered.

We shall begin by using the laws of friction enunciated above to solve some problems of the first class. Afterwards we shall consider how the directions of the friction forces are to be discovered when the system is bordering on motion.

172. *A particle is placed on a rough curve in two dimensions under the action of any forces. To find the positions of equilibrium.*

Let X, Y be the resolved forces in any position P of the particle. Let R be the reaction measured inwards of the curve on the particle, F the friction called into play measured in the direction of the arc s. Let ψ be the angle the tangent makes with the axis of x. The particle is supposed to be on the proper side of the curve, so that it is pressed against the curve by the action of the impressed forces. Taking the figure of the next article, we have, by resolving and taking moments,

$$X \cos \psi + Y \sin \psi + F = 0,$$
$$- X \sin \psi + Y \cos \psi + R = 0.$$

Now if μ be the coefficient of friction F must be numerically less than μR. The required positions of equilibrium are therefore those positions at which the expression

$$\frac{X \cos \psi + Y \sin \psi}{- X \sin \psi + Y \cos \psi}$$

is numerically less than μ. This expression is a function of the position of the particle on the curve. Let us represent it by $f(x)$.

The positions of equilibrium in which the particle borders on motion are found by solving the equations $f(x) = \pm \mu$. Since this equation may have several roots, we thus obtain several extreme positions of equilibrium. We must then examine whether equilibrium holds or fails for the intermediate positions, i.e. whether $f(x)$ is $<$ or $> \mu$ numerically.

We may sometimes determine this last point in the following manner. Suppose an extreme position, say $x = x_1$, to be determined by solving the equation $f(x) = \mu$. If equilibrium exist in the positions determined by values of x slightly less than x_1, $f(x)$ must be increasing as x increases through the value $x = x_1$. On the contrary if equilibrium fail for these values of x, $f(x)$ must be decreasing. Thus equilibrium fails or holds for values of x slightly greater than x_1 according as $f'(x)$ is positive or negative when $x = x_1$. Let us next suppose that an extreme position, say $x = x_2$, is determined by solving the equation $f(x) = - \mu$. If equilibrium exist in the positions determined by values of x slightly less than x_2, $f(x)$ must be algebraically decreasing as x increases through the value $x = x_2$, and therefore $f'(x_2)$ is negative.

If therefore any extreme position of equilibrium is determined by the value

$x = x_1$ of the independent variable, equilibrium fails or holds for values of x slightly greater than x_1 according as $f'(x_1)$ has the same sign as μ or the opposite. It is clear that this rule may also be used in the case of a rigid body whose position in space is determined by only one independent variable.

173. Cone of friction. There is another method of finding the position of equilibrium which is more convenient when we wish to use geometry. Let ϵ be the angle of friction, so that $\mu = \tan \epsilon$. At any point P draw two straight lines each making an angle ϵ with the normal at P, viz. one on each side. Let these be PA, PB. Then the resultant reaction at P (i.e. the resultant of R and F) must act between the two straight lines PA, PB. These lines may be called the extreme or bounding lines of friction.

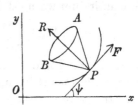

If the forces on P were not restricted to two dimensions, we should describe a right cone whose vertex is at P, whose axis is the line of action of the reaction R, and whose semi-angle is $\tan^{-1}\mu$. This cone is called the cone of limiting friction or more briefly the *cone of friction*.

Since the resultant reaction at P is equal and opposite to the resultant of the impressed forces on the particle we have the following rule. *The particle is in equilibrium at all points at which the impressed force acts within the cone of friction.* In the extreme positions of equilibrium the resultant of the impressed forces acts along the surface of the cone.

174. *A particle is placed on a rough curve in three dimensions under the action of any forces. To find the positions of equilibrium.*

Let X, Y, Z be the resolved parts of the impressed forces. Let R be their resultant, T their resolved part along the tangent to the curve at the point where the particle is placed. Since T must be less than μ times the normal pressure in any position of equilibrium we have $T^2 < \mu^2(R^2 - T^2)$. If ds be an element of the arc of the curve, this may be put into the form

$$\left(X \frac{dx}{ds} + Y \frac{dy}{ds} + Z \frac{dz}{ds} \right)^2 < \frac{\mu^2}{1 + \mu^2} (X^2 + Y^2 + Z^2).$$

Here X, Y, Z and s are functions of the coordinates x, y, z. The particle will be in equilibrium at all the points of the curve at which this inequality holds. If we change the inequality into an equality, we have an equation to find the limiting positions of equilibrium.

175. *A particle rests on a rough surface under the action of any forces. To find the positions of equilibrium.*

Let $f(x, y, z) = 0$ be the surface, let Q be the normal component of the impressed forces at the point where the particle is placed. In equilibrium we must have $R^2 - Q^2 < \mu^2 Q^2$. We have therefore

$$\frac{(Xf_x + Yf_y + Zf_z)^2}{f_x^2 + f_y^2 + f_z^2} > \frac{X^2 + Y^2 + Z^2}{1 + \mu^2}.$$

Here X, Y, Z and f are functions of the coordinates. If we change the inequality into an equality, we have a surface which cuts the given surface $f = 0$ in a curve. This curve is the boundary of the positions of equilibrium of the particle.

176. Ex. 1. A heavy bead of weight W can slide on a rough circular wire fixed in space with its plane vertical. A centre of repulsive force is situated at one extremity of the horizontal diameter, and the force on the bead when at a distance r is pr. Find the limiting positions of equilibrium.

If 2θ be the angle the radius at the bead makes with the horizon, the tangential and normal forces are $(W \cos 2\theta - pr \sin \theta)$ and $(W \sin 2\theta + pr \cos \theta)$. Putting the ratio of the first to the second equal to $\pm \tan \epsilon$, we find $\sin(\gamma \mp \epsilon - 2\theta) = \pm \cos \gamma \sin \epsilon$, where $W = pa \tan \gamma$ and a is the radius. Discuss these positions.

Ex. 2. A heavy particle rests in equilibrium on a rough cycloid placed with its axis vertical and vertex downwards. Show that the height of the particle above the vertex is less than $2a \sin^2 \epsilon$, where a is the radius of the generating circle.

Ex. 3. A rigid framework in the form of a rhombus of side a and acute angle ϕ rests on a rough peg whose coefficient of friction is μ. Prove that the distance between the two extreme positions which the point of contact of the peg with any side can have is $a\mu \sin \alpha$. See Art. 173.   [St John's Coll., 1890.]

Ex. 4. Two uniform rods AB, BC are rigidly joined at right angles at B and project over the edge of a table with AB in contact. Find the greatest length of AB that can project; and prove that if the coefficient of friction be greater than $\dfrac{AB(AB + 2BC)}{BC^2}$ the system can hang with only the end A resting on the edge.

[Math. Tripos, 1874.]

Ex. 5. Three rough particles of masses m_1, m_2, m_3, are rigidly connected by light smooth wires meeting in a point O, such that the particles are at the vertices of an equilateral triangle whose centre is O. The system is placed on an inclined plane of slope α, to which it is attached by a pivot through O; prove that it will rest in any position if the coefficient of friction for any one of the particles be not less than

$$\frac{\tan \alpha}{m_1 + m_2 + m_3} (m_1^2 + m_2^2 + m_3^2 - m_2 m_3 - m_3 m_1 - m_1 m_2)^{\frac{1}{2}}. \qquad \text{[Math. Tripos, 1877.]}$$

Ex. 6. A particle rests on the surface $xyz = c^3$ under the action of a constant force parallel to the axis of z: prove that the curve of intersection of the surface with the cone $\dfrac{1}{x^2} + \dfrac{1}{y^2} = \dfrac{\mu^2}{z^2}$ will separate the part of the surface on which equilibrium is possible from that on which it is impossible; μ being the coefficient of friction.

[Math. Tripos, 1870.]

Ex. 7. The ellipsoid $\dfrac{x^2}{a^2} + \dfrac{y^2}{b^2} + \dfrac{z^2}{c^2} = 1$ is placed with the axis of x vertical, its surface being rough. Show that a heavy particle will rest on it anywhere above its intersection with the cylinder $\dfrac{y^2}{b^2}\left(1 + \dfrac{a^2}{\mu^2 b^2}\right) + \dfrac{z^2}{c^2}\left(1 + \dfrac{a^2}{\mu^2 c^2}\right) = 1$, μ being the coefficient of friction.   [Trin. Coll., 1885.]

177. The following problem is regarded from more than one aspect to illustrate some different methods of proceeding.

Ex. 1. *A ladder is placed with one end on a rough horizontal floor and the other against a rough vertical wall, the vertical plane containing the ladder being perpendicular to the wall. Find the positions of equilibrium.*

Let AB be the ladder, $2l$ its length, w its weight acting at its middle point C. Let θ be its inclination to the horizon. See the figure of Ex. 2.

Let R, R' be the reactions at A and B acting along AD, BD respectively; μ, μ' the coefficients of friction at these points. The frictions at A and B are ξR and $\eta R'$, where ξ, η are two quantities which are numerically less than μ and μ' respectively. In many problems ξ, η may be either positive or negative. In this case however, since friction is merely a resistance and not an active force, we may assume that the frictions act along AL and LB. We may therefore regard ξ, η as positive. This limitation will also follow from the equations of equilibrium.

By resolving and taking moments we have
$$\xi R = R' \qquad\qquad \eta R' + R = w$$
$$2\eta R'l\cos\theta + 2R'l\sin\theta = wl\cos\theta.$$

Eliminating R, R' we find $\tan\theta = \dfrac{1 - \xi\eta}{2\xi}$. Any positive value of $\tan\theta$ given by this equation, where ξ, η are less than μ, μ', will indicate a possible position of equilibrium. If the roughness is so slight that $\mu\mu' < 1$, the minimum value of $\tan\theta$ is given by $\tan\theta = \dfrac{1 - \mu\mu'}{2\mu}$. If the roughness is so great that $\mu\mu' > 1$, the ladder will rest in equilibrium at all inclinations.

Ex. 2. The ladder being placed at any given inclination θ to the horizon, find what weight can be placed on a given rung that the ladder may be in equilibrium.

Let M be the rung, W the weight on it, $AM = m$. Let $\mu = \tan\epsilon$, $\mu' = \tan\epsilon'$.

Geometrical Solution. If we make the angles $DAE = \epsilon$, $DBE = \epsilon'$, the resultant reactions at A and B must lie within these angles and must meet in some point which lies within the quadrilateral $EFDH$.

Let G be the centre of gravity of the weights W and w. If the vertical line through G pass to the left of E, the weight $(W + w)$ may be supposed to act at some point P within the quadrilateral above mentioned. This weight may then be resolved obliquely into the two directions PA, PB. These may be balanced by two reactions at A and B each lying within its limiting lines.

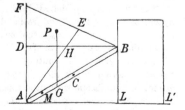

The result is that there will be equilibrium if the vertical through G passes to the left of E.

It is evident that this reasoning is of general application. We may use it to find the conditions of equilibrium of a body which can slide with a point on each of two given curves whenever the impressed forces which act on the body can be conveniently reduced to a single force. We draw the limiting lines of friction at the points of contact, and thus form a quadrilateral. *The condition of equilibrium is that the resultant impressed force shall pass through the quadrilateral area.*

The abscissæ of the points E and G measured horizontally from A to the right are easily proved to be respectively

$$x = \frac{2l\,(\mu\mu'\cos\theta + \mu\sin\theta)}{\mu\mu' + 1}, \qquad \bar{x} = \frac{(Wm + wl)\cos\theta}{W + w}.$$

If C lie to the right of the vertical through E, (i.e. $l\cos\theta > x$) there cannot be equilibrium unless the given rung lie to the left of that vertical ($m\cos\theta < x$). Also the weight W placed on the rung must be *sufficiently great* to bring the centre of gravity G to the left of that vertical ($\bar{x} < x$).

If C lie to the left of the vertical through E, ($l\cos\theta < x$) there is equilibrium whatever W may be if the given rung is also on the left of that vertical ($m\cos\theta < x$). But if the given rung is on the right of the vertical ($m\cos\theta > x$), the weight W placed on it must be *sufficiently small* not to bring the centre of gravity to the right of that vertical.

Lastly, if the vertical through E lie to the right of B, ($\tan^{-1}\mu > \frac{1}{2}\pi - \theta$) there is equilibrium whatever W may be, and on whatever rung it may be placed.

Another problem is solved on a similar principle in Jellett's treatise on friction, 1872.

Analytical solution. Following the same notation as in Ex. 1 we have by resolving and taking moments

$$\xi R = R', \quad \eta R' + R = W + w,$$
$$2\eta R'l\cos\theta + 2R'l\sin\theta = (Wm + wl)\cos\theta.$$

Eliminating R, R', we find

$$\frac{2l\,(\xi\eta\cos\theta + \xi\sin\theta)}{\xi\eta + 1} = \frac{(Wm + wl)\cos\theta}{W + w} \quad\cdots\cdots\cdots\cdots\cdots\cdots (A).$$

The condition of equilibrium is that it is possible to satisfy this equation with values of ξ, η which are less than μ, μ' respectively. By seeking the maximum value of the left-hand side we may derive from this the geometrical condition that the centre of gravity of W and w must lie to the left of a certain vertical straight line. But our object is to discuss the equation otherwise.

Let us regard ξ, η as the coordinates of some point Q referred to any rectangular axes. Then (A) is the equation to a hyperbola, one branch of which is represented in the figure by the dotted line. If this hyperbola pass within the rectangle NN' formed by $\xi = \pm\mu$, $\eta = \pm\mu'$, the conditions of equilibrium can be satisfied by values of ξ, η less than their limiting values. If the curve does not cut the rectangle, there cannot be equilibrium without the assistance of more than the available friction. The right-hand side of (A) is the quantity already

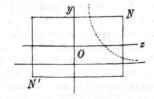

called $\bar{x}$. Let it be transferred to the left-hand side and let the equation thus altered be written $z = 0$. We notice that z is negative at the origin. In order that the hyperbola may cut the rectangle it is sufficient and necessary that z should be positive at the point N, i.e. when $\xi = \mu$, $\eta = \mu'$. The required condition of equilibrium is therefore that $\dfrac{2l\,(\mu\mu'\cos\theta + \mu\sin\theta)}{\mu\mu' + 1} - \bar{x}$ should be a positive quantity. This is virtually the same result as before and may be similarly interpreted.

Ex. 3. Let the ladder AB be placed in a given position leaning against the rough vertical face of a large box which stands on the same floor, as shown in the figure of Ex. 2. Determine the conditions of equilibrium.

We have now to take account of the equilibrium of the box BLL'. Let W' be its weight. Let R'' be the reaction between it and the floor, $\zeta R''$ the friction. We have then, in addition to the equations of Ex. 1,

$$R'' = W' + \eta R', \quad \zeta R'' = R'.$$

Eliminating R'' we find $(W' + w)\,\xi\eta\zeta + W'\zeta - w\xi = 0.$

We have also by Ex. 1, $\xi\eta + 2\xi\tan\theta - 1 = 0$..........................(A).

Eliminating η, so as to express both η and ζ in terms of one variable ξ, we find

$$2\,(W' + w)\tan\theta\,\xi\zeta + w\xi - (2W' + w)\,\zeta = 0 \quad\dots\dots\dots\dots\dots\dots \text{(B)}.$$

The conditions of equilibrium are that the two equations A and B can be simultaneously satisfied by values of ξ, η, ζ less than μ, μ', μ'' respectively.

Regarding ξ, η, ζ as the coordinates of a representative point Q, these equations represent two cylinders. These cylinders intersect in a curve. If any part of this curve lie within the rectangular solid bounded by $\xi = \pm\mu$, $\eta = \pm\mu'$, $\zeta = \pm\mu''$ the conditions of equilibrium are satisfied.

But instead of using solid geometry we may represent (A) and (B) by two hyperbolas having different ordinates η, ζ but the same abscissa ξ. The frictions being resistances, we shall assume that they act so that ξ, η, ζ are all positive. It will therefore be necessary only to draw that portion of the figure which lies in the positive quadrant. Take $OM = \mu$, $OM' = \mu'$, $OM'' = \mu''$. Let OB and AH represent the hyperbolas (B) and (A). Then we easily find

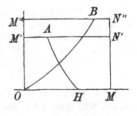

$$M'A = \frac{1}{\mu' + 2\tan\theta}, \quad M''B = \frac{(2W' + w)\,\mu''}{2\,(W' + w)\,\mu''\tan\theta + w}.$$

The condition of equilibrium is that an ordinate can be found intersecting the two hyperbolas in points Q, Q' each of which lies within the limiting rectangles. The necessary conditions are therefore found by making an ordinate travel across the figure from OM' to $N'N''$. They may be summed up as follows.

(1) The hyperbola AH must intersect the area of the rectangle ON'; the condition for this is that $M'A < \mu$.

(2) If the hyperbola OB intersect $M''N''$ on the left-hand side of N'', i.e. if $M''B < \mu$, then $M'A$ must be $< M''B$, for otherwise the ordinate QQ' would not cut both curves within the prescribed area. But this condition is included in (1) if $M''B > \mu$.

If the ladder is so placed that the inequality (2) becomes an equality while (1) is not broken, the frictions η and ζ attain their limiting values while ξ is not limiting, the ladder will therefore be on the point of slipping at its upper extremity, and the box will be just slipping along the plane.

If the ladder is so placed that the inequality (1) becomes an equality while (2) is not broken, ξ and η have their limiting values while ζ is less than its limit. The box is therefore fixed and the ladder slips at both ends.

178. Ex. 1. A ladder AB rests against a smooth wall at B and on a rough horizontal plane at A. A man whose weight is n times that of the ladder climbs up it. Prove that the frictions at A in the two extreme cases in which the man is at the two ends of the ladder are in the ratio of $2n+1$ to 1.

Ex. 2. A boy of weight w stands on a sheet of ice and pushes with his hands against the smooth vertical side of a heavy chair of weight nw. Show that he can incline his body to the horizon at any angle greater than $\cot^{-1} 2\mu$ or $\cot^{-1} 2\mu n$, according as the chair or the boy is the heavier, the coefficient of friction between the ice and boy or the ice and chair being μ.   [Queens' Coll.]

Ex. 3. Two hemispheres, of radii a and b, have their bases fixed to a horizontal plane, and a plank rests symmetrically upon them. If μ be the coefficient of friction between the plank and either hemisphere, the other being smooth, prove that, when the plank is on the point of slipping, the distance of its centre from its point of contact with the smooth hemisphere is equal to $(a \sim b)/\mu$.   [St John's Coll., 1885.]

Ex. 4. A heavy rod rests with one end on a horizontal plane and the other against a vertical wall. To a point in the rod one end of a string is tied, the other end being fastened to a point in the line of intersection of the plane and wall. The string and rod are in a vertical plane perpendicular to the wall. Show that, if the rod make with the horizon an angle a which is less than the complement of 2ϵ, then equilibrium is impossible unless the string make with the horizon an acute angle less than $a + \epsilon$, where ϵ is the angle of friction both with the wall and the plane.   [Math. Tripos, 1890.]

Ex. 5. A parabolic lamina whose centre of gravity is at its focus rests in a vertical plane upon two rough rods of the same material at right angles and in the same vertical plane; if ϕ be the inclination of the directrix to the horizon in one extreme position of equilibrium, prove that $\tan^2(a - \phi) \tan(a + \epsilon - \phi) = \tan(a - \epsilon)$; where ϵ is the angle of friction, a the inclination of one rod to the horizon.

[Trin. Coll., 1882.]

Ex. 6. Two rods AC, BC with a smooth hinge at C are placed in a given position with their extremities A and B resting on a rough horizontal plane. The plane of the rods being vertical, find the conditions of equilibrium.

Let θ, θ' be the inclinations of the rods to the horizon, W and W' their weights. Let $(R, \xi R)$, $(R', \eta R')$ be the reactions and frictions at A and B. Resolving and taking moments in the usual way, we find

$$\xi = \frac{W + W'}{W \tan \theta' + (2W + W') \tan \theta}, \qquad \eta = \frac{W' + W}{W' \tan \theta + (2W' + W) \tan \theta'}.$$

If the value of ξ thus found is $> \mu$ the system will slip at A; if $\eta > \mu$ it will slip at B. If the system slip at A only, then $\xi > \eta$; this gives $W \tan \theta < W' \tan \theta'$.

Ex. 7. A groove is cut in the surface of a flat piece of board. Show that the form of the groove may be so chosen as to satisfy this condition, that if the board will just hang in equilibrium upon a rough peg placed at any one point of the groove, it will also just hang in equilibrium when the peg is placed at any other point.   [Math. Tripos, 1859.]

Ex. 8. A lamina is suspended by three strings from a point O; if the lamina be rough, and the coefficient of friction between it and a particle P placed upon it be constant, show that the boundary of possible positions of equilibrium of the particle on the lamina is a circle.   [Math. Tripos, 1880.]

Let ON be a perpendicular on the lamina. Let D be the centre of gravity of the lamina, G that of the lamina and particle. Then in equilibrium OG is vertical and NG is the line of greatest slope. The angle NOG is equal to the inclination of the plane to the horizon and is constant because the equilibrium is limiting. The locus of G is a circle, centre N. Since $DP : DG$ is constant the locus of P is also a circle.

Ex. 9. Spheres whose weights are W, W' rest on different and differently inclined planes. The highest points of the spheres are connected by a horizontal string perpendicular to the common horizontal edge of the two planes and above it. If μ, μ' be the coefficients of friction and be such that each sphere is on the point of slipping down, then $\mu W = \mu' W'$.   [Math. Tripos.]

Consider one sphere: the resultant of T and μR balances that of W and R. By taking moments about the centre $T = \mu R$. Hence, by drawing a figure, $R = W$. Thus $T = \mu W$ and the result follows.

Ex. 10. A uniform rod passes over one peg and under another, the coefficient of friction between each peg and the rod being μ. The distance between the pegs is b, and the straight line joining them makes an angle β with the horizon. Show that equilibrium is not possible unless the length of the rod is $> b\{1 + (\tan\beta)/\mu\}$.   [Coll. Ex.]

Ex. 11. A uniform rod ACB, length $2a$, is supported against a rough wall by a string attached to its middle point C: show that the rod can rest with C at any point of a circular arc, whose extremities are distant a and $a\cos\epsilon$ from the wall, where ϵ is the angle of friction. [Take moments about C.]

Ex. 12. Two uniform and equal rods of length $2a$ have their extremities rigidly connected, and are inclined to each other at an angle $2a$. These rods rest on a fixed rough cylinder with its axis horizontal, and whose radius is $a\tan a$. Show that in the limiting position of equilibrium the inclination θ to the vertical of the line through the point of intersection of the rods perpendicular to the axis of the cylinder is given by $\sin^2 a \sin\theta = \cos(\theta - \epsilon)\sin\epsilon$, where $\tan\epsilon$ is the coefficient of friction.   [Coll. Ex.]

Ex. 13. Three equal uniform heavy rods AB, BC, CD, hinged at B and C, are suspended by a light string attached to D from a point E, and hang so that the end A is on the point of motion, towards the vertical through E, along a rough horizontal plane (coefficient of friction $\mu = \tan\epsilon$): show that

$$\frac{\cos(a - \epsilon)}{\cos a} = \frac{\cos(\beta - \epsilon)}{3\cos\beta} = \frac{\cos(\gamma - \epsilon)}{5\cos\gamma} = \frac{\mu}{6}\frac{\cos(\theta - \epsilon)}{\cos\theta},$$

where a, β, γ are the inclinations of the rods to the horizon beginning with the lowest, and θ that of the string.   [Coll. Ex., 1881.]

Take moments about B, C, D, E in succession for the rods AB, AB and BC, and so on. Subtracting each equation from the next in order, the results follow at once.

Ex. 14. A sphere rests on a rough horizontal plane, and its highest point is joined to a peg fixed in the plane by a tight cord parallel to the plane. Show that, if the plane be gradually tilted about a line in it perpendicular to the direction of the cord, the sphere will not slip until the inclination becomes equal to $\tan^{-1} 2\mu$, where μ is the coefficient of friction.   [Math. Tripos, 1886.]

Ex. 15. A uniform hemisphere, placed with its base resting on a rough inclined plane, is just on the point of sliding down. A light string, attached to the point of the hemisphere farthest from the plane, is then pulled in a direction parallel to and directly up the plane. If the tension of the string be gradually increased until the sphere begins to move, it will slide or tilt according as $13 \tan \phi$ is less or greater than 8, where ϕ is the inclination of the plane to the horizon. The centre of gravity of the hemisphere is at a distance from the centre equal to three-eighths of the radius.   [Coll. Ex., 1888.]

Ex. 16. A circular disc, of radius a, whose centre of gravity is distant c from its centre, is placed on two rough pegs in a horizontal line distant $2a \sin \alpha$ apart. Show that all positions will be possible positions of equilibrium, provided

$$a \sin \alpha \sin (\lambda_1 + \lambda_2) > c \sin (2\alpha \mp \lambda_1 \pm \lambda_2),$$

where λ_1, λ_2 are the angles of friction at the two pegs.   [St John's Coll., 1880.]

Ex. 17. A number of equally rough particles are knotted at intervals on a string, one end of which is fixed to a point on an inclined plane. Show that, all the portions of the string being tight, the lowest particle is in its highest possible position, when they are all in a straight line making an angle $\sin^{-1} (\tan \lambda / \tan \alpha)$ with the line of greatest slope, λ being the angle of friction and α the inclination of the plane to the horizon. Show also that, if any portion of the string make this angle with the line of greatest slope, all the portions below it must do so too.
                    [Math. Tripos, 1886.]

Ex. 18. A rough paraboloid of revolution, of latus rectum $4a$, and of coefficient of friction $\cot \beta$, revolves with uniform angular velocity about its axis which is vertical: prove that for any given angular velocity greater than $(g/2a)^{\frac{1}{2}} \cot \frac{1}{2}\beta$ or less than $(g/2a)^{\frac{1}{2}} \tan \frac{1}{2}\beta$ a particle can rest anywhere on the surface except within a certain belt, but that for any intermediate angular velocity equilibrium is possible at every point of the surface.   [Math. Tripos, 1871.]

Let mg be the weight of the particle. It is known by dynamics that we may treat the paraboloid as if it were fixed in space, provided we regard the particle as acted on by a force $m\omega^2 r$ tending directly from the axis, where r is the distance of the particle from the axis, and ω the angular velocity of the paraboloid.

We may prove that the ordinates in the limiting positions of equilibrium are given by $\mu\omega^2 y^2 - (2a\omega^2 - g) y + 2a\mu g = 0$. That a belt may exist, the roots of this quadratic must be real.

Ex. 19. A rod rests partly within and partly without a box in the shape of a rectangular parallelepiped, presses with one end against the rough vertical side of the box, and rests in contact with the opposite smooth edge. The weight of the box being four times that of the rod, show that, if the rod be about to slip and the box about to tumble at the same instant, the angle the rod makes with the vertical is $\frac{1}{2}\lambda + \frac{1}{2}\cos^{-1}(\frac{1}{2}\cos \lambda)$, where λ is the angle of friction.   [Math. Tripos, 1880.]

Ex. 20. A glass rod is balanced partly in and partly out of a cylindrical tumbler with the lower end resting against the vertical side of the tumbler. If α and β are the greatest and least angles which the rod can make with the vertical, prove that the angle of friction is $\frac{1}{2} \tan^{-1} \dfrac{\sin^3 a - \sin^3 \beta}{\sin^2 a \cos a + \sin^2 \beta \cos \beta}$.   [Math. Tripos, 1875.]

Ex. 21. A heavy rod, of length $2l$, rests horizontally on the inside rough surface of a hollow circular cone, the axis of which is vertical and the vertex downwards. If $2a$ is the vertical angle of the cone, and if the coefficient of friction μ is less than $\cot a$, prove that the greatest height of the rod, when in equilibrium, above the vertex of the cone is $l \cot a \left\{ \dfrac{1 + \cos^2 a + \sin a \sqrt{(\sin^2 a + 4\mu^2)}}{2(1 - \mu^2 \tan^2 a)} \right\}^{\frac{1}{2}}$.

[Math. Tripos, 1885.]

Ex. 22. A heavy uniform rod AB is placed inside a rough curve in the form of a parabola whose focus is S and axis vertical. Prove that, when it is on the point of slipping downwards, the angle of friction is $\frac{1}{2}(SAB - SBA)$.   [Coll. Ex., 1889.]

Ex. 23. A rod MN rests with its ends in two fixed straight rough grooves OA, OB, in the same vertical plane, which makes angles a and β with the horizon: prove that, when the end M is on the point of slipping down AO, the tangent of the inclination of MN to the horizon is $\dfrac{\sin(a - \beta - 2\epsilon)}{2 \sin(\beta + \epsilon) \sin(a - \epsilon)}$.   [Math. Tripos, 1876.]

Ex. 24. A uniform rectangular board $ABCD$ rests with the corner A against a rough vertical wall and its side BC on a smooth peg, the plane of the board being vertical and perpendicular to that of the wall. Show that, without disturbing the equilibrium, the peg may be moved through a space $\mu \cos a (a \cos a + b \sin a)$ along the side with which it is in contact, provided the coefficient of friction (μ) lie between certain limits; a being the angle BC makes with the wall, and a, b the lengths of AB, BC respectively. Also find the limits of μ.   [Math. T., 1880.]

Ex. 25. An elliptical cylinder, placed in contact with a vertical wall and a horizontal plane, is just on the point of motion when its major axis is inclined at an angle a to the horizon. Determine the relation between the coefficients of friction of the wall and plane; and show from your result that, if the wall be smooth, and a be equal to 45^0, the coefficient of friction between the plane and cylinder will be equal to $\frac{1}{2} e^2$, where e is the eccentricity of the transverse section of the cylinder.

[Math. Tripos, 1883.]

Ex. 26. A rough elliptic cylinder rests, with its axis horizontal, upon the ground and against a vertical wall, the ground and the wall being equally rough; show that the cylinder will be on the point of slipping when its major axis plane is inclined at an angle of $\pi/4$ to the vertical if the square of the eccentricity of its principal section be $2 \sin \epsilon (\sin \epsilon + \cos \epsilon)$, where ϵ is the angle of friction.   [Coll. Ex., 1885.]

Ex. 27. Three uniform rods of lengths a, b, c are rigidly connected to form a triangle ABC, which is hung over a rough peg so that the side BC may rest in contact with it; find the length of the portion of the rod over which the peg may range, showing that, if $\mu > \dfrac{a(a + b + c)}{b(b + c)} \operatorname{cosec} C + \tan \frac{1}{2}(C - B)$, where $C > B$, the triangle will rest in any position.   [Math. Tripos, 1887.]

Ex. 28. A waggon, with four equal wheels on smooth axles whose plane contains the centre of gravity, rests on the rough surface of a fixed horizontal circular cylinder, the axles being parallel to the axis of the cylinder; investigate the pressures on the wheels, and prove that the inclination to the horizontal of the plane containing the axles is $\tan^{-1} \{\tan a (w - w')/W\}$, where w, w' are the weights on the two axles, W that of the whole waggon, and $2a$ is the angle between the tangent planes at the points of contact.   [Math. Tripos, 1888.]

Ex. 29. Three circular cylinders A, B, C, alike in all respects, are placed with their axes horizontal and their centres of gravity in a vertical plane; A is fixed, B is at the same level, and C at a lower level touches them both, the common tangent planes being inclined at 45^0 to the vertical. B and C are supported by a perfectly rough endless strap of suitable length passing round the cylinders in the plane containing the centres of gravity. Show that equilibrium can be secured by making the strap tight enough, provided that the coefficient of friction between the cylinders is greater than $1 - 1/\sqrt{2}$; and find how slipping will first occur if the strap is not quite tight enough.   [Math. Tripos, 1888.]

Ex. 30. Two uniform rods AB, BC, of equal length, are jointed at B. They are at rest in a vertical plane, equally inclined to the horizon, with their lower ends in contact with a rough horizontal plane. Prove that, if they be on the point of slipping both at A and C, the frictional couple at the joint is $Wa\,(\sin a - 2\mu \cos a)$, where W is the weight of each rod, a the inclination of each rod to the horizon, $2a$ the length of each rod, and μ the coefficient of friction. [St John's Coll., 1890.]

Ex. 31. Six uniform rods, each of length $2a$, are joined end to end by five smooth hinges, and they stand on a rough horizontal plane in equilibrium in the form of a symmetrical arch, three on each side; prove that the span cannot be greater than $2a\sqrt{2}\,(1 + \sqrt{\tfrac{1}{5}} + \sqrt{\tfrac{1}{13}})$, if the coefficient of friction of the rods and plane be $\tfrac{1}{5}$.   [Coll. Ex., 1886.]

Consider only half the arch. The reaction at the highest point is horizontal, and equal to half the weight of one rod. Take moments (1) for the upper, (2) for the two upper, (3) for all three rods. We find that their inclinations to the vertical are $\tfrac{1}{4}\pi$, $\tan^{-1}\tfrac{1}{3}$, $\tan^{-1}\tfrac{1}{5}$. The result follows easily.

179. Friction between wheel and axle. *Ex. 1. A gig is so constructed that when the shafts are horizontal the centre of gravity of the gig and the shafts is over the axle of the wheels. The gig in this position rests on a perfectly rough ground. Find the direction and magnitude of the least force which, acting at the extremity of the shaft, will just move the gig.*

When an axle is made to fit the nave of a wheel, the relative sizes of the axle and hole are so arranged that the wheel can turn easily round the axle. The axle is therefore just a little smaller than the hole. Thus the two cylinders touch along some generating line and the pressures act at points in this line. Even if the axle were somewhat tightly clasped at first, yet by continued use it would be worn away so that it would become a little smaller than the hole.

It is possible that the axle may be so large that it has to be forced into the hole. When this is the case, besides the pressures produced by the weight of the gig, there will be pressures due to the compression of the axle. These last will act on every element of the surface of the axle and their magnitudes will depend on how much the axle has to be compressed to get it into the hole. If the axle and hole are not perfectly circular, these pressures may be unequally distributed over the surface of the axle. When these circumstances of the problem are not given, the pressures on the axle are indeterminate.

Let X, Y be the required horizontal and vertical components of the force applied at the extremity S of the shaft.

Consider the equilibrium of the wheel. Since it touches a perfectly rough ground
at *A*, the friction at this point
cannot be limiting. Let *R* and *F*
be the reaction and friction. It is
evident that the friction *F* must act
to the left, if it is to balance the
force *X* which is taken as acting
to the right.

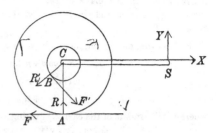

The axle will touch the circular
hole in which it works at *some one
point B*. At this point there will
be a reaction *R'* and a friction *F'*,
which is limiting when the gig is on the point of motion. Thus $F'=\mu R'$. The
resultant of *R'* and $\mu R'$ must balance the resultant of *R* and *F* and the weight of
the wheel. It therefore follows that the point *B* is on the left of *C*, i.e. *behind the
axle*. Let θ be the angle *ACB*, let *a* and *b* be the radii of the wheel and axle.
Taking moments about *A* we have

$$R'a \sin \theta = \mu R' (a \cos \theta - b).$$

Putting $\mu = \tan \epsilon$, this gives $\sin (\epsilon - \theta) = \dfrac{b}{a} \sin \epsilon.$

Since *b* is less than *a*, we see that θ is positive and less than ϵ.

Consider next the equilibrium of the gig. The forces *R'* and $\mu R'$ act on the gig
in directions opposite to those indicated in the figure. Let *W* be the weight of the
gig, then resolving and taking moments about *C* we have

$$X = - R' \sin \theta + \mu R' \cos \theta,$$
$$Y = - R' \cos \theta - \mu R' \sin \theta + W,$$
$$Yl = \mu R'b,$$

where *l* is the length of the shaft. These equations give *X* and *Y*.

Ex. 2. A light string, supporting two weights *W* and *W'*, is placed over a wheel
which can turn round a fixed rough axle. Supposing the string not to slip on the
wheel, find the condition that the wheel may be on the point of turning round the
axle. If *a*, *b* be the radii of the wheel and axle, and $\mu = \tan \epsilon$, prove that

$$(W - W') a = (W + W') b \sin \epsilon.$$

Ex. 3. A solid body, pierced with a cylindrical cavity, is free to turn about
a fixed axle which just fits the cavity, and the whole figure is symmetrical about
a certain plane perpendicular to the axle. The axle being rough, and the body
acted on by forces in the plane of symmetry, find the least coefficient of friction
that the body may be in equilibrium.

The circular sections of the cavity and axle are drawn in the figure as if they
were of different sizes. This has been
done to show that the reaction and
friction act at a definite point, but in
the geometrical part of the investigation
they should be regarded as equal.

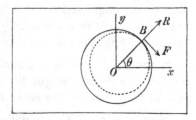

Let the plane of symmetry be taken
as the plane of *xy*, and let its intersection
O with the axis be the origin. Let *X*, *Y*, *G*
be the components of the forces, and let
these urge the body to turn round the
axis in a direction opposite to that of the hands of a watch.

The axle will touch the cavity along a generating line, let B be its point of intersection with the plane of xy. Let θ be the angle BOx. Let R and F be the normal reaction and the friction at B; when the body borders on motion we have $F = \mu R$.

By resolving and taking moments we find

$$R (\cos \theta + \mu \sin \theta) + X = 0,$$
$$R (\sin \theta - \mu \cos \theta) + Y = 0,$$
$$-\mu R a + G = 0,$$

where a is the radius of the cavity. Putting $\mu = \tan \epsilon$, we deduce from these equations

$$\tan (\theta - \epsilon) = Y/X, \qquad R^2 = (X^2 + Y^2) \cos^2 \epsilon.$$

These determine the point B and the reaction R. The least value of the coefficient of friction is then given by

$$(X^2 + Y^2) a^2 \sin^2 \epsilon = G^2.$$

180. Lemma. *If a lamina be moved from any one position to any other in its own plane, there is one point rigidly connected to the lamina whose position in space is unchanged. The lamina may therefore be brought from its first to its last position by fixing this point and rotating the lamina about it through the proper angle.*

Let A, B be any two points in the lamina in its first position, A', B' their positions in the last position. Then if A, B can be brought into the positions A', B' by rotation about some point I, fixed in space, the whole lamina will be brought from its first to its last position. Bisect AA', BB' at right angles by the straight lines LI, MI. Then $IA = IA'$, and $IB = IB'$. Also, since AB is unaltered in length by its motion, the sides of the triangles AIB, $A'IB'$ are equal, each to each. It follows that the angles AIB, $A'IB'$ are

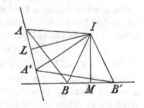

equal, and therefore that the angles AIA' and BIB' are equal. If then we turn the lamina round I, as a point fixed in space, through an angle equal to AIA', A will take the position A', and B will take the position B'. Thus the whole body has been transferred from the one position to the other.

If the body be simply translated, so that every point moves parallel to a given straight line, the bisecting lines LI, MI are parallel, and therefore the point I is infinitely distant.

If the angle AIA' is indefinitely small, the fixed point I of the lamina is called the *instantaneous centre of rotation*.

181. Frictions in unknown directions. We are now prepared to make a step towards the generalization of the laws of friction. Let us suppose a heavy body to rest on a rough horizontal table on n supports. Let these points be $A_1, A_2, ...A_n$, and let the pressures at these points be $P_1, P_2, ...P_n$. We shall also suppose the body to be acted on by a couple and a force applied at some convenient base of reference, the forces being all parallel to the table. To resist these forces a frictional force is called into play at each point of support. The directions and magnitudes of these frictional forces are unknown, except that the magnitude of each is less than the limiting friction, and the direction is opposed to the resultant of all the external and molecular forces which act on that point of support. If the pressures $P_1, ...P_n$ are known, there are thus $2n$ unknown quantities, and there are only three equations of equilibrium. The frictions at the points of support are therefore generally indeterminate.

By calling the frictions indeterminate we mean that there are different ways of arranging forces at the points of support which could balance the given forces and which *might* be frictional forces. Which of these is the true arrangement of the frictional forces depends on the manner in which the body, regarded as partially elastic, begins to yield to the forces. Suppose, for example, a force Q to act at a point B of the body, and to be gradually increased in magnitude. The frictions on the points of support nearest to B will at first be sufficient to balance the force, but, as Q gradually increases, the frictions at these points may attain their limiting values. As soon as they begin to yield, the frictions at the neighbouring points will be called into play, and so on throughout the body.

When the external forces are insufficient to move the body as a whole, the directions and magnitudes of the frictions at the points of support depend on the manner in which the body yields, however slight that yielding may be. Even if the external forces were absent, the body could be placed in a state of constraint and might be maintained in that state by the frictions. Thus the frictions depend on the *initial state of constraint* as well as on the external forces. It is also possible that the body, though apparently at rest, may be performing small oscillations about some position of stable equilibrium. This might cause other changes in the frictions.

182. Limiting Equilibrium. Let us now suppose that the external forces have been gradually increased according to some given law until the whole body is on the point of motion. By this we mean that the least diminution of roughness or the least increase of the forces will cause the body to move. We may enquire what is the condition that these forces may be just great enough to move the body, or just small enough not to move it.

When the body is just beginning to move, the arrangement of the frictional forces is somewhat simplified. We suppose the body to be so nearly rigid that the distances between the several particles do not sensibly change. Thus their motions are not independent, but are sensibly governed by the law proved in the lemma of Art. 180. The directions of the frictions, also, being opposite to the directions of the motions, are governed by the same law.

It will be seen from what follows that, when a rigid body turns round an instantaneous axis, the friction at every point of support acts in the direction which is most effective to prevent motion. If, therefore, the frictional forces thus arranged are insufficient to prevent motion, there is no other arrangement by which they can effect that result.

If the body move on a horizontal plane, no matter how slightly, it must be turning about some vertical axis; let this vertical axis intersect the plane in the point I. There are then two cases to be considered, (1) the point I may not coincide with any one of the points of support, and (2) it may coincide with some one of them.

Let us take these cases in order. The position of I is unknown; let its coordinates be ξ, η referred to any axes in the plane of the table. The points $A_1,\dots A_n$ are all beginning to move each perpendicular to the straight line which joins it to the point I. The frictions at these points will therefore be known when I is known. Their directions are perpendicular to IA_1, IA_2, &c., and they all act the same way round I. Their magnitudes are $\mu_1 P_1$, $\mu_2 P_2$, &c., if μ_1, μ_2, &c. are the coefficients of friction. Since the impressed forces only just overbalance the frictions, we may regard the whole as in equilibrium. Forming then the three equations of equilibrium, we have sufficient equations to find both ξ, η and the condition that the body should be on the point of motion. It may be that these equations do not give any available values of ξ, η, and in such a case the point I cannot lie away from one of the points of support.

183.　Let us consider next the case in which I coincides with one of the points of support, say A_1. The coordinates ξ, η of I are now known. Just as before the frictions at $A_2, \ldots A_n$ are all known, their directions are perpendicular to A_1A_2, A_1A_3, &c. and their magnitudes are μ_2P_2, &c. Since A_1 does not move, the friction at A_1 is not necessarily limiting friction. It may be only just sufficient to prevent A_1 from moving. Let the components of this friction parallel to the axes x and y be F_1 and F_1'. Forming as before the three equations of equilibrium, we have sufficient equations to find F_1, F_1' and the required condition that the body may be on the point of motion. If, however, the values of F_1, F_1' thus found are such that $F_1^2 + F_1'^2$ is greater than $\mu_1^2P_1^2$, the friction required to prevent A_1 from moving is greater than the limiting friction. It is then impossible that the body could begin to turn round A_1 as an instantaneous centre. We can determine by a similar process whether the body could begin to turn round A_2, and so on for all the points of support.

184.　We shall now form the Cartesian equations from which the coordinates ξ, η and the condition of limiting equilibrium are to be found. These however are rather complicated, and in most cases it will be found more convenient to find the position of I by some geometrical method of expressing the conditions of equilibrium.

Let the impressed forces be represented by a couple L together with the components X and Y acting at the origin. Let the coordinates of A_1, A_2 &c. be (x_1y_1), (x_2y_2), &c. Let the coordinates of I be $(\xi\eta)$. Let the distances IA_1, IA_2 &c. be r_1, r_2 &c. Let the direction of rotation of the body be opposite to that of the hands of a watch. Then since the frictions tend to prevent motion, they act in the opposite direction round I.

The resolution of these frictions parallel to the axes will be facilitated if we turn each round its point of application through an angle equal to a right angle. We then have the frictions acting along the straight lines IA_1, IA_2 &c., all towards or all from the point I. Taking the latter supposition, their resolved parts are to be in equilibrium with X acting along the positive direction of the axis of y and Y along the negative direction of x.

We find by resolution

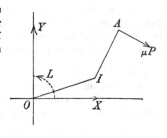

$$\Sigma\mu P\,\frac{\xi - x}{r} + Y = 0$$
$$\left.\begin{array}{l}\\ \\\end{array}\right\}\ \ldots\ldots\text{:(1).}$$
$$\Sigma\mu P\,\frac{\eta - y}{r} - X = 0$$

The equation of moments must be formed without changing the directions of the frictions. Taking moments about I, we have

$$\Sigma \mu P r + Y\xi - X\eta - L = 0 \ldots\ldots\ldots\ldots\ldots\ldots\ldots\ldots\ldots\ldots(2).$$

If the instantaneous centre I coincide with A_1, the equations are only slightly altered. We write $(x_1 y_1)$ for $(\xi\eta)$, F_1 and $-F_1'$ for $\mu_1 P_1 \dfrac{y_1 - \eta}{r_1}$ and $\mu_1 P_1 \dfrac{x_1 - \xi}{r_1}$, and finally omit the term $\mu_1 P_1 r_1$ in the moment.

185. The Minimum Method. There is another way of discussing these equations which will more clearly explain the connection between the two cases. If the body is just beginning to turn about some instantaneous axis, it would begin to turn about that axis if it were fixed in space. Let then I be any point on the plane of xy and let us enquire whether the body can begin to turn about the vertical through I as an axis fixed in space. Supposing all the friction to be called into play, the moment of the forces round I, measured in the direction in which the frictions act, is

$$u = \Sigma \mu P r + Y\xi - X\eta - L.$$

If, in any position of I, u is negative, the moment of the forces is more powerful than that of the frictions; the body will therefore begin to move. If on the other hand u is positive, the moment of the frictions is more powerful than that of the forces, and the body could be kept at rest by less than the limiting frictions. Let us find the position of I which makes u a minimum. If in this position u is positive or zero, there is no point I about which the body can begin to turn.

To make u a minimum we equate to zero the differential coefficients of u with regard to ξ, η. Since $r^2 = (x - \xi)^2 + (y - \eta)^2$, the equations thus formed are exactly the equations (1) already written down in Art. 184.

The statical meaning of these equations is that the pressures on the axis which has been fixed in space are zero when that axis has been so chosen that u is a minimum. If this is not evident, let R_x and R_y be the resolved pressures on the axis. The resolved parts parallel to the axes of the impressed forces and the frictions together with R_x and R_y must then be zero. But the equations (1) express the fact that these resolved parts without R_x and R_y are zero. It evidently follows that both R_x and R_y are zero.

That this position of I makes u a minimum and not a maximum may be shown analytically by finding the second differential coefficients of u with regard to ξ and η. The terms of the second order are then found to be

$$\Sigma \mu P \{(\eta - y)\, d\xi - (\xi - x)\, d\eta\}^2 / 2r^3,$$

where the Σ implies summation for all the points A_1, A_2, &c. Since each of these squares is positive, u must be a minimum.

It appears therefore that *the axis about which the body will begin to turn may be found by making the moment (viz. u) of the forces about that axis a minimum; and the condition that the forces are only just sufficient to move the body is found by equating to zero the least value thus found.*

186. The quantities r_1, r_2, &c. are necessarily positive, and therefore not capable of unlimited decrease. Besides the minima found by the rules of the differential calculus, other maxima or minima may be found by making some one of the quantities r_1, r_2, &c. equal to zero.

Suppose u to be a minimum when $r_1 = 0$, i.e. when the point I coincides with A_1. Take A_1 as the origin of coordinates. Let I receive a small displacement from the position A_1, and let its coordinates become $\xi = r_1 \cos \theta_1$, $\eta = r_1 \sin \theta_1$. Let the coordinates of A_2, &c. be $(r_2 \theta_2)$, &c. The value of u, when the first power only of the small quantity r_1 is retained, becomes

$$u = \mu_1 P_1 r_1 + \mu_2 P_2 \{ r_2 - r_1 \cos (\theta_1 - \theta_2) \} + \&c. + Y r_1 \cos \theta_1 - X r_1 \sin \theta_1 - L.$$

The condition that u should be a minimum is that the increment of u should be positive for all small displacements of I. This will be the case if the coefficient of r_1, viz. $\qquad \mu_1 P_1 - \mu_2 P_2 \cos (\theta_1 - \theta_2) - \&c. + Y \cos \theta_1 - X \sin \theta_1,$

is positive for all values of θ_1. We may write this in the form

$$\mu_1 P_1 + A \cos \theta_1 + B \sin \theta_1,$$

where A and B are quantities independent of θ_1. It is clear that if this is positive for all values of θ_1, $\mu_1 P_1$ must be numerically greater than $(A^2 + B^2)^{\frac{1}{2}}$.

We notice that since $\qquad A = - \mu_2 P_2 \cos \theta_2 - \&c. + Y,$
$$B = - \mu_2 P_2 \sin \theta_2 - \&c. - X,$$

the quantities A and $-B$ are the resolved parts parallel to the axes of the external forces and of all the frictional forces except that at A_1. If F be the friction at the point A_1, the resultant pressure on the axis will be $(A^2 + B^2)^{\frac{1}{2}} + F$. This can be made to vanish by assigning to the friction F a value less than the limiting friction. See Art. 183.

It appears therefore that, *if we include all the positions of I which make the moment u a minimum, viz. those which do, as well as those which do not coincide with a point of support, that position in which u is least is the position of the instantaneous axis.*

187. It will be observed that, if the lamina is displaced round the axis through I through any small angle $d\theta$, the *work done by the forces and the frictions* is $u d\theta$, where $d\theta$ is measured in the direction in which the frictions act. *To make u a minimum is the same thing as to make this work a minimum for a given angle of displacement.*

188. Ex. 1. *A triangular table with a point of support at each corner A, B, C is placed on a rough horizontal floor. Find the least couple which will move the table.*

It may be shown that the pressure on each point of support is equal to one third of the weight of the triangle. The limiting frictional forces at A, B, C are therefore each equal to $\frac{1}{3} \mu W$.

Let the triangle begin to turn about some point I not at a corner. Since the frictions balance a couple, these frictions when rotated through a right angle so as to act along AI, BI, CI must be in equilibrium. Hence I must lie within the triangle. Also, the frictions being equal, each of the angles AIB, BIC, CIA must be $= 120°$. If then *no angle of the triangle is so great as 120°, the point I is the intersection of the arcs described on any two sides of the triangle to contain 120°.* The least couple which will move the triangle is therefore $\frac{1}{3} \mu W (AI + BI + CI)$.

The triangle might also begin to turn about one of its corners. Suppose I to coincide with the corner C. Rotating the frictions as before, the magnitude of the friction at C must be just sufficient to balance two forces, each equal to $\frac{1}{3} \mu W$, acting along AC and BC. The resultant of these is clearly $\frac{1}{3} \mu W . 2 \cos \dfrac{C}{2}$. Unless the angle C is $> 120°$ this resultant is $> \frac{1}{3} \mu W$ and is therefore inadmissible. Thus *the table cannot turn round an axis at any corner unless the angle at that corner is greater than* 120°. If the corner is C, the magnitude of the least couple is $\frac{1}{3}\mu W (CA + CB)$.

This statical problem might also be solved by finding the position of a point I such that the sum of its distances AI, BI, CI (all multiplied by the constant $\frac{1}{3}\mu W$) from the corners is an absolute minimum.

Ex. 2. Four equal heavy particles A, B, C, D are connected together so as to form a rigid quadrilateral and placed on a rough horizontal plane. Supposing the pressures at the four particles are equal, find the least couple which will move the system.

The instantaneous centre I is the intersection of the diagonals or one of the corners according as that intersection lies inside or outside the quadrilateral.

Ex. 3. A heavy rod is placed in any manner resting on two points A and B of a rough horizontal curve, and a string attached to the middle point C of the chord is pulled in any direction so that the rod is on the point of motion. Prove that the locus of the intersection of the string with the directions of the frictions at the points of support is an arc of a circle and a part of a straight line. Find also how the force must be applied that its intersection with the frictions may trace out the remainder of the circle.

Firstly let the rod be on the point of slipping at both A and B, and let F, F' be the frictions at the two points. Then F, F' are both known, and depend only on the weight and on the position of the centre of gravity of the rod. Supposing the centre of gravity to be nearer B than A, the limiting friction at B will be greater than that at A. Since there is equilibrium, the two frictions and the tension must meet in one point; let this be P. Then since $AC = CB$, it is evident that CP is half the diagonal of the parallelogram whose sides are AP, BP. Hence, by the triangle of forces, AP, BP and $2PC$ will represent the forces in those directions. Hence $AP : PB :: F : F'$, and thus the ratio $AP : PB$ is constant for all directions of the string. The locus of P is therefore a circle.

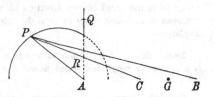

Let the point C be pulled in the direction PC, so that the line CP in the figure represents the produced direction of the string.

The string CP cuts the circle in two points, but the forces can meet in only one of these. It is evident that the rod must be on the point of turning round some one point I. This point is the intersection of the perpendiculars drawn to PA, PB at A and B. Now the frictions, in order to balance the tension, must act *towards* P, and therefore the directions of motion of A and B must be *from* P. This clearly cannot be the case unless the point I is on the same side of the line AB as P. Therefore the angle PAB is greater than a right angle. Thus the point I cannot lie on the dotted part of the circle.

Secondly. Let the rod be on the point of slipping at one point of support only. Supposing as before that the centre of gravity is nearer B than A, the rod will slip at A and turn round B as a fixed point. Thus the friction acts along QA and the locus of P is the fixed straight line QA.

But P cannot lie on the dotted part of the straight line, for if possible let it be at R. Then if AR represent F, RB must be less than F', because there is no slipping at B. But, because R lies within the circle, the ratio $AR : RB$ is less than the ratio $AP : PB$, i.e. is less than $F : F'$, and therefore RB is greater than F'. But this is contrary to supposition.

Thus the string being produced will always cut the arc of the circle and the part of the straight line in one point and one point only. The frictions always tend to that point when the rod is on the point of motion.

In order that the locus of P may be the dotted part of the circle it is necessary that the frictions should tend one from P and the other to P and the tension must therefore act in the angle between PA and PB produced. By the triangle of forces APB we see that the tension must act parallel to AB, and be proportional to it.

Ex. 4. A lamina rests on three small supports A, B, C placed on a horizontal table; one of these, viz. C, is smooth and the other two, A and B, are rough. A string attached to any point D, fixed in the lamina, is pulled horizontally so that the lamina is on the point of motion. If the position of the centre of gravity and the coefficients of friction are such that the limiting frictions F and F' at A and B are in the ratio $BD : AD$, prove that the locus of the intersection P of the string and the frictions F, F' is (1) a portion of the circle circumscribing ABD, (2) a portion of a rectangular hyperbola having its centre at the middle point of AB and also circumscribing ABD, (3) a portion of two straight lines.

Let $\qquad AD=b,\ BD=a,\ $ then $\ Fb=F'a$.

Draw LAL', HBH' perpendiculars to AB. If the lamina slip at one point only of the supports A, B, the point P lies on these perpendiculars.

If the lamina slip at both A and B, we find, by taking moments about D, that sin $PAD=$ sin PBD. The angles PAD and PBD are therefore either supplementary or equal. The locus of P is therefore the circle circumscribing the triangle ABD, and a rectangular hyperbola also circumscribing ABD. The first locus follows also from the triangle of astatic forces considered in Art. 71. The second locus may be found by taking AB as axis of x and equating the tangents of the angles PBA and $PAB - \gamma$, where γ is the difference of the angles DAB and DBA.

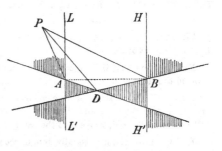

To determine the branches of these two curves which form the true locus of P we consider the relative positions of P and the instantaneous centre I. These two points lie at opposite ends of a diameter of a circle drawn round ABP. Hence, if P lie outside the perpendiculars LL', HH', I also must lie outside. The frictions cannot then balance the tension T unless the straight line PD passes *inside* the angle APB. Similarly, if P lie between the perpendiculars, PD must be *outside* the angle APB.

The straight lines LL', HH', DA, DB divide space into ten compartments. Several of these compartments are excluded from the locus of P by the rules just given. It will be convenient to mark (by shading or otherwise) the compartments in which P can lie. We then sketch the circle and the hyperbola and take only those branches which lie on a marked compartment. The figures are different according as D lies between or outside the lines LL', HH'.

Ex. 5. If in the last example the limiting frictions are in any ratio, the locus of the intersection of the string and frictions is a portion of a curve of the fourth degree and of two straight lines. The proper portions, as before, are those branches which lie in the marked compartments.

189. Ex. 1. *A uniform straight rod AB is placed on a rough table, and all its elements are equally supported by the table. Find the least force which, acting at one extremity A perpendicular to the rod, will move it.*

Let l be the length of the rod, w its weight per unit of length. Each element dx of the rod presses on the table with a weight wdx. The limiting friction at this element is therefore μwdx. If I be the centre of instantaneous rotation, the friction at each element acts perpendicular to the straight line joining it to I, and all these are in equilibrium with the impressed force P at A.

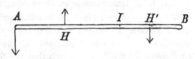

The point I must lie in the length of the rod. For suppose it were on one side of the rod, then, rotating (as already explained) the frictions through a right angle so that they all act towards I, these should be in equilibrium with a force P acting parallel to the rod. But this is impossible unless I lie in the length of the rod.

Next, let I be on the rod, and let $AI = z$. The friction at any element H or H' acts perpendicular to the rod in the direction shown in the figure. The resultant frictions on AI and BI are therefore μwz and $\mu w (l - z)$. These act at the centres of gravity of AI and BI. Resolving and taking moments about A, we have

$$\mu wz - \mu w (l - z) = P, \qquad \mu wz^2 = \mu w (l^2 - z^2).$$

The last equation gives $z\sqrt{2} = l$, and the first shows that $P = \mu W (\sqrt{2} - 1)$, where W is the weight of the rod.

Ex. 2. Show that the rod could not begin to turn about a point I on the left of A or on the right of B.

Ex. 3. If the pressure of an element on the table vary as its distance from the extremity A of the rod; and P, Q be the forces applied at A, B respectively which will just move the rod, prove that the ratio of P to Q is $2 (\sqrt[3]{2} - 1)$.

Ex. 4. Two uniform equally rough rods AB, BC, smoothly hinged together at B, are placed in the same straight line on a rough horizontal table, and the extremity A is acted on by a force P in a direction perpendicular to the rods. If P is gradually increased until motion begins, show that the rod AB begins to move before BC or both begin to move together according as $2 (\sqrt{2} - 1) W'$ is greater or less than W, where W, W' are the weights of the rods AB, BC respectively. If both rods begin to move together, prove that the instantaneous centre of rotation of AB is at a distance z from A where $\dfrac{2z^2}{l^2} = 1 + 2 (\sqrt{2} - 1) \dfrac{W'}{W}$ and l is the length of AB.

Ex. 5. A heavy rod AB placed on a rough horizontal table is acted on at some point C in its length by a force P, in a direction making an angle a with the rod, and the force is just sufficient to produce motion. If the instantaneous centre lie in a straight line drawn through B perpendicular to the rod and be a distance from A equal to twice the length AB, prove that $\tan \alpha = 2\,(2 - \sqrt{3})/\sqrt{3}\log 3$.   Find the position of C.

Ex. 6. A hoop is laid upon a rough horizontal plane, and a string fastened to it at any point is pulled in the direction of the tangent line at the point. Prove that the hoop will begin to move about the other end of the diameter through the point.   [Math. Tripos, 1873.]

Let A be the point, AB the diameter through A. If we rotate each force round its point of application through a right angle the frictional forces will act towards the centre I of rotation Art. 184. The point I is therefore so situated that the resultant of the frictional forces (regarded as acting towards I from the elements of the hoop) is parallel to the diameter AB. It easily follows that I must lie on the diameter AB.

Let us next consider the equation of moments. The point I must be so situated in the diameter AB that the moment about A of the frictions at all the elements of the hoop is zero. This condition is satisfied if I is at the end B of the diameter AB, for then the line of action of the friction at every element passes through A.

It is, perhaps, unnecessary to prove that no point, other than B, will satisfy this condition. It may however be shown in the following manner. If possible let I lie on AB within the circle. Whatever point P is taken on the hoop the angle IPA is less than a right angle. Since the friction at P acts in a direction at right angles to IP, it will become evident by drawing a figure that the friction at every element tends to produce rotation round A in the same direction. The moment therefore of the frictions about A could not be zero. In the same way we can prove that I cannot lie outside the circle.

Ex. 7. A uniform semicircular wire, of weight W, rests with its plane horizontal on a rough table, AB is the diameter joining its ends, and C is the middle point of the arc; a string tied to C is pulled gently in the direction CA, and the tension increased until the wire begins to move. Show that the tension at this instant is equal to $2\sqrt{2}\mu W/\pi$. [The instantaneous axis is at B.]   [St John's Coll., 1886.]

Ex. 8. A uniform piece of wire, in the form of a portion of an equiangular spiral, rests on a rough horizontal plane; show that the single force which, applied to a point rigidly connected with it, will cause it to be on the point of moving about the pole as instantaneous centre, is equal to the weight of a straight wire of length equal to the distance between the ends of the spiral, multiplied by the coefficient of friction. Show how to find the point.   [Math. Tripos, 1888.]

Ex. 9. Three equal weights, occupying the angles A, B, C of an equilateral triangle, are rigidly connected and placed upon a rough inclined plane with the base AB of the triangle along the line of greatest slope, and the highest weight A is attached by a string to a point O in the line of the base produced upwards; if the system be on the point of moving, prove that the tangent of the inclination of the plane is $(2 + \sqrt{3})\,\mu/\sqrt{3}$, where μ is the coefficient of friction.   [Math. Tripos, 1870.]

Suppose I not at a corner, the three frictions are then equal. Since A can only move perpendicular to OA, I must lie in OAB. Unless I lie between A and B and at the foot of the perpendicular from C on AB, the three frictions will have a component perpendicular to AB. Taking moments about I, we find the result given in the question. Next suppose I to be at the corner A. The frictions at B and C when resolved perpendicular to AB are then too great for the limiting friction at A. This supposition is therefore impossible.

Ex. 10. A three-legged stool stands on a horizontal plane, the coefficient of friction being the same for the three feet; a small horizontal force is applied to one of the feet in a given direction, and is gradually increased until the stool begins to move; show that this force will be greatest when its direction intersects the vertical through the centre of gravity of the stool.

Show also that if the force when equal to twice the whole friction of the foot on which it acts, applied in a direction whose normal at the foot passes between the two other feet, causes the foot to begin to move in its own direction, the centre of gravity of the stool is vertically above the centre of the circle inscribed in the triangle formed by the feet.   [Math. Tripos.]

Ex. 11. A flat circular heavy disc lies on a rough inclined plane and can turn about a pin in its circumference; show that it will rest in any position if $32\mu > 9\pi \tan i$, where i is the inclination of the plane to the horizon. The weight is supposed to be equally distributed over its area.   [Pet. Coll., 1857.]

Let W be the weight of the disc. The origin being at the pin the friction at any element $rd\theta dr$ is $\mu W \cos i \cdot rd\theta dr/\pi a^2$. Taking moments about the pin the result follows by integration.

Ex. 12. A right cone, of weight W and angle $2a$, is placed in a circular hole cut in a horizontal table with its vertex downwards. Show that the least couple which will move it is μWr cosec a, where r is the radius of the hole.

The pressure Rds on each element ds of the hole acts normally to the surface of the cone, hence, resolving vertically, $\int Rds \sin a = W$. The limiting friction on each element is μRds, hence, taking moments about the axis of the cone, the result follows.

Ex. 13. A heavy particle is placed on a rough inclined plane, whose inclination is equal to the limiting angle of friction; a thread is attached to the particle and passed through a hole in the plane, which is lower than the particle but not in the line of greatest slope; show that, if the thread be *very slowly* drawn through the hole, the particle will describe a straight line and a semicircle in succession.

[Maxwell's problem, Math. Tripos, 1866.]

Let W be the weight resolved along the line of greatest slope, F the friction, then $F=W$. As the particle moves very slowly, the forces F, W and the tension T are always in equilibrium. As long as the hole O is lower than the particle, T is infinitely small and just disturbs the equilibrium. The particle therefore descends along the line of greatest slope. When the particle P passes the horizontal line through O, T becomes finite. Hence T bisects the angle between F and W. The path is therefore such that the radius vector OP makes the same angle with the tangent (i.e. F) that it makes with the line of greatest slope. This, by a differential equation, obviously gives a semicircle having O for one extremity of its horizontal diameter.

Ex. 14. If, on a table on which the friction varies inversely as the distance from a straight line on it, a particle is moved from one given point to another, so that the work done is a minimum, the path described is a circle.   [Trin. Coll.]

This result follows at once from Lagrange's rule in the Calculus of Variations.

190. Ex. 1. Two heavy particles A, A', placed on a rough table, are connected by a string without tension and very slightly elastic. The particle A is acted on by a force P in a given direction AC making with $A'A$ produced an angle β less than a right angle. As P is gradually increased from zero, will A move first or will both move together?

Let F, F' be the limiting frictions at A, A'. Suppose P to increase from zero: while P is less than F it is entirely balanced by the friction at A. The string, however nearly inelastic it may be, has no tension until A has moved. Let P be a little greater than F; take AL to represent P and draw LMM' parallel to AA'; with centre A and radius F describe a circle cutting LMM' in M and M', then LM represents the tension of the string. Of the two intersections M, M', the nearest to L is chosen, for this makes the friction at A act opposite to P when $P = F$.

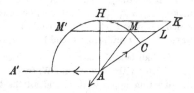

As P gradually increases M travels along the arc CH. The equilibrium of the particle A becomes impossible when LMM' does not cut the circle, i.e. when M reaches H. The particle A' borders on motion when the tension LM becomes equal to F'. Now $HK = F \cot \beta$. Hence the particle A moves alone if $F \cot \beta < F'$ but both move together if $F \cot \beta > F'$.

When the limiting frictions F, F' are equal, and β is less than half a right angle, both particles move together. One friction acts along AA' and the other makes an angle β with the force P. Also $P = 2F \cos \beta$.

In this solution the point M' has been excluded by the principle of continuity, though statically A would be in equilibrium under the forces represented by AL, LM', $M'A$. If the string AA' had a proper initial tension, but balanced by frictions at A and A' together with an initial force P along AC, then M' would be the proper intersection to take.

Ex. 2. Two weights A and B are connected by a string and placed on a horizontal table whose coefficient of friction is μ. A force P, which is less than $\mu A + \mu B$, is applied to A in the direction BA, and its direction is gradually turned round an angle θ in the horizontal plane. Show that if P be greater than $\mu \sqrt{A^2 + B^2}$, then both A and B will slip when $\cos \theta = \{\mu^2 (B^2 - A^2) + P^2\}/2\mu BP$, but if P be less than $\mu \sqrt{A^2 + B^2}$ and greater than μA, then A alone will slip when $\sin \theta = \mu A / P$.   [Math. Tripos.]

Ex. 3. The n particles A_0, A_1, ..., A_{n-1}, of equal weights, are connected together, each to the next in order, by $n-1$ strings of equal length and very slightly elastic. These are placed on a rough horizontal plane with the strings just stretched but without tension, and are arranged along an arc of a circle less than a quadrant. The particle A_{n-1} is now acted on by a force P in the direction $A_{n-1}A_n$, where A_n is an imaginary $(n+1)$th particle. Supposing P to be gradually increased from zero, find its magnitude when the system begins to move.

Let us suppose that any two consecutive particles A_m and A_{m+1} both border on motion. Let ϕ_m be the angle the friction at A_m makes with the chord $A_{m+1}A_m$. Let T_m be the tension of the string A_mA_{m+1}. Let β be the angle between any string and the next in order. Let F be the limiting friction at any particle.

Resolving the forces on the particles A_m and A_{m+1} perpendicularly to $A_{m-1}A_m$ and $A_{m+1}A_{m+2}$ respectively, we find

$$T_m \sin \beta = F \sin (\phi_m + \beta), \qquad T_m \sin \beta = F \sin \phi_{m+1}.$$

Resolving the same forces perpendicularly to the frictions on the two particles, we have $\qquad T_m \sin \phi_m = T_{m-1} \sin (\phi_m + \beta), \qquad T_{m+1} \sin \phi_{m+1} = T_m \sin (\phi_{m+1} + \beta).$

Comparing the first two equations, we see that $\phi_m + \beta$ and ϕ_{m+1} are either equal or supplementary. The other two equations show that the second alternative makes $T_{m+1} = T_{m-1}$. Both these alternatives are statically possible, and thus forces which might be friction forces could be arranged at the several particles in many ways so that equilibrium would be preserved.

We shall take the alternative which agrees with the supposition that the strings are initially without tension. When P is less than F the friction at A_{n-1} acts in the direction opposite to P, and all the tensions are zero. When P has become greater than F, the string $A_{n-2} A_{n-1}$ is slightly stretched and the tension $A_{n-2} A_{n-1}$ is called into play. The friction at A_{n-2} acts opposite to this tension, and all the other tensions are zero. Thus, as P continually increases, the tensions and frictions are one by one called into play. Supposing the tensions to be initially zero, we shall assume that the tensions produced by P are such that their magnitudes continually increase from the string with zero tension up to the string $A_{n-1} A_n$. Any other supposition would lead to the result that by pulling a string at one end we could produce, after overcoming the resistances, a greater tension at the other end. Since then T_{m+1} must be greater than T_{m-1}, we have $\phi_{m+1} = \phi_m + \beta$.

Suppose that all the particles from A_p to A_{n-1} border on motion and that $T_{p-1} = 0$; we have then $\phi_p = 0$, $\phi_{p+1} = \beta$, and in general

$$\phi_{p+\kappa} = \kappa\beta, \qquad T_{p+\kappa} \sin \beta = F \sin (\kappa + 1) \beta.$$

Since $T_{n-1} = P$, we see that the force P required to make all the particles from A_p to A_{n-1} border on motion is

$$P = F \sin (n - p) \beta \cdot \operatorname{cosec} \beta.$$

When P becomes greater than the value given by this equation, a tension in the string $A_{p-1} A_p$ will be called into play. The tension of $A_p A_{p+1}$ required to move A_p without A_{p-1} is $F \operatorname{cosec} \beta$, while that required to move both is $F \sin 2\beta \cdot \operatorname{cosec} \beta$. Since the latter is less than the former tension, the friction at A_{p-1} will become limiting before A_p begins to move. Thus we see that, as P continues to increase, the successive particles border on motion, but no one begins to move without the others.

If $n\beta$ be less than a right angle, we conclude that all the particles begin to move together, and that the force required to move them is $P = F \sin n\beta \operatorname{cosec} \beta$.

If $n\beta$ be greater than a right angle, we have shown that, without destroying the equilibrium, P can increase up to $F \sin p\beta \cdot \operatorname{cosec} \beta$, where $p\beta$ is less and $(p+1)\beta$ greater than a right angle. We have then $T_{n-p-1} = 0$. When P becomes greater than this value, the particle A_{n-1} will begin to move alone. For the tension required to move A_{n-1} is $F \operatorname{cosec} \beta$, and the tension T_{n-2} is then $F \cot \beta$. Since this is less than $F \sin p\beta \operatorname{cosec} \beta$, the system A_{n-2}, A_{n-3}, &c. is not bordering on motion.

CHAPTER VI.

THE PRINCIPLE OF VIRTUAL WORK.

191. In a former chapter the principle of virtual work has been established for forces which act on a particle. It is now proposed to consider this principle more fully, and to apply it to a system of bodies in two and three dimensions.

The principle itself may be enunciated as follows. *Let any number of forces P_1, P_2 &c. act at the points A_1, A_2 &c. of a system of bodies. These bodies are connected together in any manner so as either to allow or exclude relative motion, and they therefore exert mutual actions and reactions on each other. Let the system be slightly displaced so that the points A_1, A_2 &c. assume the neighbouring positions A_1', A_2' &c. Let dp_1, dp_2 &c. be the projections of the displacements A_1A_1', A_2A_2' &c. on the directions of the forces P_1, P_2 &c. respectively, and let $dW = P_1 dp_1 + P_2 dp_2 + $ &c. Then the system is in equilibrium if $dW = 0$ for all displacements consistent with the geometrical connexions between the bodies of the system.*

Also the system is not in equilibrium if one or more displacements can be found for which dW is not equal to zero.

Strictly speaking we should say, not that dW is zero, but that dW, in the language of the differential calculus, is a small quantity of the second order. This will be understood in what follows.

192. These displacements are to be regarded as imaginary motions which the system might, but does not necessarily, take. The principle of virtual work supplies a test, whether a given position of the system is one of equilibrium or not. We first consider what are the possible ways in which the system could begin to move out of the given position. If for any one of these the sum ΣPdp is zero, then the system will not begin to move in that mode of displacement. In this way all the possible displacements are examined, and if ΣPdp is zero for each and every one, the given position is one of equilibrium.

These small tentative displacements of the system are called *virtual displacements*. The product Pdp is called, sometimes the *virtual moment*, and sometimes the *virtual work* of the force P. The sum ΣPdp is called the virtual moment or virtual work of all the forces.

193. A proof of the principle of virtual work for forces acting on a single particle has been already given in Chap. II. No satisfactory method has yet been found by which the principle for a system of bodies can be deduced directly from the elementary axioms of statics. Lagrange has made a brilliant attempt which will be discussed a little further on.

There is another line of argument which may be adopted. The system is regarded as composed of simpler bodies, each acted on by some of the forces, and connected together by mutual actions and reactions. Thus Poisson regards the system as a collection of points in equilibrium connected together as if by flexible strings or inflexible rods without weight. To avoid making any assumptions concerning the molecular structure of bodies, we shall regard the system as made up of rigid bodies of such size that the elementary laws of statics may be applied to them.

The principle will first be proved for the simpler body, assuming the composition and resolution of forces. The principle will therefore be true for the general system, provided we include amongst the forces P_1, P_2 &c. all the mutual actions and reactions of the bodies of the system.

Lastly, these actions and reactions are examined, and it will be proved that they do not put in an appearance in the general equation of virtual work. It follows that the principle may be used as if P_1, P_2 &c. were the only forces acting on the system.

The chief objection to this mode of proof is that the mutual actions and reactions must be sufficiently known to enable us to prove that their separate virtual works are either zero or cancel each other.

In this mode of proof we have in part followed the lead of Fourier. See *Journal Polytechnique*, Tome II.

To prove the converse theorem we shall examine how a system could begin to move from a position of rest. We shall show that every such displacement is barred if for that displacement the virtual work of the forces is zero.

194. Proof of the principle for a free rigid body. We begin by proving that *the virtual work of any system of finite forces P_1, P_2 &c. is equal to that of their resultants provided the points of application of all the forces are connected by invariable relations.* See Art. 19.

The general process by which these resultants are found may be separated into three steps; (1) we may combine or resolve forces acting at a point by the parallelogram of forces; (2) we may transfer a force from one point A of its line of action to another B; (3) we may remove from or add to the system, equal and opposite forces. By the repeated action of these steps we have been able in the preceding chapters to change one set of forces into another simpler set, which we called their resultant. See Art. 117.

It has been proved in Art. 66 that the virtual work is not altered by the first of these processes. We shall now show that the virtual work of a force is not altered by the second process. It follows that the sum of the virtual works of two equal and opposite forces introduced by the third process is zero, and cannot affect the general virtual work of all the forces.

Let $A'B'$ be the displaced position of AB. Draw $A'M$, $B'N$ perpendiculars on AB. Let F be the force whose point of application is to be transferred from A to B. Before and after the

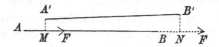

transference its virtual works are $F.AM$ and $F.BN$ respectively. Since $A'B'$ makes with AB an infinitely small angle whose cosine may be regarded as unity, we have MN equal to $A'B'$. Hence, *if the distance between the two points of application remain unaltered,* i.e. $AB = A'B'$, we have $BN = AM$. It immediately follows that $F.AM = F.BN$.

Thus in all changes of forces into other forces consistent with the principles of statics, the work of the forces due to any given small displacement is unaltered.

195. We may now apply this result to a system of forces P_1, P_2 &c. acting on a free rigid body.

All these forces can be reduced to a force R acting at an arbitrary point O, and a couple G, Art. 105. By what precedes the virtual work of the forces P_1, P_2 &c. due to any displacement is equal to the virtual work of R and G.

If the forces P_1, P_2 &c. are in equilibrium, both R and G are zero, Art. 109. Hence the virtual work of P_1, P_2 &c. for any displacement is zero.

Conversely, if the virtual work of P_1, P_2 &c. is zero for all displacements, then the virtual work of R and G is zero. We shall now show that this requires that R and G should each be zero. First let the body be moved parallel to itself through any small space δr in the direction in which R acts. The virtual work of the force R is $R\delta r$. Let AB be the arm of the couple and let the forces act at A and B. Since equal and parallel displacements AA', BB' are given to A and B, while the forces acting at A and B are equal and opposite, it is evident that the works due to the two forces cancel each other. The work of the couple G is therefore zero. Hence the sum of the works of R and G cannot vanish unless $R = 0$.

Next let the body be turned through a small angle $\delta\omega$ round a perpendicular drawn through O to the plane of the couple, and let this rotation be in the direction in which the couple urges the body. Let O bisect the arm AB and let the forces of the couple be $\pm Q$. Each of the points A and B receives a displacement equal to $\frac{1}{2}AB\delta\omega$ in the direction of the force acting at that point. The sum of the works due to these two forces is therefore $AB \cdot Q\delta\omega$, i.e. $G\delta\omega$. Since the point of application of R is not displaced, the virtual work of R (even if R were not zero) is zero. Hence the sum of the virtual works of R and G cannot vanish unless $G = 0$. It immediately follows that the body is in equilibrium.

196. On the forces which do not put in an appearance in the equation of virtual work. When the body is not free but can move either under the guidance of fixed constraints or under the action of other rigid bodies it becomes necessary (as explained in Art. 193) to determine what actions and reactions do not appear in the general equation of virtual work. We cannot make an exhaustive list, but we may make one which will include those cases which commonly occur.

I. *Let two particles A, B of the system act on each other by means of forces along AB, then if the distance AB remain invariable for any displacement, the virtual works of the action and the reaction destroy each other.* For example, if the points A, B are connected by an inelastic string, the tension does not appear in the equation of virtual work.

This follows at once from Art. 194, for the force at A may be transferred to B. The two equal and opposite forces acting at B have then the same displacement. Hence their virtual works are equal and opposite.

II. *If any body of the system is constrained to turn round a point or an axis fixed in space, the virtual work of the reaction at this point or axis is zero.* This is evidently true, for the displacement of the point of application of the force is zero.

III. *Let any point A of a body be constrained to slide on a surface fixed in space.*

If the surface is smooth, the action R on the point A of the body is normal to the surface. Let A move to a neighbouring point A', then AA' is at right angles to the force. The work by Art. 68 is therefore zero.

If the surface is rough, let F be the friction. This force acts along $A'A$, and its work is $-F \cdot AA'$. This is not generally zero.

IV. *If any body of the system roll without sliding on a fixed surface, the work of the reaction is zero.*

If this is not evident, it may be proved as follows. In the figure the body DAE rolls on the fixed surface $MABN$ and takes a neighbouring position $D'BE'$. The plane of the paper represents a section of the surfaces drawn through their common normal at A, and contains the elementary arc AB of rolling. In this displacement the point A of the body begins to move along the common normal and arrives at A'. If we replace the curves DAE, MAB by their circles of curvature, we know (since the arcs AB, $A'B$ are equal) that $AA' : AB^2$ is half the sum of the opposite curvatures. Assuming

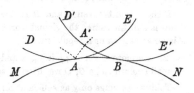

these curvatures to be finite, it follows that AA' is of the same order of small quantities as AB^2, i.e. AA' is of the second order of small quantities. Hence, when we retain only terms of the first order, as in the principle of virtual work, we may treat the rolling body as if it were turning round a point A fixed (for the instant) in space. It follows therefore from the result of the last article that, when a body rolls on a fixed surface, which may be either rough or smooth, the virtual work of the reaction is zero.

V. If the surface on which the body rolls is another body of the system, the surface is moveable. But we may show that, *if both bodies are included in the same equation of virtual work, the mutual action does not appear in that equation.*

To prove this we notice that we may construct any such displacement of the two bodies (1) by moving the two bodies together until the body $MABN$ assumes its position in the given displacement, and then (2) rolling the body DAE on the body $MABN$, now considered as fixed, until DAE also reaches its final position. During the first of these displacements the action and reaction at A are equal and opposite, while their common point of application A has the same displacement for each body. Their virtual works are therefore equal and opposite, and their sum is zero. During the second displacement the body DAE rolls on a fixed surface, and the virtual work of its reaction is zero. See Art. 65.

197. Work of a bent elastic string. If the points A, B are connected by an elastic string, it may be necessary to know what the work of the tension is when the length is increased from l to $l + dl$. We shall show that, whether *the string connecting A and B is straight, or bent by passing through smooth rings fixed or moveable or over a smooth surface, the work is* $-Tdl$.

For the sake of greater clearness we shall consider the cases separately.

(1) Let the string be straight. Referring to the figure of Art. 194, the virtual work of the tension at A is $+T.AM$. The positive sign is given because the tension acts at A in the direction AB and the displacement AM is in the same direction, Art. 62. The work of the tension at B is $-T.BN$. The sum of these two is $-T(A'B' - AB)$ i.e. $-Tdl$.

If the action between A and B is a push R instead of a pull T, the same argument will apply but we must write $-R$ for T, so that the virtual work is Rdl.

If the action between A and B is due to an attractive or repulsive force F the result is still the same; the virtual works are $-Fdl$ or $+Fdl$ according as the force F is an attraction or a repulsion.

(2) Suppose the string joining A and B is bent by passing through any number of small smooth rings C, D &c. fixed in space.

Taking two rings only as sufficient for our argument, let these be C and D. Let A, B be displaced to A', B', and let $A'M$, $B'N$ be perpendiculars on AC and DB. The whole length l of the string is lengthened by BN and shortened by AM, hence $dl = BN - AM$. The tension T being the same throughout the string, the work at A

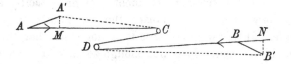

is $T \cdot AM$, that at B is $- T \cdot BN$. Exactly as before, the whole work is the sum of these two, i.e. $- Tdl$.

(3) Suppose the rings C, D &c., through which the string passes, are attached to other bodies of the system. The rings themselves will now be also moveable.

Supposing all these bodies to be included in the same equation of virtual work, the system is acted on by the following forces, viz. T at A along AC, T at C along CA, T at C along CD, T at D along DC and so on. By what has just been proved, the work of the first and second of these taken together is $- Td (AC)$, the work of the third and fourth is $- Td (CD)$ and so on. Hence, if l be the whole length of the string, viz. $AC + CD + $&c., the whole work is $- Tdl$.

In all these cases we see that, if the length of the string is unaltered by the displacement, the tension does not appear in the equation of virtual work.

(4) Let the string joining A and B pass over any smooth surface, which either is fixed in space, or is one of the bodies to be included in the equation of virtual work. Each elementary arc of the string may be treated in the manner just explained. The work done by the tension is therefore as before equal to $- Tdl$.

In order not to interrupt the argument, we have assumed that *the tension of a string is unaltered by passing over a smooth pulley or surface*. To prove this, let us suppose the string to pass over any arc BC of a smooth surface. Any element PP' of the string is in equilibrium under the action of the tensions at P, P' and the normal reaction of the smooth surface. The resolved part of these forces along the tangent at P must therefore be zero. Let T, T' be the tensions at P, P', $d\psi$ the angle between the tangents at these points, and let ds be the length of PP'. Supposing the pressure per unit of length of the string on the surface to be finite and equal to R, the pressure on the arc PP' is Rds. The resolved part of this along the tangent at P is less than $Rds \sin d\psi$, and is therefore of the second order of small quantities. The difference of the resolved parts of the tensions is $T - T' \cos d\psi$, which, when small quantities of the second order are neglected, reduces to $T - T'$. Since this must be zero, we have $T = T'$. Taking a series of elements of the string, viz. PP', $P'P''$ &c., it immediately follows that the tensions at P, P', P'' &c. are all equal, i.e. the tension of the string is the same throughout its length. If the surface were rough, this result would not follow, for the frictions must then be included in the equation of equilibrium formed by resolving along the tangent. *We may also prove the equality of the tensions by applying the principle of virtual work to the string BC.* Sliding the string without change of length along the surface, we have $T \cdot BB' = T' \cdot CC'$. Hence $T = T'$.

When the surface is a rough circular pulley which can turn freely about a smooth axis, and the string lies in a plane perpendicular to the axis, we can prove the equality of the tensions by taking moments about the axis. Let the string be $ABCD$ and let it touch the cylinder along the arc BC. Let T, T' be the tensions of AB, CD, r the radius of the cylinder. Taking moments about the axis, we have $Tr = T'r$. This gives $T = T'$.

198. In the preceding arguments we have tacitly assumed that the pressures which replace the constraints are finite in magnitude. If this were not true it is not clear that the virtual work would be zero. It is not enough to make a product $P \cdot dp$ vanish that one factor viz. dp should be zero, if the other factor P is infinite. Such cases sometimes occur in our examples when we treat the body under consideration as an unyielding rigid mass. But in nature the changes of structure of the body cannot be neglected when the forces acting on it become very great. The displacements are therefore different from those of a rigid body.

199. Converse of the principle of virtual work. We shall now prove the converse principle of virtual work for a system of bodies. *The system being placed at rest in some position, it is given that the work of the external forces is zero for all small displacements which do not infringe on the constraints. It is required to prove that the system is in equilibrium.*

If the system is not in equilibrium it will begin to move. Let us then examine all the ways in which the system could begin to move from its position of rest. Some one way having been selected, it is clear that by introducing a sufficient number of smooth constraining curves we can so restrain the system that it cannot move in any other way. Thus if any point of one of the bodies would freely describe a curve in space, we can imagine that point attached to a small ring which can slide along a rigid smooth wire, whose form is the curve which the point would freely describe. The point is thus prevented from moving in any other way. The reaction of this smooth curve has been proved to have no virtual work. It is also clear that these constraining curves in no way alter the work of the external forces during the displacement of the body.

In order to prevent the system from moving from its initial position it will now only be necessary to apply some force F to some one point A in a direction opposite to that in which A would move if F did not act. The forces of the system are now in equilibrium with F. Let the system receive an arbitrary virtual displacement along the only path open to it. In this displacement let the point A come to A'. Then the work of the forces plus the work of F is zero. But it is given that the work of the forces is zero for every such displacement, hence the work of F is zero. But this work is $-F \cdot AA'$, and since AA' is arbitrary it immediately follows that F must be zero. Thus no force is required to prevent the system from moving from its place of rest along any selected path. The system is therefore in equilibrium. *Treatise on Natural Philosophy*, Thomson and Tait, 1879, Art. 290.

200. Initial motion. Let us imagine a system to be placed at rest, and yet not to be in equilibrium under the action of the given external forces. We shall show that *the system will so begin to move* * *that the work of the forces in the initial displacement is positive.*

The proof of this is really a repetition of the argument already given in Art. 199. If the system begin to move from the position of rest in any given way, we constrain it to move only in that way. If F be the force acting at A which will prevent motion, we find as before that the work of the forces plus that of F is zero. But F must act opposite to the direction in which A would move if F were not applied, hence its work is negative; and the work of the impressed forces in this displacement is therefore positive.

201. It follows from this result, that *it is sufficient to ensure equilibrium that the work of the forces should be negative instead of zero for all displacements,* for then there is no displacement which the system could take from its state of rest. If however the work of the forces is negative for any one displacement, it must be positive for an equal and opposite displacement, i.e. one in which the direction of motion of every particle is reversed. To exclude therefore all displacements which make the work positive, it is in general necessary that the work should be zero for all displacements.

In some special cases of constraint it may happen that one displacement is possible while the opposite is impossible. *It is then not necessary that the work should be zero for this displacement.* For example, a heavy particle placed inside a cone with the axis vertical is clearly in equilibrium, yet the work done in any displacement is negative and not zero.

202. Method of using the principle. Let us suppose that points A_1, A_2, &c. of a system are constrained to move on fixed surfaces. We have then two objects, (1) to form those equations of equilibrium which do not contain the reactions, (2) to find the reactions. To effect the former purpose we give the system all necessary displacements which do not separate A_1, A_2, &c. from the constraining surfaces, and equate the sum of the

---

* *Dynamical proof.* When a system starts from a position of rest, it is proved in dynamics that the semi vis viva after a displacement is equal to the work done by the external forces. Now the vis viva cannot be negative, because it is the sum of the masses of the several particles multiplied by the squares of their velocities. It is therefore clear that the system cannot begin to move in any way which makes the virtual work of the forces negative.

virtual moments for each displacement to zero. To effect the
latter purpose we give the system a series of displacements such
that each of the points A_1, A_2, &c. in turn is alone moved off the
surface on which it rests. Including the work of the corresponding reaction and still equating the sum of the virtual works to
zero we have an equation to find that reaction.

203. *To deduce the equations of equilibrium from the principle
of work.*

The equations of equilibrium of a system are really equivalent
to two statements, (1) the sum of the resolved parts of the forces
in any direction for each body or collection of bodies in the system
is zero, (2) the sum of the moments about any or every straight
line is zero.

The equations of equilibrium of a system in one plane have been obtained
in Chap. IV., Arts. 109—111. The corresponding equations of a system in space
will be given at length in a later chapter. But to avoid repetition they are included
in the following reasoning. See also Arts. 105 and 113.

We have now to deduce these two results from the principle of
work. As before, let P_1, P_2 &c. be the forces, A_1, A_2 &c. their
points of application, $(\alpha_1, \beta_1, \gamma_1)$, $(\alpha_2, \beta_2, \gamma_2)$ &c. their direction
angles. Let the body or collection of bodies receive a linear
displacement parallel to the axis of x through a small space dx.

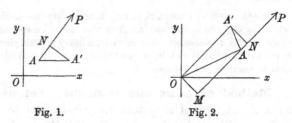

Fig. 1.   Fig. 2.

Then if A be moved to A', $AA' = dx$, (Fig. 1), and the projection
AN on the line of action of P is $dx \cos \alpha$. Hence, by the principle
of work, $P_1 \cos \alpha_1\, dx + P_2 \cos \alpha_2\, dx + \ldots = 0.$
Dividing by dx, this gives the equation of resolution, viz.

$$P_1 \cos \alpha_1 + P_2 \cos \alpha_2 + \ldots = 0.$$

In this equation all the reactions on the special body considered
due to the other bodies are to be included.

To find the sum of the moments of the forces about any straight line, say the axis of z, let us displace the special body considered round that axis through an angle $d\omega$.

First let the forces act in the plane of xy, and let p_1, p_2 &c. be the perpendiculars from the origin on their respective lines of actions. Thus in Fig. 2, $OM = p$. The displacement AA' of A due to the rotation is $OA \cdot d\omega$. The projection of this on the line of action of P is $OA\, d\omega \sin OAM$, i.e. $p\,d\omega$. Hence by the principle of work $P_1 p_1\, d\omega + P_2 p_2\, d\omega + \dots = 0.$

Dividing by $d\omega$, we have the equation of moments, viz.

$$P_1 p_1 + P_2 p_2 + \dots = 0.$$

Next, let the forces act in space. We first resolve each force parallel and perpendicular to the axis about which we take moments. The resolved parts of P are respectively $P \cos \gamma$ and $P \sin \gamma$. The displacement AA' of its point of application due to a rotation round z is perpendicular to the axis of z. The work of the first of these components is therefore zero. The second component is parallel to the plane of xy, and its work is found in exactly the same way as if it acted in the plane of xy. If p be the length of the perpendicular from O on the projection on xy of its line of action, the work is $P \sin \gamma\, p\,d\omega$. We therefore find as before

$$P_1 \sin \gamma_1 p_1 + P_2 \sin \gamma_2 p_2 + \dots = 0,$$

which is the usual equation of moments.

204. Combination of equations. The equations of equilibrium of each of the bodies forming a system having been found by resolving and taking moments, we can combine these equations at pleasure in any linear manner. For example we might multiply by λ an equation obtained by resolving parallel to some straight line x, and multiply by μ another equation obtained by taking moments about some straight line z. Adding the results, we get a new equation which may be more suited to our purpose than either of the original ones.

We shall now show that this derived equation might be obtained directly from the principle of work by a suitable displacement. Suppose both the equations combined as above to be equations of equilibrium of the same body. Let these be written in the form $\Sigma P \cos a = 0, \qquad \Sigma P p = 0.$

If we displace the body parallel to x through a small space dx and rotate it round z through an angle $d\omega$, the work of any force P due to the whole displacement is, by Art. 65, equal to the sum of the works of P due to each displacement. The equation of work obtained by this displacement is therefore

$$(\Sigma P \cos a)\, dx + (\Sigma P p)\, d\omega = 0.$$

If then we take $dx : d\omega$ in the ratio $\lambda : \mu$, the derived equation follows at once.

If the equations to be combined are equations of equilibrium of different bodies, these different bodies are to be displaced, a linear displacement corresponding always to a resolution and an angular displacement to a moment. If several equations are combined together the corresponding displacements are to be taken in any order, and the resulting displacement regarded as the single displacement which gives the corresponding work equation.

As in forming the equations of equilibrium by resolving and taking moments we suppose the constraints removed and replaced by corresponding reactions, so in forming these work equations the same supposition must be made.

It further appears that, if we can eliminate any unknown reactions from the equations of equilibrium by choosing the multipliers λ, μ &c. properly and adding the equations, then the same resulting equation can always be obtained (equally free from the same reactions) from the principle of work by giving the system a suitable displacement or series of displacements.

205. Examples on Virtual Work. Ex. 1. *A flat semicircular board with its plane vertical and curved edge upwards rests on a smooth horizontal plane, and is pressed at two given points of its circumference by two beams which slide in smooth vertical tubes. Find the ratio of the weights of the beams that the board may be in equilibrium.* [Math. Tripos, 1853.]

Let W, W' be the weights of the beams AB, $A'B'$; ϕ, ϕ' the angles which the radii CA, CA' make with the horizontal diameter Cx. Let a be the radius of the sphere, b the distance between the tubes. If y, y' be the altitudes above Cx of the centres of gravity of the rods, we have by the principle of work, $-Wdy - W'dy' = 0$.

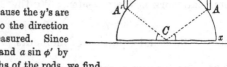

The negative sign is used because the y's are measured upwards *opposite* to the direction in which the weights are measured. Since y and y' differ from $a \sin \phi$ and $a \sin \phi'$ by constants, viz. half the lengths of the rods, we find

$$W \cos \phi d\phi + W' \cos \phi' d\phi' = 0.$$

But by geometry $a \cos \phi + a \cos \phi' = b$.

Differentiating the latter equation, and eliminating $d\phi : d\phi'$, we find

$$W \cot \phi = W' \cot \phi',$$

which gives the required ratio.

Ex. 2. Three heavy rods, which can slide freely through three vertical tubes fixed in space, rest with one extremity of each on a smooth hemisphere. The hemisphere rests with its plane face on a smooth horizontal plane. If Cx be any horizontal line through the centre C, θ_1, θ_2, θ_3 the angles which the planes through Cx and the lower extremities of the rods make with a horizontal plane, and W_1, W_2, W_3 the weights of the rods, prove that in equilibrium $\Sigma W \cot \theta = 0$.

Ex. 3. Eight rods perfectly similar and uniform are jointed together in the form of an octahedron, and being suspended from one of the angles are supported by a string fastened to the opposite angle, the string being elastic and such that the weight of all the rods together would stretch it to double its natural length, viz. that of one of the rods. Prove that in the position of equilibrium the rods will be inclined to the vertical at an angle $\cos^{-1} \frac{3}{4}$.   [Coll. Ex., 1889.]

Let the eight rods be AE, BE, CE, DE; AF, BF, CF, DF and let EF be the elastic string. Let W be the weight of any rod, $2a$ its length, and θ the inclination to the vertical. The octahedron being in its position of equilibrium, let the system receive a symmetrical displacement so that the angle θ is increased by $d\theta$. Taking E for origin, the depth of the centre of gravity of any one of the four upper rods is $a \cos \theta$, the virtual work of the weights of these rods is therefore $4Wd (a \cos \theta)$. The depth of the centre of gravity of any one of the four lower rods is $3a \cos \theta$, the virtual work of their weights is $4Wd (3a \cos \theta)$.

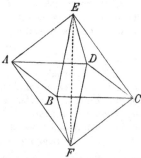

Since the unstretched length of the string is $2a$ and its stretched length is $EF = 4a \cos \theta$, the tension is, by Hooke's law, $T = E (4a \cos \theta - 2a)/2a$, where E is the weight which would stretch the string to twice its natural length, i.e. $E = 8W$. The virtual work is $-Td (4a \cos \theta)$, Art. 197. Adding all these several virtual works together we have $16 Wd (a \cos \theta) - Td (4a \cos \theta) = 0$. Substituting for T we easily find that $\cos \theta = \frac{3}{4}$.

Ex. 4. Show that the force necessary to move a cylinder of radius r and weight W up a plane inclined at angle a to the horizon by a crowbar of length l, inclined at β to the horizon, is $\dfrac{Wr}{l} \cdot \dfrac{\sin a}{1 + \cos (a + \beta)}$.   [Math. Tripos, 1874.]

Ex. 5. A smooth rod passes through a smooth ring at the focus of an ellipse whose major axis is horizontal, and rests with its lower end on the quadrant of the curve which is furthest removed from the focus. Show that its length must be at least $\frac{3}{4}a + \frac{1}{4}a \sqrt{(1 + 8e^2)}$, where a is the semi-major axis and e the eccentricity.   [Math. Tripos, 1883.]

Ex. 6. An isosceles triangular lamina with its plane vertical rests vertex downwards between two smooth pegs in the same horizontal line; show that there will be equilibrium if the base make an angle $\sin^{-1} (\cos^2 a)$ with the vertical; $2a$ being the vertical angle of the lamina, and the length of the base being three times the distance between the pegs.   [Math. Tripos, 1881.]

Ex. 7. Three rigid rods AB, BC, CD, each of length $2a$, are smoothly jointed at B, C. The system is placed so that the rods AB, CD are in contact with two smooth pegs distant $2c$ apart in the same horizontal line, and the rods AB, CD make equal angles a with the horizon. Prove that the tension of a string in AD which will maintain this configuration is $\frac{1}{4}W \operatorname{cosec} a \sec^2 a \{3c/a - (3 + 2 \cos^3 a)\}$, where W is the weight of either rod.   [St John's Coll., 1890.]

Ex. 8. Four rods, equal and uniform, rest in a vertical plane in the form of a square with a diagonal vertical and the two upper rods resting on two smooth pegs in a horizontal line. Show that the pegs must be at the middle points of the rods, and find the actions at the hinges.   [Coll. Ex., 1884.]

Ex. 9. Three equal and similar uniform heavy rods AB, BC, CD, freely jointed at B and C, have small smooth weightless rings attached to them at A and D: the rings slide on a smooth parabolic wire, whose axis is vertical and vertex upwards, and whose latus rectum is half the sum of the lengths of the three rods: prove that in the position of equilibrium the inclination θ of AB or CD to the vertical is given by the equation $\cos \theta - \sin \theta + \sin 2\theta = 0$.   [Coll. Ex., 1881.]

Ex. 10. A smooth hemispherical bowl of radius r is fixed with its rim horizontal. A uniform heavy rectangle $ABCD$ rests with two points A, B on the internal surface of the bowl, and its sides AD, BC resting on, and reaching beyond, the edge of the bowl. If θ be its inclination to the horizontal, show that

$$4 (r^2 - b^2) \cos^2 2\theta - a^2 \cos^2 \theta = 0,$$

where $AB = 2b$, $BC = 2a$.   [Coll. Ex., 1891.]

Ex. 11. n equal uniform rods, each of weight W' and length l, are jointed so as to form symmetrical generators of a cone whose semi-vertical angle is a, the joint being at the vertex of the cone. The rods are placed with their other ends in contact with the interior of a sphere whose radius is r, so that the axis of the cone is vertical, and a weight W is hung on at the joint. Show that

$$l^2 (3n^2 W'^2 + 4nW'W) \cos^2 a = (r^2 - l^2) (nW' + 2W)^2,$$

and find the action at the joint on each rod.   [Coll. Ex., 1884.]

Ex. 12. A conical tent resting on a smooth floor is made of an indefinitely great number of equal isosceles triangular elements hinged at the vertex, and kept in shape by a heavy circular ring placed on it as a necklace. Show that in equilibrium the semi-vertical angle of the cone is $\sin^{-1} \left\{ \dfrac{r}{h} \left(\dfrac{3W'}{W + 3W'} \right) \right\}^{\frac{1}{2}}$, where W, W' are respectively the weights of the cone and the ring and r, h are in like manner the radius of the ring and the slant side of the cone.   [St John's Coll., 1885.]

Ex. 13. A smooth fixed sphere supports a zone of very small equal smooth spherical particles, and the whole is prevented from slipping off the sphere by an elastic ring occupying a horizontal circle of angular radius a. Show that in the position of equilibrium the tension of the band is T, where $2\pi T = W \tan a$, and W is the whole weight of the ring and particles together.   [St John's Coll., 1885.]

It may be assumed that the centre of gravity of such a zone is half way between the bounding planes.

The work function.

206. Coordinates of a system. Our general object in statics is to find the positions of equilibrium of a system. To solve this problem we require some quantities which when given will determine the position of the system in space. Thus the position of a particle in geometry of two dimensions is defined when we know its coordinates x, y. In the same way if a body is free to move in the plane of xy, its position is fixed when we know the coordinates x, y of some point in it and also the angle θ some straight line fixed in the body makes with the axis of x. These three quantities, viz. x, y and θ, are called the coordinates of the body.

If the body is in space we define its position by giving (1) the coordinates x, y, z of some point A fixed in the body, (2) the two angles some straight line AB fixed in the body makes with the axes of x and y. If no more than this is given, the position of the body is not fixed, for it could be turned round AB as an axis. We therefore require (3) the angle some plane drawn through AB and fixed in the body makes with some plane fixed in space. These six quantities, or any other six which fix the place of the body, are called its coordinates.

If the body be under constraint the case is a little altered. Thus suppose the extremities of a rod of given length are constrained to rest on two given curves in a vertical plane; its position is defined simply by its inclination to the horizon or by the abscissa of one extremity. Either of these, or any other quantity which defines the position of the rod, is called its coordinate.

207. In the general case of a system of bodies, *any quantities which, when given, determine the positions of all the members of the system are called the coordinates of that system.* Just as the Cartesian coordinates of a point are connected by one or more equations when the point is constrained to lie on a given surface or curve, so the coordinates of a system are connected by equations when the system is subject to constraints. By help of these equations we can eliminate as many coordinates as there are equations, and thus make the position of the system depend on a smaller number of coordinates. There being now no equations of constraint, these remaining coordinates are independent of each other.

Let us suppose that the system is referred to independent coordinates. Since each may be varied without altering the others, there are as many ways of moving the system as there are co-ordinates. Any *small* displacement, indicated by varying simultaneously several coordinates, may be *constructed* by varying first one of the coordinates and then another, and so on. *The number of independent coordinates is therefore called the number of degrees of freedom of the system.*

208. The work function. Let a system of bodies be placed in any position, and let it receive any indefinitely small displacement which the constraints imposed on the system permit it to take. Let X, Y, Z be the components of any force P, and let (xyz) be the rectangular Cartesian coordinates of its point of application. The work of P is the same as that of its components, so that the general expression for the work is

$$\Sigma Pdp = \Sigma\,(Xdx + Ydy + Zdz)\ldots\ldots\ldots\ldots(1),$$

where the Σ implies summation for all the forces of the system.

Let the independent coordinates of the system be θ, ϕ, ψ &c. Then since these determine its position, the coordinates x, y, z of every point of each body can be expressed in terms of θ, ϕ &c. Thus x, y, z and X, Y, Z are all known functions of θ, ϕ &c. Substituting, the equation (1) takes the form

$$\Sigma Pdp = \Theta d\theta + \Phi d\phi + \&c. \quad\quad\dots\dots\dots\dots(2),$$

where Θ, Φ &c. are all known functions of the coordinates θ, ϕ &c.

209. *The coefficients* Θ, Φ, *&c. have sometimes an elementary statical meaning.* Suppose for example that the change in the coordinate θ (the others remaining constant) had the effect of turning the body about some straight line through the angle $d\theta$. Then $\Theta d\theta$ is the work of the forces when this displacement is given to the body. But, by Art. 203, this work is $Md\theta$, where M is the moment. It follows that Θ is the moment of the forces about the straight line.

Again, suppose that the change of some abscissa ϕ had the effect of moving the body parallel to the axis of x, then by the same article, ϕ is the resolved part of the forces parallel to that axis.

210. In most cases *the expression for the work is found to be a perfect differential of some quantity which we may call W.* For example, suppose the force P which acts on the point (xyz) to be due to the repulsion of some centre of force C, i.e. let P be a force whose line of action always passes through a point C fixed in space. If r be the distance from C to the point of application, the work of such a force for any small displacement is Pdr. If then the magnitude of P is some function of the distance r, the part contributed by such a central force to the expression ΣPdp is a perfect differential.

To take another case, let a force T acting between two points A, A' which move with the system be caused by such an elastic string as that described in Art. 197 or in any other way, so only that the force is some function of the distance between A and A'. The work of such a force is $\pm Tdr$, and as T is a function of r, this again is a perfect differential.

The system may be under the action of a variety of central forces, attracting many points of the system; or again there may be any number of actions between different sets of points, yet in all these cases *the share contributed by each force to the virtual work is a perfect differential.*

These two typical cases represent the forces which in most cases act on the system. The external forces are generally central forces, and the internal forces either do not appear in the equation of virtual work or appear as forces between one point and another such as those just described.

211. Since the expression (2) in Art. 208 represents the work of the forces due to any general small displacement, the integral of that expression when taken between any limits is the work of the forces as the system makes a finite displacement, i.e. as the system moves from any position I. to another II. The lower limit of the integral is found by giving the coordinates θ, ϕ &c. their values in the position I., and the upper limit by giving the same coordinates their values in the position II.

When the expression (2) is a perfect differential, this integration can be effected without knowing the route by which the system travels from the one position to the other. The integral W is a function of the upper and lower limits, and will thus depend on the initial and final position of the system and not on any intermediate position. It follows that *the work due to a displacement from one given position to another is the same, whatever route is taken by the system, provided always none of the geometrical constraints are violated.*

When the forces are such that the expression $\Sigma P dp$ is a perfect differential, they are said to form a conservative system.

Suppose we select any one position of the system of bodies as a standard, and let this position be defined by the values of the coordinates $\theta = \theta_1$, $\phi = \phi_1$, &c. Then *taking this standard position as the lower limit of the integral* and any general position as the upper limit, we have

$$W = \int \Sigma P dp = F(\theta, \phi, \&c.) - F(\theta_1, \phi_1, \&c.);$$

when it is not necessary to make an immediate choice of a standard position we write the integral in its indefinite form, viz.

$$W = F(\theta, \phi, \&c.) + C.$$

The function W, particularly when used in the indefinite form, is often called the *force function*, or *work function*.

Sometimes *the upper limit is made the standard position* and the general position the lower limit. If this standard is determined by the values $\theta = \theta_2$, $\phi = \phi_2$, &c.; the integral becomes

$$V = F(\theta_2, \phi_2, \&c.) - F(\theta, \phi, \&c.).$$

This is usually called *the potential energy of the forces with reference to the position defined by* $\theta = \theta_2$, $\phi = \phi_2$, &c.

If the two standards of reference were identical, we should have $W = -V$. But both these standards are seldom used in the same problem. In every case that standard of reference is generally chosen which is most suitable to the particular problem under discussion. We notice that $W + V$ is the work of the forces as the system moves along any route from the position $(\theta_1, \phi_1, \&c.)$ to the position $(\theta_2, \phi_2, \&c.)$, and these being fixed, the sum is constant for all positions of the system of bodies.

212. Maximum and Minimum. Suppose the system to be in a position of equilibrium. We then have $dW = 0$ for every virtual displacement, so that W is a maximum, a minimum, or stationary. The last alternative represents the case in which the evanescence of the first differential coefficients does not indicate a true maximum or minimum.

We have therefore another method of finding the positions of equilibrium of a system. We regard the work function as a known function of the coordinates, θ, ϕ &c. of the system, say

$$W = F(\theta, \phi, \ldots) + C.$$

To find the positions of equilibrium we use any of the rules given in the differential calculus to find the values of θ, ϕ &c. which make W a maximum or minimum.

213. If the coordinates θ, ϕ, &c. are all independent, we make the differential coefficient of W with regard to each of the variables equal to zero. This is equivalent to giving the system the geometrical displacements indicated by varying θ, ϕ, &c. in turn, and equating the virtual work in each case to zero. *But the process is analytical instead of geometrical*, and this has sometimes great advantages.

When we cannot express the position of the system by independent coordinates, *we may yet reduce the problem to the solution of equations* by using Lagrange's method of *indeterminate multipliers*. Let the n coordinates θ_1, θ_2, &c. be connected by the m geometrical relations

$$f_1(\theta_1, \theta_2, \&c.) = 0, \qquad f_2(\theta_1, \theta_2, \&c.) = 0, \qquad \&c. = 0,$$

so that $n - m$ of the coordinates are independent. Differentiating and using the m multipliers λ_1, λ_2, &c. we have

$$\Sigma\left(\frac{dW}{d\theta} + \lambda_1 \frac{df_1}{d\theta} + \lambda_2 \frac{df_2}{d\theta} + \ldots\right) d\theta = 0,$$

where Σ implies summation for θ_1, θ_2, &c. Since there are m multipliers at our disposal we choose these so that the coefficients of the differentials of the dependent coordinates are zero. The remaining θ's being independent we can make each vary separately and it then follows from the equation that the corresponding coefficient is zero. The coefficient of every $d\theta$ being zero, we obtain n equations of the form

$$\frac{dW}{d\theta} + \lambda_1 \frac{df_1}{d\theta} + \lambda_2 \frac{df_2}{d\theta} + \ldots = 0.$$

Joining these to the m given geometrical relations we have $m + n$ equations to find the n coordinates and the m multipliers.

214. Stable and Unstable equilibrium. It should be noticed that it is necessary and sufficient for equilibrium that the work function W is a maximum, a minimum, or stationary. There is however an important distinction between these cases.

Suppose the system is in equilibrium in such a position that W is a true maximum, i.e. W is decreased if the system is moved into any neighbouring position which is consistent with the constraints. Let the system be actually placed at rest in any one of these neighbouring positions. Not being in equilibrium in this new position it will begin to move. By Art. 200 it must so move that the initial work of the forces is positive, i.e. it must so move that W increases. The system therefore tends to approach closer to its original position of equilibrium. *The original position is therefore said to be stable.*

Suppose next the system is in equilibrium in such a position that W is a true minimum, i.e. W is increased if the system is moved into any neighbouring position. Let the system be placed at rest in one of these neighbouring positions, then, by the same reasoning as before, it will begin to move on some path which will take it further off from its original position of equilibrium. *The equilibrium is then said to be unstable.*

Lastly, *suppose the system is in equilibrium in such a position that W is neither a true maximum nor a true minimum,* i.e. W is decreased when the system is moved into some neighbouring positions and increased when the system is moved into some others. By the same reasoning as in the two preceding cases the equilibrium is stable for some displacements and unstable for others. According to the definition given in Art. 75 *this state of equilibrium is to be regarded as on the whole unstable.*

215. We have only considered how the system *begins to move,* and not whether it may afterwards approach or recede from the position of equilibrium. As explained in Art. 75, this is a dynamical problem. The general result however agrees with what has been proved above.

216. Instead of using the work function we may use the potential energy. Since their sum $W + V$ is constant, the general results are just reversed. When the system is placed at rest in any position other than one of equilibrium, it *begins to move so* *that the potential energy decreases.* In a position of equilibrium the potential energy is a maximum, a minimum, or stationary. The equilibrium is *stable or unstable according as the potential energy is a true minimum or maximum.*

217. We have supposed in what precedes that none of the neighbouring positions are also positions of equilibrium. It is of course possible that W should be constant for two consecutive positions of the system of bodies, and yet (say) greater than when the system is moved into any other neighbouring position. In such a case the equilibrium is *neutral* for the displacement from one of the consecutive positions to the other and stable for all other displacements. Various cases may occur. For example, the equilibrium may be neutral for more than one or for all displacements from a given position of equilibrium; or again W may be constant for all positions defined by some relations between the coordinates, and yet (say) a maximum for all displacements from this locus. We then have a locus of positions of equilibrium, each of which is stable for all displacements which do not move the system along the locus.

In a system with two coordinates θ, ϕ, we could regard W as the ordinate of a surface whose x and y coordinates are θ and ϕ. Every geometrical peculiarity connected with the maximum and minimum ordinates of such a surface has a corresponding statical peculiarity in the positions of equilibrium of the system.

218. Altitude of the centre of gravity a maximum or minimum. There is one important application of the theorem on virtual work of which much use is made. Let gravity be the only external force acting on the system. Let z_1, z_2 &c. be the altitudes above any fixed horizontal plane of the several heavy particles, and $\bar{z}$ the altitude of their centre of gravity. If m_1, m_2 &c. be the masses of these particles, we have $\bar{z}\Sigma m = \Sigma mz$. If g be a constant, so that mg represents the weight of the mass m, the virtual work of the weights is $dW = -\Sigma mgdz = -g\Sigma md\bar{z}.$

The work function is therefore $W = -\bar{z}g\Sigma m + C.$

This is a true maximum or a true minimum, according as $\bar{z}$ is at the least or greatest height.

We deduce the following theorem. *Let a system of bodies be under the influence of no forces but their weights, together with such mutual reactions as do not appear in the equation of virtual work, and let it be supported by frictionless reactions with other fixed surfaces, or in some other way by forces which do not appear in the equation of virtual work; the possible positions of equilibrium may be found by making the altitude of the centre of gravity of the system above any fixed horizontal plane a maximum, a minimum, or stationary. The equilibrium will be stable or unstable according as the altitude of the centre of gravity is or is not a true minimum.*

219. *Alternation of stable and unstable positions.* Suppose the constraints are such that *the system moves with one degree of freedom.* Then as the system moves through space the centre of gravity will describe some definite curve. The positions in which the ordinate is a true maximum and a true minimum must evidently occur alternately. It follows that the truly stable and truly unstable positions of equilibrium occur alternately.

220. Analytical method of determining the stability of a system. To show how this theorem may be used to find positions of equilibrium in an analytical manner, let us suppose, as an example, that the system has one degree of freedom. We first choose some convenient quantity by which the position of the system is fixed, and which is therefore called its coordinate. Let this be called θ. Then the value of θ when the system is in equilibrium is the quantity to be found. Let $\bar{z}$ be the altitude of the centre of gravity of the system above some fixed horizontal plane. From the geometry of the question we now express $\bar{z}$ in terms of θ. The required value of θ is then found by making $d\bar{z}/d\theta = 0$. To determine whether the equilibrium is stable or unstable, we differentiate again and find $d^2\bar{z}/d\theta^2$. If this second differential coefficient is positive, when θ has the value just found, the equilibrium is stable. If negative, the equilibrium is unstable. If zero we must examine the third and higher differential coefficients of $\bar{z}$, following the rules given in the differential calculus to discriminate whether a function of one independent variable is a maximum or minimum.

If the coordinate θ cannot vary from $\theta = -\infty$ to $\theta = +\infty$, it may itself have maxima and minima. It must be remembered that *these values of θ may lead to maxima and minima values of $\bar{z}$ other than those given by the ordinary theory in the differential calculus.*

221. Examples. Ex. 1. *A uniform heavy rod AB rests against a smooth vertical wall and over a smooth peg C. Find the position of equilibrium, and determine whether it is stable or unstable.*

Let the length of the rod be $2a$ and let the distance of C from the wall be b. Let the inclination of the rod to the wall be θ. Taking the horizontal through C for the axis of x, we find for the altitude z of the centre of gravity

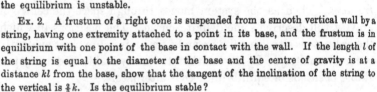

$$z = a \cos \theta - b \cot \theta,$$
$$dz/d\theta = - a \sin \theta + b (\sin \theta)^{-2},$$
$$d^2z/d\theta^2 = - a \cos \theta - 2b (\sin \theta)^{-3} \cos \theta.$$

Putting $dz/d\theta = 0$, we find that in the position of equilibrium $\sin^3 \theta = b/a$. Since $d^2z/d\theta^2$ is negative the equilibrium is unstable.

Ex. 2. A frustum of a right cone is suspended from a smooth vertical wall by a string, having one extremity attached to a point in its base, and the frustum is in equilibrium with one point of the base in contact with the wall. If the length l of the string is equal to the diameter of the base and the centre of gravity is at a distance kl from the base, show that the tangent of the inclination of the string to the vertical is $\frac{2}{3}k$. Is the equilibrium stable?

Ex. 3. A body is kept in equilibrium by three forces P, Q, R acting at certain points A, B, C in it. When the body is disturbed the forces continue to act at these points parallel to directions fixed in space and their magnitudes are unaltered. If a, b, c be the distances of A, B, C from O, the point of intersection of the three lines of action when the body is in equilibrium, show that the equilibrium is stable, neutral, or unstable, for displacements in the plane of the forces, according as $Pa + Qb + Rc$ is positive, zero, or negative; a, b, c being counted positive if drawn from O in the directions of the forces. [Coll. Ex., 1892.]

An elementary solution of this problem has been given in Art. 77. To use the test given by the principle of work we turn the body round O through an angle θ and place it at rest in this new position. The work done in returning to its old position is X versin θ where $X = Pa + Qb + Rc$. If X is positive, the equilibrium is stable by Art. 200 or 214.

222. Ex. *A heavy body can move in a vertical plane in such a manner that two of its points, viz. A and B, are constrained to slide, one on each of two equal and similar smooth curves whose equations are respectively $x = f(y)$ and $x = -f(y)$, y being vertical. The perpendicular on the chord AB drawn from the centre of gravity G bisects AB in E. Show how to find the positions of equilibrium, and determine whether the position in which AB is horizontal is stable or not.*

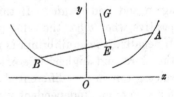

Let $AB = 2a$, $GE = h$. Let θ be the inclination of AB to the horizon and (xy) the coordinates of G. Then since the points A, B lie on the given curves we find

$$\left. \begin{array}{l} x + h \sin \theta + a \cos \theta = f (y - h \cos \theta + a \sin \theta) \\ x + h \sin \theta - a \cos \theta = -f (y - h \cos \theta - a \sin \theta) \end{array} \right\} \quad \dots\dots\dots(1).$$

Eliminating x, we have

$$2a \cos \theta = f (y - h \cos \theta + a \sin \theta) + f (y - h \cos \theta - a \sin \theta) \dots\dots\dots (2).$$

Differentiating this and putting $dy/d\theta = 0$, we find

$$\left.\begin{array}{l} -2a \sin \theta = f' \, (y-h \cos \theta + a \sin \theta) \, (h \sin \theta + a \cos \theta) \\ + f' \, (y-h \cos \theta - a \sin \theta) \, (h \sin \theta - a \cos \theta) \end{array}\right\} \dots\dots\dots(3).$$

Joining this equation to (1) and (2) we have three equations to find x, y, θ. It is clear that (3) is satisfied by $\theta = 0$, this therefore is one position of equilibrium.

To determine if this horizontal position is stable, we differentiate (2) *twice* to find $d^2y/d\theta^2$. We easily find after reduction

$$-\frac{d^2y}{d\theta^2} = \frac{a + a^2 f'' \, (y-h)}{f' \, (y-h)} + h \dots\dots\dots\dots\dots\dots(4).$$

The position of equilibrium is stable or unstable according as the right hand side is negative or positive.

We may obtain a geometrical interpretation for the equation (4) in the following manner. The straight line AB being in its horizontal position, let n be the length of the normal to the curve at either A or B intercepted between the curve and the axis of y. Let ρ be the radius of curvature at A or B, estimated positive when measured from the curve in the direction of n, and let ψ be the inclination of the tangent at A or B to the axis of y. We know by the differential calculus that if $x = f(y)$ be the equation to a curve, $\tan \psi = f'(y)$, while n and ρ are given by

$$n = x \, \{1 + (f'(y))^2\}^{\frac{1}{2}}, \quad \rho = \frac{\{1 + (f'(y))^2\}^{\frac{3}{2}}}{-f''(y)} ;$$

remembering that a and $y - h$ are the equilibrium coordinates of A we find

$$\frac{d^2y}{d\theta^2} = \frac{n^3 - a^2 \rho}{a\rho \tan \psi} - h \dots\dots\dots\dots\dots\dots\dots(5).$$

The horizontal position of equilibrium is therefore stable or unstable according as the right hand side of this equation is positive or negative.

If in the position of equilibrium $d^2y/d\theta^2$ should be zero, the equilibrium is said to be neutral to a first approximation. We must then continue our differentiations of (2) to ascertain if y is a true maximum or minimum, or neither. We find that $d^3y/d\theta^3 = 0$, and

$$-\frac{d^4y}{d\theta^4} = \frac{-a + (3h^2 - 4a^2) \, f'' \, (y-h) + 6a^2 h f''' \, (y-h) + a^4 f''''' \, (y-h)}{f' \, (y-h)} - h.$$

The equilibrium is therefore stable or unstable according as the right hand side is negative or positive. If this again vanish we proceed to higher differential coefficients.

223. Ex. 1. A prism whose cross section is an equilateral triangle rests with two edges on smooth planes inclined at angles a, β to the horizon. If θ be the angle which the plane containing these edges makes with the vertical, show that

$$\tan \theta = \frac{2 \sqrt{3} \sin a \sin \beta + \sin (a + \beta)}{\sqrt{3} \sin (a \sim \beta)}. \qquad \text{[Coll. Ex., 1889.]}$$

Ex. 2. The form of a bowl of revolution is such that every rod resting horizontally in it is in neutral equilibrium to a first approximation. Show that the differential equation to the generating curve is $(dx/dy)^2 = 2 \log a/x$ where y is vertical. Show also that the equilibrium is stable or unstable according as the length of the rod is less or greater than $2a/e^{\frac{1}{2}}$, where e is the base of Napier's logarithms.

Ex. 3. A uniform square board is capable of motion in a vertical plane about a hinge at one of its angular points; a string attached to one of the nearest angular points, and passing over a pulley vertically above the hinge at a distance from it equal to a side of the square supports a weight whose ratio to the weight of the board is $1 : \surd 2$. Find the positions of equilibrium, and determine whether they are respectively stable or unstable.   [Math. Tripos, 1855.]

Ex. 4.   The extremities of a rod without weight are capable of sliding on a smooth fixed vertical wire bent into the form of a circle. A weight is suspended from the extremities of the rod by two strings, which pass through a small smooth fixed ring, vertically below the centre of the circle. Show that the weight will be in stable equilibrium when the rod passes through the middle point of the polar of the ring with respect to the circle.   [Math. Tripos, 1859.]

Ex. 5.   A uniform regular tetrahedron has three corners in contact with the interior of a fixed hemispherical bowl of such magnitude that the completed sphere would circumscribe the tetrahedron; prove that every position is one of equilibrium. If P, Q, R be the pressures on the bowl, and W the weight of the tetrahedron, prove that $3(P^2 + Q^2 + R^2) - 2(QR + RP + PQ) = 3W^2$.   [Math. Tripos, 1869.]

Ex. 6.   A right cone rests with its curved surface in contact with two smooth equal cylinders whose axes are parallel, in the same horizontal plane, and distant d apart, and whose cross sections are circles of radii a. Show that the cone can rest in equilibrium with its axis in a plane perpendicular to the axes of the cylinders and inclined at an angle θ to the vertical given by $4d \cos \theta = 3r \cos^2 a + 4a \cos a$, where $2a$ is the vertical angle of the cone and r is the radius of its base; and determine whether the position is one of stable equilibrium.   [Math. Tripos, 1890.]

Ex. 7.   A conical plug of height h and semi-vertical angle a is at rest in a circular hole of radius a. Show that the vertical position of equilibrium is one of stability or of instability according as $16a$ is greater or less than $3h \sin 2a$.
[St John's Coll., 1887.]

224.   Ex.   *One end A of a straight beam AB rests against a smooth vertical wall, and the other B rests on an unknown curve. If l be the length of the beam, h the altitude of the centre of gravity, find the form of the curve that the relation $4ch - l^2 = c^2$ may hold in the position of equilibrium whatever values l and h may have.*   [Boole's problem.]

Let $(0, y')$ (x, y) be the coordinates of A and B.   Then
$$2h = y + y' \quad \ldots\ldots (1), \qquad x^2 + 4(y - h)^2 = l^2 \quad \ldots\ldots (2).$$

We notice that a curve could be found such that a rod of given length l could rest on it in equilibrium in the manner described in the question. Such a curve is found by making the altitude h constant.

The curve is therefore the ellipse (2) where h and l have any constant values which satisfy the given relation. The envelope of all these ellipses must also satisfy the mechanical problem, because the envelope touches every ellipse and the reaction will suit either curve. The envelope found in the usual way is the parabola $x^2 = 4cy$.

We might find this parabola without using the theory of envelopes. Since in equilibrium $dh = 0$ when l is constant, we have by differentiating (2)
$$x dx + 4(y - h) dy = 0.$$
But (2) is satisfied when h and l both vary; $\therefore x dx + 4(y - h)(dy - dh) = l dl$,
also since $4ch - l^2 = c^2$, $2cdh = l dl$.

Eliminating the differentials we find $2(h - y) = c$. Joining this to the given relation we can express h and l in terms of y. Substituting these in (2) the required relation between x and y is found. It reduces to the parabola already found.

225. Ex. A heavy body can move in a vertical plane in such a manner that
two straight lines CA, CB fixed in it are
constrained to slide on two equal and
similar curves fixed in space. The equa-
tions to the curve are $p = f(\omega)$ and $q = f(\omega')$,
where p, q are the perpendiculars drawn
from the origin on the tangents, and ω, ω'
are the angles which these perpendiculars
make with opposite sides of the axis of x, y
being vertical as before. The centre of
gravity G lies in the bisector of the angle C
at a distance h from either of the straight
lines CA, CB. Show how to find the inclination of CG to the vertical when the body

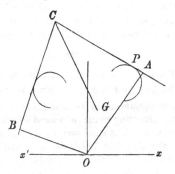

is in equilibrium, and determine whether the position in which CG is vertical is
stable or unstable.

Let a be the angle CG makes with either CA or CB, and θ the inclination of CG
to the vertical. Let y be the altitude of G. We first show by geometrical considerations that $y \sin 2a = (p - h) \cos(\theta - a) + (q - h) \cos(\theta + a)$.

Remembering that $p = f(\theta + a)$ and $q = f(a - \theta)$ we have, by equating $dy/d\theta$ to zero,
an equation to find θ.

In the position in which CG is vertical $\theta = 0$, hence $p = q$. Differentiating a
second time, we have

$$\frac{\sin 2a}{2} \frac{d^2y}{d\theta^2} = \left(h - p + \frac{d^2p}{d\theta^2} \right) \cos a + 2 \frac{dp}{d\theta} \sin a.$$

We may obtain a geometrical interpretation of this value of $d^2y/d\theta^2$. The body
being in the position in which CG is vertical, the straight line CA will touch one
of the curves in some point P. Let ρ be the radius of curvature of the curve at P,
ξ the horizontal abscissa of P. We may then show that

$$\sin a \frac{d^2y}{d\theta^2} = h + \rho - 2\xi \sec a.$$

The equilibrium is stable or unstable according as the value of $d^2y/d\theta^2$ is positive or
negative. If the value is zero, we must differentiate a second time.

226. Examples of atoms. Some good examples of the method of using the
work function to determine questions of stability are supplied by Boscovich's theory
of atoms. Almost all the following results are enunciated by Sir W. Thomson in
an interesting paper contributed to *Nature*, October 1889.

It is enough for our present purpose to say that Boscovich supposed matter to
consist of atoms or points between which there is repulsion at the smallest distance,
attraction at greater distances, repulsion at still greater distances, and so on, ending
with attraction according to the Newtonian law for all distances for which this law
has been proved. Boscovich suggested numerous transitions from attraction to
repulsion and vice versa, but for the sake of simplicity, we shall here consider
problems which involve only one change from repulsion to attraction.

Suppose then that the mutual force between two atoms is repulsive when the
distance between them is less than p, zero when it is equal to p, and attractive when
greater than p. With this supposition we shall consider the stability of the equilibrium of some groups of atoms.

227. Ex. 1. Three particles, whose masses are m, m', m'' repel each other so that the force between m and m' is $F = -mm'(r-p)^{n-1}$ where n is an even integer. The particles are in equilibrium when placed at the corners of an equilateral triangle each of whose sides is equal to p. Show that the equilibrium is stable.

The term of the work function W corresponding to F is $\int F dr = -\dfrac{mm'}{n}(r-p)^n$.

When the atoms are displaced, let the three sides of the triangle be $p+x$, $p+y$, $p+z$. We have by Art. 211, $n(C-W) = m'm''x^n + m''my^n + mm'z^n$.

The equilibrium is stable or unstable according as W is a maximum or a minimum, i.e. according as the right hand side is a minimum or a maximum. But, since n is even, the right hand side is a minimum when x, y, z are each zero; for these values make the right hand side zero and all others make it greater than zero. The equilibrium is therefore stable.

We have taken the law of force to be a single power of $r-p$, but it is clear that the same reasoning will apply if the law of force is expressed by several terms with different odd powers. Even greater generality may be given to the law, for it is sufficient that the lowest power should be odd.

In just the same way we may prove that a group of four particles placed at the corners of a regular tetrahedron, each of whose edges is equal to p, is a stable arrangement.

Ex. 2. Three equal atoms A, B, C are placed in equilibrium in a straight line. Supposing the force of repulsion to be $F = -\mu(r-p)^{n-1}$, where n is even, determine if the configuration is stable or unstable.

It is clear that in the position of equilibrium the distances AB, BC are each less than the critical distance p, while AC is greater than p. Let AB and BC be each equal to a. As we are only concerned with relative displacements, let A be fixed. Let B', C' be the displaced positions of B, C; let (xy) be the coordinates of B' referred to B, and $(x'y')$ those of C' referred to C. If $r = AB'$, we have

$$r = \{(a+x)^2 + y^2\}^{\frac{1}{2}} = a + x + \frac{y^2}{2a} + \&c.$$

$$\therefore\ (r-p)^n = (a-p)^n + n(a-p)^{n-1}\left(x + \frac{y^2}{2a}\right) + n\frac{n-1}{2}(a-p)^{n-2}x^2 + \&c.$$

If we replace (xy) by $(x'-x,\ y'-y)$, this expression gives the value of $(r''-p)^n$ where $r'' = B'C'$. If instead we replace (xy) by $(x'y')$ and write $2a$ for a, the expression gives the value of $(r'-p)^n$, where $r' = AC'$.

Taking all these expressions, we have as before

$$\frac{n}{\mu}(C-W) = (r-p)^n + (r'-p)^n + (r''-p)^n$$

$$= n(a-p)^{n-1}\left\{x' + \frac{(y-y')^2 + y^2}{2a}\right\} + n\frac{n-1}{2}(a-p)^{n-2}\{x^2 + (x'-x)^2\}$$

$$+ n(2a-p)^{n-1}\left\{x' + \frac{y'^2}{4a}\right\} + n\frac{n-1}{2}(2a-p)^{n-2}x'^2 + \&c.,$$

where all the constant terms have been absorbed into one constant, viz. C.

To find the position of equilibrium, we make W a maximum or a minimum, i.e. we put $\dfrac{dW}{dx} = 0$, $\dfrac{dW}{dx'} = 0$, $\dfrac{dW}{dy} = 0$, $\dfrac{dW}{dy'} = 0$. These give $(a-p)^{n-1} + (2a-p)^{n-1} = 0$.

Hence, since $n-1$ is odd and p lies between a and $2a$, we find $-(a-p)=2a-p$ and therefore $a=\frac{2}{3}p$. This result might have been more simply obtained by equating the forces on the particle A due to the repulsion of B and the attraction of C.

To distinguish whether W is a maximum or a minimum, we examine the terms of the second order. We find that those on the right-hand side are

$$- n\,(p-a)^{n-1}\,\frac{(2y-y')^2}{4a} + n\,\frac{n-1}{2}\,(p-a)^{n-2}\,\{x^2+x'^2+(x'-x)^2\}.$$

It is clear that this expression cannot keep one sign for all values of $x,\ y,\ x',\ y'$ for the terms with $(y,\ y')$ are negative and those with $(x,\ x')$ positive. We therefore infer that W is neither a maximum nor a minimum. The equilibrium is stable for all displacements in which the particles remain in the original straight line. It is unstable for all displacements in which they are moved perpendicular to that straight line. On the whole the equilibrium is unstable.

This method of solution has been adopted in order to show how the rules of the differential calculus may be used in making W a maximum or minimum. The result may be more simply obtained by displacing one particle perpendicularly to the straight line ABC and calculating the normal force of repulsion on it. The equilibrium is then seen to be unstable for this displacement.

Ex. 3. Show that the following configurations of four equal atoms are unstable.

(1) Three atoms at the corners of an equilateral triangle and one at the centre. (2) The four atoms at the corners of a square. (3) The four atoms in one straight line.

Ex. 4. Three equal particles repelling each other according to the nth power of the distance are connected together by three equal elastic strings. Find the position of equilibrium and show that it is stable if $n<p/(p-a)$, where a is the unstretched, and p the stretched length of any string.

228. Ex. Three fine rigid bars, coinciding with the diagonals of a regular hexagon, are each freely moveable about their common centre in the plane of the hexagon; six equal particles at the extremities of the bars repel one another with a force varying inversely as any power of the distance. Show that the equilibrium of the system is stable.   [Math. Tripos, 1859.]

229. On Frameworks. The determination of the forces which act along the rods of a framework supply some good examples of the use of the theory of work. The general method of proceeding may be described as follows. If we remove such of the connecting rods as we may choose, and replace these by forces acting at their extremities, we so loosen the constraints that the framework admits of displacement. The principle of work then gives equations connecting the forces which act on the system but omitting all those reactions which act between the rods not removed. We thus form equations to find the reactions on any one or more rods we choose to select.

230. Ex. A framework, consisting of any number of rods, not necessarily in one plane, is acted on by forces at the corners. If R be the reaction along any rod regarded as positive when in a state of thrust, r the length of that rod, and if

X, Y, Z be the components of the forces at that corner whose coordinates are
x, y, z, prove that $\Sigma Rr + \Sigma (Xx + Yy + Zz) = 0$,
where the Σ implies summation over the whole framework.   Maxwell, *Edinburgh
Transactions*, 1872, Vol. 26, p. 14.

Let us remove all the rods and apply the corresponding reactions at particles
placed at the corners. We now displace the system by giving it a slight enlarge-
ment, so that the displaced figure is similar to the original one. The principle of
work gives $\Sigma Rdr + \Sigma (Xdx + Ydy + Zdz) = 0$. But, since the figures are similar,
$dr/r = dx/x = \&c$. Substituting, the result follows at once. As an example of this
theorem see Art. 130, Ex. 5.

231. When we apply the principle of work to a frame, we
have to displace the corners. It will be found convenient to
distinguish these displacements by different names.

If the frame is not stiffened by the proper number of rods
(Art. 151) the angles may receive finite changes of magnitude
without altering the length of any side. When this is the case
any change is called a *normal or ordinary deformation*. The
actual displacement given may be infinitely small, but in a
normal deformation the change of angle may be increased until
it becomes finite.

If the framework is stiffened by the proper number of rods,
the connecting rods may possibly be so arranged that the angles
can receive infinitely small changes in magnitude, but not finite
changes, without altering the length of any side (Art. 151). Such
a displacement is called an *abnormal or singular deformation*.
This is an imaginary displacement, which could be a real one only
when small quantities of the second order are neglected.

If the frame is stiffened by only just the proper number of
rods so that there are no relations between the lengths of the
rods, any side of the frame can be increased in length without
breaking its connection with the others. Such a frame is said to
be *simply stiff* or *freely dilatable*.

If there are more rods than are necessary to stiffen the frame,
so that there are relations between the lengths of the sides, one
rod cannot be altered in length without altering some of the others.
Such a frame is said to be *indilatable* or *dilatable under one or
more conditions*.

These names are due partly to Maxwell, *Phil. Mag.* 1864, and partly to
M. Lévy, *Statique Graphique*.

232. *A simply stiff frame of rods connected by smooth hinges at
the corners A_1, A_2 &c. is in equilibrium under the action of any forces.*

It is required to find the stress along any side A_1A_2 which is not acted on by the external forces.

Let R_{12} be the reaction along this rod, and let it be regarded as positive when the rod is in a state of thrust. Let l_{12} be the length of the side.

Since the external forces are in equilibrium the work due to any virtual displacement of the frame which does not alter the length of any side is zero. Let us remove the rod A_1A_2 from the frame and replace its effects by applying to the particles at its extremities forces each equal to R_{12}. If we now fix in space any other side, say the adjoining side A_1A_n, the polygon will have one degree of freedom. It may be deformed, and each corner will describe a curve fixed in space. Supposing a small deformation given, let the length l_{12} be increased by dl_{12}, and let dW be the work of the external forces. Then, since the other reactions do not put in any appearance in the equation of work, we have

$$R_{12}dl_{12} + dW = 0 \dots\dots\dots\dots\dots\dots\dots(1).$$

If in addition to this deformation we give the side A_1A_n any virtual displacement, the frame moving with it as a whole, the work dW is not altered. We see therefore that the mode of displacement is immaterial. It is not even necessary to remove the side l_{12}, we simply let its length increase by dl_{12}. If dW be the resulting work of the forces, the reaction R_{12} is given by

$$R_{12} = -\frac{dW}{dl_{12}} \dots\dots\dots\dots\dots\dots(2).$$

It appears that, *if the length of any rod, not acted on by the external forces, can be increased without undoing the frame the reaction along that rod is determinate.* For example, if there are no external forces acting on the frame, the reaction along any such side is zero.

233. If the rod A_1A_2 is acted on by some of the external forces the reactions at the corners A_1, A_2 do not necessarily act along the length of the rod. We may reduce this case to the one already considered in the last article by replacing each of these forces by two parallel forces, one acting at each extremity of the rod. This method has been explained in Art. 134. We may also find the reactions by a more direct process.

Let R_{12}, S_{12} be the components of the action at the corner A_1 of the rod A_1A_2, resolved along and perpendicular to the length of the rod. In the same way R_{21}, S_{21} are the components at the corner A_2 of the same rod. Let us remove the rod A_1A_2 and replace its effects on the rest of the frame by applying at its extremities the forces R_{12}, S_{12} and R_{21}, S_{21}. Let R_{12}, R_{21} be regarded as positive when the rod is in a state of thrust.

Let the system be so deformed that the length of the side A_1A_2 is increased by dl_{12}, while the corner A_2 and the direction in space of that side are unaltered. The virtual work of the reactions R_{21}, S_{21} and S_{12} in this displacement is evidently zero. Let dW be the virtual work of the external forces which act on the system, excluding the rod A_1A_2, then

$$R_{12}dl_{12} + dW = 0.$$

To find the reaction S_{12} a different displacement must be given to the system. The external forces which act on the rod A_1A_2 having been removed, the remaining external forces are not in equilibrium. Their virtual work for a displacement of the frame as a whole is not necessarily zero. Keeping A_2 as before fixed in space and not altering the length l_{12}, let us turn the frame round an axis perpendicular to the plane containing A_2 and the force S_{12}. If $d\theta$ be the angle of displacement and dW the work of the forces, we have

$$S_{12}d\theta + dW = 0.$$

By giving the frame these two deformations the reactions R_{12} and S_{12} at the corner A_1 can be found. If the frame be perfectly free, the deformation necessary to find S_{12} can always be given. The deformation necessary to find R_{12} requires that the length of the rod can be altered. It follows that *both these reactions are determinate if the length of the rod A_1A_2 can be altered without destroying the connections of the frame.*

If the frame is subject to any external constraints, these may be replaced by pressures at the points of constraint. When the magnitudes of these pressures have been deduced from the general equations of equilibrium, we may regard the frame as perfectly free and acted on by known forces. The reactions at any corner may then be found as if the frame were free.

It is not meant that in every case exactly these displacements must be given to the system, for these may not suit the geometrical conditions of the problem. Other displacements may recommend themselves by their symmetry or by the ease with which the virtual work due to those displacements can be found. Any two displacements which introduce only R_{12} and S_{12} into the equations of virtual work will supply two equations from which these two components may be found.

If the system be in three dimensions, the direction of S_{12} may be unknown as well as its magnitude. In this case the components of S_{12} in two convenient directions may be used instead of S_{12}. Three displacements to supply three equations of virtual work will then be necessary.

234. Examples. Ex. 1. Six equal heavy rods, freely hinged at the ends, form a regular hexagon $ABCDEF$, which when hung up by the point A is kept from altering its shape by two light rods BF, CE. Prove that the thrusts of the rods BF, CE are as 5 to 1, and find their magnitudes. [Math. T., 1874.]

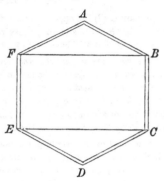

Let the length of any side be $2a$, and let θ be the angle which either of the upper sides makes with the vertical.

To find the thrust T of BF, we suppose the length of BF to be slightly increased. The inclinations of AB and AF to the vertical are therefore increased by $d\theta$. The work of the thrust T is $Td\,(4a\sin\theta)$. The work of the weights of the two upper rods is $2Wd\,(a\cos\theta)$. The centre of gravity of each of the four other rods is slightly raised, and the work of their weights is $4Wd\,(2a\cos\theta)$. We have therefore

$$Td\,(4a\sin\theta)+2Wd\,(a\cos\theta)+4Wd\,(2a\cos\theta)=0, \quad \therefore\ 2T=5W\tan\theta.$$

To find the thrust T' of the rod CE, we suppose the length of CE to be slightly altered. No work is done by the weights of the four upper rods. The centres of gravity of the two lower rods are however slightly raised. If θ be the angle either of the lower rods makes with the vertical, we easily find

$$T'd\,(4a\sin\theta)+2Wd\,(a\cos\theta)=0, \quad \therefore\ 2T'=W\tan\theta.$$

The result given in the question follows at once.

Ex. 2. A tetrahedron, formed of six equal uniform heavy rods, freely jointed at their extremities, is suspended from a fixed point by a string attached to the middle point of one of its edges. It is required to find the reactions at the corners.

The tetrahedron is regular, hence the upper and lower rods, viz. AB and CD, are horizontal. Let L and M be their middle points, then LM is vertical; let $LM=z$. Let P, P' be the thrusts along these rods and w the weight of any rod.

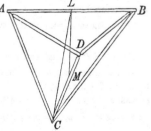

Without altering the direction in space of the upper rod, or the position of its middle point, let us increase its length by dr. Since the transverse reactions at its extremities will do no work in this displacement, the equation of virtual work is

$$P\,dr+4w\cdot\tfrac{1}{2}dz+w\,dz=0\ldots\ldots(1).$$

In the same way, if we increase the length of the lower bar by dr without altering its direction in space or the position of its middle point, the equation of virtual work is

$$P'dr - 4w \cdot \tfrac{1}{2}dz - wdz + Tdz = 0 \quad\text{.........................(2)},$$

where T is the tension of the string. Since $T = 6w$, and the ratio $dr : dz$ is the same for each rod, these two equations give at once $P = P'$.

To find the relation between dr and dz we require some geometrical considerations. From the right-angled triangles BLC, LCM we have

$$BC^2 - BL^2 = CL^2 = CM^2 + z^2 \quad\text{..........................(3)}.$$

In obtaining equation (1), the half side BL is altered by $\tfrac{1}{2}dr$, the other lengths CM and BC being unaltered; we therefore have

$$- BL \cdot dBL = zdz, \quad \therefore dr = -2\sqrt{2}dz.$$

In obtaining (2) the opposite half side is altered by $\tfrac{1}{2}dr$, we therefore have as before $dr = -2\sqrt{2}dz$. Substituting these values of dr in (1) and (2) we find that each of the thrusts P and P' is equal to $\tfrac{3}{4}\sqrt{2}w$.

We have now to find the other reactions. Since three rods meet at each corner, it is necessary to specify the arrangement of the hinges. We assume that each of the rods which meet at any corner is freely hinged to a weightless particle situated at that corner. Since this particle may afterwards be considered as joined to the extremity of any one of the three rods, we thus include the case in which two of the rods at any corner are hinged to the third.

The reaction between a particle and any one of the rods which meet it will be a single force. By taking moments for the rod about a vertical drawn through one end, we may show that the reaction at the other end lies in the vertical plane through the rod. The reaction may therefore be obliquely resolved into a force acting along that rod and a vertical force. Let Q and Z be the components at A on either of the rods AC, AD, Q being positive when it compresses the rod and Z when acting upwards. In the same way Q' and Z' will represent the components on either of these rods at their lower extremities.

Let us now lengthen each of the four inclined rods by $d\rho$, keeping the upper rod fixed. The equation of virtual work for the lower bar together with the two particles at each end is then

$$4Q'd\rho + 4Z'dz + wdz = 0 \quad\text{................................(4)}.$$

Since the rod CD has here received simply a vertical displacement, this equation might have been obtained by resolving vertically the forces on the rod and equating the sum to zero, Art. 204.

To find the relation between $d\rho$ and dz we recur to (3). In obtaining the equation (4), BC is altered by $d\rho$ while BL and CM are unaltered, hence

$$BC \cdot dBC = zdz, \quad \therefore dz = \sqrt{2}d\rho.$$

We therefore have

$$2\sqrt{2}Q' + 4Z' + w = 0 \quad\text{...................................(5)}.$$

Resolving the forces on the particle at C in the direction CD, we find

$$- P' = 2Q' \cos 60^\circ \quad\text{....................................(6)}.$$

The value of P' having been already found, we have $Q' = -\tfrac{3}{4}\sqrt{2}w$, $Z' = \tfrac{1}{2}w$.

In the same way, if we lengthen each of the inclined rods by $d\rho$ keeping the lower rod fixed, the equation of virtual work for the upper rod and the two particles at each end becomes

$$- 4Zdz + 4Qd\rho - wdz + Tdz = 0 \quad\text{.........................(7)}.$$

Resolving the forces on the particle at A along AB, we have

$$- P = 2Q \cos 60^\circ = Q, \quad \therefore Q = -\tfrac{3}{4}\sqrt{2}w, \quad Z = \tfrac{1}{2}w.$$

Ex. 3.　Two rods CA, CB, freely jointed at C, are placed in a vertical plane, and rest with the points A, B on a smooth horizontal table, A and B being connected by a weightless string $AQPB$ passing through smooth rings at P and Q, the middle points of CA, CB. Prove that the tension T of the string is given by

$$T \cdot AB \cdot \left(\frac{1}{BP} + \frac{1}{AQ} + \frac{1}{AB} \right) = W \cos A \cos B \operatorname{cosec} C,$$

where W is the weight of the two rods.　　　　　　　　　　[Coll. Exam., 1890.]

Ex. 4.　A frame $ABCD$ is formed of four light rods, each of length a, freely jointed together; it rests with AC vertical and the rods BC, CD in contact with fixed frictionless supports E, F in the same horizontal line at a distance c apart, the joints B, D being kept apart by a light rod of length b. Show that, when a weight W is placed on the highest joint A, it produces in BD a thrust of magnitude R, where $Rb^2 (4a^2 - b^2)^{\frac{1}{2}} = W (2a^2c - b^3)$. Examine the case when $b = (2a^2c)^{\frac{1}{3}}$. [Math. T., 1886.]

Ex. 5.　Four equal rods ARB, CRD, ESB, FSD form with each other a rhombus $RBSD$; A and C are fixed hinges at a distance a from R; R, B, S and D are free hinges, and at E and F forces, each equal to P, are applied perpendicular to the rods. If a be the angle which the reactions at A and C make with AC, 2θ the angle ARC, and b a side of the rhombus, show that $a \cot a = 2 (a + b) \tan \theta + a \cot \theta$.
　　　　　　　　　　　　　　　　　　　　　　　[Coll. Exam., 1889.]

Take AC as axis of x, its middle point as origin. Let X, Y be the reactions at A; $x = a \sin \theta$, $y = 2 (a + b) \cos \theta$ the coordinates of E. Increasing the length of AC without altering its direction in space, or the position of its middle point, we have, by the principle of virtual work, $Xd (a \sin \theta) + P \sin \theta dy - P \cos \theta dx = 0$. Also by resolution $Y + P \sin \theta = 0$. The result follows at once.

Ex. 6.　Four equal rods AB, BC, CD, DA are freely jointed at the ends so as to form a square and are suspended by the corner A. The rods are kept apart by a single string without weight joining the middle points of AB, BC. Show that the tension of the string and the reaction at the lowest point C are respectively $4W$ and $\frac{1}{2} W \sqrt{5}$, where W is the weight of any rod.

Ex. 7.　A succession of n rhombus figures of equal sides, each being b, are placed having equal diagonals in a straight line and one angular point common to two successive figures, and the extreme sides of the first and last rhombus are produced through equal lengths a in opposite directions to points A, B, C, D respectively. Consider now all the straight lines in the figure to be rods hinged freely where they intersect and having fixed hinges at C and D. At A and B, the free ends, are applied equal forces perpendicular to the rods; show that the reactions at C and D make an angle ϕ with CD, where $a \cot \phi = 2 (a + nb) \tan \theta + a \cot \theta$, θ being the angle which the common diagonal makes with any side.　　　　[Coll. Exam., 1889.]

Ex. 8.　A tripod stand is constructed of three equal uniform rods connected by means of a universal joint at one extremity of each; the whole rests on a smooth floor, and is prevented from collapsing through having the lower extremities connected by strings equal in length to the rods. Find the tensions of the strings. In particular, if a weight W equal to that of each rod be suspended from this joint, then the tension is $\frac{5}{36} \sqrt{6} W$.　　　　　　　　　　[St John's Coll., 1882.]

Ex. 9.　Six uniform rods, each of weight W, are jointed together to form a regular hexagon, which is hung up from a corner. The two middle rods are connected by a light horizontal rod. Show that, if they rest vertically, the horizontal rod divides them in a ratio which is independent of its length. If the horizontal rod be heavy, and uniform in length and material with the others, show that the ratio is 6 : 1, and that the stress in the horizontal rod is $\frac{4}{5}W_\sim/3$. Find also the stresses at the joints. [Coll. Exam., 1888.]

235. Abnormal deformations. Referring to the general theorem considered in Art. 232 we notice that there is a peculiar case of exception. Let us suppose that the forces which act on the frame are applied at the corners so that the reactions act along the sides of the polygon.

The side A_1A_2 being removed, the polygon may be deformed; the principle of virtual work then gives

$$R_{12}dl_{12} + dW = 0 \dots\dots\dots\dots\dots\dots(1).$$

Supposing the side $A_n A_1$ to be fixed in space, it is possible, when the frame is deformed, that the corner A_2 may begin to move perpendicularly to the side A_1A_2. In this case $dl_{12} = 0$. If the side $A_n A_1$ is also displaced in any manner, by the frame moving as a whole, the quantity dl_{12} is unaltered and is therefore still zero. When the rod A_1A_2 is replaced, it is now possible to give the frame a small deformation without altering the length of any side, provided we neglect small quantities of the second order. Since the frame is now stiff, this deformation is of the kind called *abnormal*. Art. 231.

The external forces acting on the frame are in equilibrium, hence their virtual work for every displacement of the frame as a whole is zero. If it be not zero for this abnormal deformation also, the reaction R_{12} must be infinite. But if it be zero the equation (1) becomes nugatory, since both dl_{12} and dW are zero. The reaction R_{12} may now be finite.

In order, then, to deform the frame so that the reaction R_{12} may do work, we must remove, or lengthen, *two or more sides*. Let these be the given side l_{12} and any other say l_{23}. We now have

$$R_{12}dl_{12} + R_{23}dl_{23} + dW = 0 \dots\dots\dots\dots(2).$$

To use this equation we must know the ratio between the corresponding increments of any two sides. The equation (2) will then give the relation between the corresponding reactions. Thus *the reactions are indeterminate; one is arbitrary* but the others may be found in terms of this one.

236. In most cases the relation between the increments of any two sides may be found by inspection or by differentiating some known relations between the sides of the polygon. In more difficult cases we may proceed in the following manner. See Lévy, *Statique Graphique*.

Regarding the stiff framework as a general polygon with undetermined sides, we can find as many angles as may be convenient in terms of the sides. Let us suppose, as an example, that two equations have been found connecting, say, the two angles θ_1, θ_2 with the sides. Let these be

$$\left.\begin{array}{l} f_1\left(\cos\theta_1,\ \cos\theta_2,\ l_{12},\ l_{23},\ \&c.\right)=0 \\ f_2\left(\cos\theta_1,\ \cos\theta_2,\ l_{12},\ l_{23},\ \&c.\right)=0 \end{array}\right\} \quad \dots\dots\dots\dots\dots\dots (3).$$

Since this particular polygon can have a slight deformation without altering the sides we must have

$$\frac{df_1}{d\theta_1}d\theta_1 + \frac{df_1}{d\theta_2}d\theta_2 = 0, \qquad \frac{df_2}{d\theta_1}d\theta_1 + \frac{df_2}{d\theta_2}d\theta_2 = 0 \dots\dots\dots\dots (4).$$

These give $d\theta_1 = 0$ and $d\theta_2 = 0$, unless the special polygon under consideration is such that the determinant $\quad J = \left| \begin{array}{cc} df_1/d\theta_1 & df_1/d\theta_2 \\ df_2/d\theta_2 & df_2/d\theta_2 \end{array} \right| = 0 \dots\dots\dots\dots\dots(5).$

If we vary the lengths of the rods, the corresponding changes of the angles θ_1, θ_2 are given by

$$\left.\begin{array}{l} \dfrac{df_1}{d\theta_1}d\theta_1 + \dfrac{df_1}{d\theta_2}d\theta_2 = -\Sigma\dfrac{df_1}{dl}dl \\ \dfrac{df_2}{d\theta_1}d\theta_1 + \dfrac{df_2}{d\theta_2}d\theta_2 = -\Sigma\dfrac{df_2}{dl}dl \end{array}\right\} \dots\dots\dots\dots\dots\dots(6).$$

Multiplying these equations by the minors of the first row of the determinant J, and adding the results, the left-hand side will vanish. We thus obtain a relation between the increments of length of the rods of the form

$$P_{12}dl_{12} + P_{23}dl_{23} + \dots = 0.$$

This relation must be satisfied by any assumed changes of length of the rods.

237. Indeterminate tensions. It is generally more convenient to consider these indeterminate reactions apart from any external forces. To make this point clear, let us suppose that two sets of external forces in all respects the same can produce two different sets of internal stress when they act separately on the frame. Then, reversing one set of the external forces and making them act simultaneously, we have the frame in a self-strained state with no external forces. If then we can find all the internal stresses when no forces act, we can superimpose them on any one set of stress produced by a given set of forces, to find all the states of stress consistent with those forces.

In the tenth volume of the *Proceedings of the Mathematical Society*, 1878, Mr Crofton discusses some cases of hexagons and octagons in a state of self-strain. His theory was afterwards enlarged by M. Lévy in 1888 in his *Statique Graphique*.

238. Ex. 1. *A plane framework, having an even number n of corners, has for its bars the n sides joining these corners and the $\frac{1}{2}n$ diagonals joining the opposite corners. Show that it may be in a state of stress without any external forces if the $\frac{1}{2}n$ points of intersection of opposite sides lie in one straight line.*   [*Lévy's theorem.*]

The following proof applies generally though the figure is drawn for a hexagon.
To fix the ideas, let the sides be in a state
of thrust and the diagonals in tension.

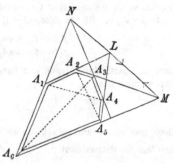

First. If the reactions R_{12} &c. are in
equilibrium, the forces R_{12}, R_{32} balance R_{25}
and are therefore equivalent to R_{54} and
R_{56}. Hence by transposition R_{12} and R_{45}
are equivalent to R_{23} and R_{56}. Each pair
by symmetry is equivalent to R_{34} and R_{61}.
The resultants of these act respectively at
L, M, N, and are equivalent. Hence L,
M, N, i.e. the intersections of opposite
sides of the hexagon, lie in a straight line.

Conversely. If L, M, N lie in a straight
line, apply two opposite forces, each equal to an arbitrary force F, at L and M. Let
the components along the sides which meet in L and M be (R_{12}, R_{45}) and (R_{32}, R_{65})
respectively. Then these four forces are in equilibrium, i.e. R_{12} and R_{32} acting at
A_2 are in equilibrium with R_{45} and R_{65} acting at A_5. Hence the two forces on A_2
have a resultant acting along $A_2 A_5$, and the two forces on A_5 have a resultant along
$A_5 A_2$ and these two resultants are equal. The other diagonals may be treated in
the same way. It follows that the forces at each corner are in equilibrium. Also
the ratio of each reaction to the arbitrary force F has been found. Another proof
will be indicated in the chapter on graphical statics.

This theorem is the more remarkable because the number of connecting rods
viz. $\frac{3}{2}n$ (being less than $2n - 3$ when n is greater than 6) is not sufficient to define
the figure, Art. 151.

By making one side infinitely small we obtain the corresponding theorem for a
framework with an odd number of corners.

Ex. 2. The bars of a framework are the sides of a hexagon and the diagonals
joining the opposite corners, prove that it may be in a state of internal stress if it is
inscribed in a conic. Find also the ratio of the reactions.   [*Crofton's theorem.*]

Ex. 3. The bars of a frame are the sides of a hexagon $A_1...A_6$, a diagonal $A_1 A_4$
and the lines $A_2 A_6$, $A_3 A_5$. Show that it may be in stress if corresponding bars on
each side of the diagonal $A_1 A_4$ intersect two and two on that diagonal. [Crofton.]

**239. Geometrical method of determining the stability
of a body.** When a body moves in any way in two dimensions,
the motion or displacement during a time dt may be constructed
by turning the body round some point I through an infinitesimal
angle; see Art. 180. The position of this point is continually
changing, so that it describes (1) a curve fixed in space, and
(2) a curve fixed in the body. Let a series of infinitesimal
arcs II', $I'I''$ &c. be taken on the first curve, and let equal
arcs IJ', $J'J''$ &c. be measured off on the second curve. After
the body has rotated round I through some angle $d\theta$, the point

J' has come into the position I'. This point then becomes the instantaneous centre, and the displacement during the next element of time may be similarly constructed by turning the body round I'. Let the arc $II' = ds$.

Since the angle between the tangents II', IJ' to the two curves is infinitely small, these curves touch each other at the point I. *The motion of the body may therefore be constructed by making the second curve roll without sliding on the first, carrying the body with it.* It is also clear that $ds : d\theta$ *is the ratio of the velocity with which the instantaneous centre describes either curve to the angular velocity of the body.*

At the beginning of the first element of time let P be the position of any point of the body, then since P begins to move in a direction perpendicular to PI, PI is a normal to the path of P. Let P' be the position in space of P at the end of the time dt; then the angle $PIP' = d\theta$. Since the body now begins to turn round I', $P'I'$ is a consecutive normal to the path of P.

If then P be so placed that the angle $IP'I'$ is also equal to $d\theta$, two consecutive normals to the path of P will be parallel, and hence the radius of curvature of the path of P will be infinite. If therefore we describe a circle passing through I and I', so as to contain an angle equal to $d\theta$, then *every point on the circumference of this circle is at a point of its path at which the radius of curvature is infinite.* For statical purposes we shall refer to this circle as the *circle of stability*. To construct this circle, we draw

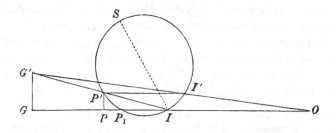

a normal at the instantaneous centre of rotation I to the path of I in space and measure along this normal a length $IS = ds/d\theta$. The circle described on IS as diameter is the circle of stability.

240. *A body moves in any manner in one plane, and in any position the circle of stability is known. To find the radius of curvature R of the path of any point attached to the moving body.*

Let G be any point of the body not on the circle of stability, and let P be that point in the straight line IG, at which the radius of curvature is infinite. As before GPI is a normal both to the locus of G and to that of P. See the figure of the last article. If we now turn the body round I through an angle $d\theta$, the points G and P will assume the positions G' and P' where the angles GIG' and PIP' are each equal to $d\theta$, and $I'P'$ is parallel to IPG. Also $G'I'$ is the consecutive normal to the locus of G; and if $G'I'$ intersect GI in O, O will be the required centre of curvature. We have by similar triangles

$$GP : GI = G'P' : G'I = G'I' : G'O.$$

In the limit the three points, P, P', and the intersection P_1 of the circle with GO, coincide. We then have $R \cdot GP_1 = GI^2$.

We have therefore the following rule[*]; *to find the radius of curvature R of the path of G, let GI intersect the circle of stability in P_1; then $R \cdot GP_1 = GI^2$.*

In the standard figure, lines drawn from G towards I have been taken as positive; it follows that R is positive or negative according as GP is positive or negative. We therefore infer that *the path of every point G is concave or convex towards I according as G lies without or within the circle of stability.*

241. Statical rule. In a position of equilibrium the tangent to the path of the centre of gravity G is horizontal, hence the position of equilibrium is such that IG is vertical. The equilibrium is stable or unstable according as the altitude of the centre of gravity is a minimum or a maximum, i.e. according as the concavity of the path is upwards or downwards. But this point is settled at once by the rule that the path of G is concave towards I except when G lies within the circle of stability.

242. Ex. 1. Two points A, B of a moving body describe known curves. Show how to find (1) the position of the instantaneous centre I, (2) the circle of stability.

---

[*] This formula for R is practically equivalent to that given by Abel Transon in *Liouville's Journal*, 1845, x. p. 148, though he uses the diameter IS of the circle instead of the circle itself. His object is to find the radius of curvature of a roulette. See also a paper by Chasles on the radius of curvature of the envelope of a roulette in the same volume.

The normals at A and B to the two curves meet in I; hence I is found. Art. 180.
If ρ_1, ρ_2 be the radii of curvatures of the curves at A and B, measure along AI and
BI respectively the lengths $AP_1 = AI^2/\rho_1$ and $BP_2 = BI^2/\rho_2$; the circle circumscribing
the triangle IP_1P_2 is the circle of stability.

Ex. 2. A body moves in one plane and the instantaneous centre of rotation is
known. Show that a straight line attached to the moving body touches its envelope
in a point G which is found by drawing a perpendicular IG on the straight line.

Since GI is normal to the locus of G, an element GG' of the path of G lies on
the straight line. Thus the straight line intersects its consecutive position in G',
i.e. G' or G is a point on the envelope.   [Roberval's rule.]

Ex. 3. A body moves in one plane and the instantaneous position of the circle
of stability is known. Prove the following construction to find the radius of

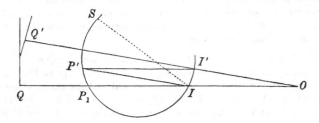

curvature of the envelope of a straight line attached to the moving body: draw a
perpendicular IQ on the straight line from the instantaneous centre I and let it cut
the circle of stability in P_1. Take $IO = IP_1$ on QP_1I produced if necessary, then O
is the required centre of curvature.

By the last example, IO is a normal at Q to the envelope. If we now turn the
body and the attached straight line round I through an angle $d\theta$, and draw from I'
a perpendicular $I'Q'$ on the straight line thus displaced, it is clear that $Q'I'$ is the
consecutive normal to the envelope. Let $Q'I'$ intersect QI in O, then O is the
required centre of curvature.

Since IO and $I'O$ are perpendiculars to two consecutive positions of the same
straight line, the angle IOI' is equal to $d\theta$. Draw $I'P'$ parallel to IP_1 to intersect
the circle of stability in P', then as in Art. 239 the angle $P'IP_1$ is also equal to $d\theta$.
Thus $I'O$ is parallel to $P'I$ and $P'O$ is a parallelogram. Therefore IO is equal
to $I'P'$, and in the limit IO and IP_1 are equal.

Ex. 4. The corners of a triangle ABC move along three curves, the normals
at A, B, C meet in I and α, β, γ are the angles at I subtended by the sides. If
ρ_1, ρ_2, ρ_3 be the radii of curvature of the curves, prove that

$$\frac{AI^2 \sin \alpha}{\rho_1} + \frac{BI^2 \sin \beta}{\rho_2} + \frac{CI^2 \sin \gamma}{\rho_3} = AI \sin \alpha + BI \sin \beta + CI \sin \gamma.$$

243. Ex. 1. *A homogeneous rod AB, of length $2l$, rests in a horizontal position
inside a bowl formed by a surface of revolution with its axis vertical. Show that the
equilibrium is stable or unstable according as $l^2\rho$ is less or greater than n^3, where ρ is
the radius of curvature at A or B and n is the length of the normal.* [See Art. 222.]

The normals at A and B meet in a point I on the axis of revolution. Take AL and BM so that each is equal to AI^2/ρ. The circle described about ILM is the circle of stability. Let the circle drawn through I touching the rod at G cut AI in a point H, then $AH . AI = AG^2$. The equilibrium is unstable if G is within the circle ILM, i.e. if AL is less than AH, i.e. if n^2/ρ is less than l^2/n.

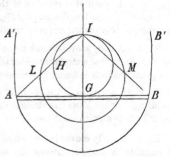

If the extremities of the rod terminate in small smooth rings which slide on a curve symmetrical about the vertical axis, the position $A'B'$, in which the normals at $A'B'$ meet in a point I *below the rod*, is also a position of equilibrium. Following the same reasoning the concavity of the path of G is turned towards I when $l^2\rho < n^3$. *The conditions of stability are therefore reversed*, the equilibrium is therefore stable or unstable according as $l^2\rho$ is $>$ or $< n^3$.

Ex. 2. The extremities of a rod are constrained by small rings to be in contact with a smooth elliptic wire. If the *major axis is vertical* prove that the lower horizontal position is unstable and the upper stable if the length of the rod is greater than the latus rectum. These conditions are reversed if the length is less than the latus rectum. If the *minor axis is vertical* the lower horizontal position is stable and the upper unstable.

In an ellipse $\rho (b^2/a)^2 = n^3$, where $2a$ and $2b$ are respectively the vertical and horizontal axes. Using this property, the results follow from those of Ex. 1.

It has been shown in Art. 126, that when the major axis of the ellipse is vertical the rod is in equilibrium only when it is *horizontal or passes through one focus*. The condition of stability in the latter case follows easily from the principle that the altitude of the centre of gravity must be a minimum. Let the rod AB be in any position and let S be the lower focus. Let AM, BN be perpendiculars on the lower directrix. The altitude of the centre of gravity above the lower directrix is

$$\tfrac{1}{2}(AM + BN) = \frac{1}{2e}(SA + SB).$$ Since SA and SB are two sides of the triangle SAB, this altitude is a minimum when S lies on the rod AB. In the same way if S is the upper focus, the depth of the centre of gravity below the upper directrix is represented by the same expression. *When therefore the rod passes through the lower focus the equilibrium is stable, when it passes through the upper focus the equilibrium is unstable.*

Ex. 3. The extremities A, B of a rod are constrained by two fine rings to slide one on each of two equal and opposite catenaries having a common vertical directrix and a common horizontal axis. Prove that the lower horizontal position of the rod is stable, see Art. 126, Ex. 5.

By drawing a figure it will be seen that the paths of A and B are convex to I. Hence A and B lie inside the circle of stability. Hence G also lies inside the circle and its path also is convex to I. The equilibrium is therefore stable.

Ex. 4. A rod rests in a horizontal position with its extremities on a cycloid with its axis vertical. Prove that the equilibrium is stable.

244. Rocking Stones. *A perfectly rough heavy body rests in equilibrium on a fixed surface: it is required to determine whether the equilibrium is stable or unstable. We shall first suppose the body to be displaced in a plane of symmetry so that the problem may be considered to be one in two dimensions.*

The geometrical method explained in Art. 241 supplies in most cases an easy solution. Let I be the point of contact of the two bodies, then I is the centre of instantaneous rotation. Let $C'IC$ be the common normal in the position of equilibrium, C, C' the centres of curvature. We shall suppose these curvatures positive when measured in opposite directions. If the upper body is slightly displaced so that I' becomes the new point of contact, the angle viz. $d\theta$ turned round by the body is equal to the angle between the normals CJ' and $C'I'$, and this is evidently equal to the sum of the angles $J'CI$, $I'C'I$. We therefore have

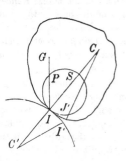

$$\frac{ds}{\rho} + \frac{ds}{\rho'} = d\theta,$$

where $II' = IJ' = ds$ as before. See also Salmon's *Higher Plane Curves*, Art. 312, or Besant's *Roulettes and Glissettes*, Art. 33.

To construct the circle of stability we measure along the common normal IC in the position of equilibrium a length $IS = ds/d\theta$. Writing z for this length, we see that $\dfrac{1}{z} = \dfrac{1}{\rho} + \dfrac{1}{\rho'}$. The circle described on IS as *diameter* is the circle of stability. Let IG cut this circle in P.

If the centre of gravity G lie without this circle, the concavity of its path is turned towards I. Hence *the equilibrium is stable or unstable according as G is below or above the point P*. If G coincide with P the equilibrium is neutral to a first approximation.

The critical altitude IP which separates stability and instability is clearly $IP = z \cos \alpha = \dfrac{\rho\rho' \cos \alpha}{\rho + \rho'}$, where α is the inclination to the vertical of the common normal in the position of equilibrium.

245. Ex. 1. A solid hemisphere (radius ρ) rests on the summit of a fixed sphere (radius ρ') with the curved surfaces in contact. If the centre of gravity of the hemisphere is at a distance $\frac{3}{8}\rho$ from the centre, prove that the equilibrium is stable or unstable according as ρ is less or greater than $\frac{3}{8}\rho'$.

In this example $a=0$, and therefore IG i.e. $\frac{3}{8}\rho$ must be less than z if the equilibrium is to be stable.

Ex. 2. A solid hemisphere rests on a rough plane inclined to the horizon at an angle β. Find the inclination of the plane base to the horizon and show that the equilibrium is stable.

The centre of gravity must lie in the vertical through I, and CG is also perpendicular to the base. Hence the required inclination of the base is the supplement of the angle CGI. The vertical through I cannot pass through G if $CI \sin \beta$ is greater than CG. Since $CG = \frac{3}{8}\rho$, it is necessary for equilibrium that $\sin \beta < \frac{3}{8}$.

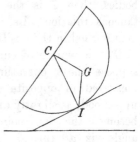

To find the circle of stability we notice that $\rho' = \infty$, and therefore $z = \rho$. The circle described on IC is therefore the circle of stability. Since the angle CGI is greater than a right angle, it is obvious that G lies inside the circle. The concavity of the path of G is therefore upwards, and the equilibrium is stable.

Ex. 3. A solid homogeneous hemisphere, of radius a and weight W, rests in apparently neutral equilibrium on the top of a fixed sphere of radius b. Prove that $5a = 3b$. A weight P is now fastened to a point in the rim of the hemisphere. Prove that, if $55P = 18\ W$, it still can rest in apparently neutral equilibrium on the top of the sphere.   [Math. Tripos, 1869.]

Ex. 4. A heavy hemispherical bowl, of radius a, containing water, rests on a rough inclined plane of angle a; prove that the ratio of the weight of the bowl to that of the water cannot be less than $\dfrac{2 \sin a}{\sin \phi - 2 \sin a}$, where $\pi a^2 \cos^2 \phi$ is the area of the surface of the water.   [Math. Tripos, 1877.]

When the bowl is displaced the water is supposed to move in the bowl so as to be always in a position of equilibrium. Its statical effect is therefore the same as if it were collected into a particle and placed at the centre of the bowl. The weight of the bowl may be collected at its centre of gravity, i.e. at the middle point of the middle radius.

Ex. 5. A parabolical cup, the weight of which is W, standing on a horizontal table, contains a quantity of water, the weight of which is nW: if h be the height of the centre of gravity of the cup and the contained water, the equilibrium will be stable provided the latus rectum of the parabola be $> 2\ (n+1)\ h$.

    [Math. Tripos, 1859.]

Let H be the centre of gravity of the water when the axis of the cup is vertical. Let the cup and the contained water be placed at rest in a neighbouring position with the surface of the water horizontal; Art. 215. It may be shown that the vertical through the centre of gravity H' of the displaced water intersects the axis of the paraboloid in a point M, where HM is half the latus rectum. The point M is called the *metacentre*. As in the last example the weight of the fluid may be collected into a particle and placed at the metacentre. The weight of the cup may be collected at the centre of gravity G of the cup. The equilibrium is stable if the altitude of the common centre of gravity of the two weights at M and G satisfies the criterion given in Art. 244.

246. When a cylindrical body rests on a fixed horizontal plane, it easily follows from what precedes that the equilibrium is stable or unstable according as the centre of gravity of the body is below or above the centre of curvature at the point of contact.

There is one case however which requires a little further consideration. Let us suppose that the evolute has a cusp O which points vertically downwards when the point of contact is at some point A. Let us also suppose that the centre of gravity G of the body is at a very little distance above O. The position of the body is unstable, but a stable position exists in immediate proximity on each side in which the tangent from G to the evolute is vertical. That these positions are stable is clear, for since the cusp points downwards either tangent from G will touch the evolute at a point L or M which is above G when that 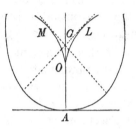 tangent is vertical. When G moves down to O these two flanking stable positions come nearer to the unstable position and finally come up to it. When therefore the centre of gravity is at the cusp of the evolute, the equilibrium is stable.

In the same way, if the cusp O point upwards and G be situated at a very short distance below O, the equilibrium is stable with a near position of instability on each side. In the limit when G coincides with O, the equilibrium becomes unstable. The reader may consult a paper by J. Larmor on *Critical Equilibrium* in *the fourth volume of the Proceedings of the Cambridge Philosophical Society*, 1883.

247. Spherical bodies, second approximation. When the equilibrium is neutral it is necessary to examine the higher differential coefficients to settle the stability or instability of the equilibrium. The geometrical method is not very convenient for this purpose. When both surfaces are spherical we can investigate all the conditions of equilibrium by the method of Art. 220.

Let the body, as represented in the figure of Art. 244, be displaced so that J' comes into the position I'. The position of the body is then represented in the adjoining figure, where J represents that point of the upper body which in equilibrium coincided with I. Let $JG = r$. Let $\psi' = IC'I'$, $\psi = JCI'$, then $\rho'\psi' = \rho\psi$. Let y be the altitude of G above C'. The inclinations to the vertical of $C'C$, CJ and JG are respectively $\alpha + \psi'$, $\alpha + \psi + \psi'$ and $\psi + \psi'$. Projecting these three lines on the vertical, we have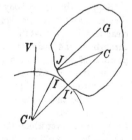

$$y = (\rho + \rho')\cos(\alpha + \psi') - \rho\cos(\alpha + \psi + \psi') + r\cos(\psi + \psi').$$

We now substitute for ψ its value $\rho'\psi'/\rho$ and expand the expression in powers of ψ'. The coefficients of ψ', $\frac{1}{2}\psi'^2$ &c. are the successive differential coefficients of y, hence the stability is determined to any degree of approximation by the rule of Art. 220.

The coefficient of ψ' is zero, that of $\frac{1}{2}\psi'^2$ is $(z\cos\alpha - r)\rho'^2/z^2$, where z has the same meaning as before. The equilibrium is stable or unstable according as this coefficient is positive or negative, i.e. according as r is less or greater than $z\cos\alpha$.

If this coefficient also vanish the equilibrium is neutral to a first approximation. We then examine the coefficient of ψ'^3. Unless this also vanishes the equilibrium is stable for displacements on one side of the position of equilibrium and unstable for displacements on the other. Supposing however that the coefficient of ψ'^3 does vanish, we examine the terms of the fourth order. The equilibrium is then stable or unstable according as the coefficient of ψ'^4 is positive or negative.

248. Ex. 1. A spherical surface rests on the summit of a fixed spherical surface, the centre of gravity being at such a height above the point of contact that the equilibrium is neutral to a first approximation. If the lower surface is convex upwards as in the diagram, prove that, whether the upper body has its convexity upwards or downwards, the equilibrium is unstable. If the lower surface has its concavity upwards, the equilibrium is stable or unstable according as the radius of curvature of the lower body is greater or less than twice that of the upper body.

The coefficient of ψ'^2 is here zero. The coefficient of ψ'^4 after elimination of r reduces to $-\rho'(\rho'+2\rho)(\rho'+\rho)/24\rho^2$. Since the equilibrium is therefore stable or unstable according as this coefficient is positive or negative, the results follow at once.

Ex. 2. A body, whose lower portion is bounded by a spherical surface, rests in apparently neutral equilibrium within a fixed spherical bowl with the point of contact at the lowest point. If the radius of one surface is twice that of the other, show that the equilibrium is really neutral.

249. Non-spherical bodies, second approximation. If the boundaries of the bodies in contact are not spherical we may adopt the following method.

Suppose the upper body has rolled away from its position of equilibrium into that represented in the figure of Art. 247. Then it is clear that, if G in that figure is to the right of the vertical through I', the body will roll further away from the position of equilibrium, but if G is on the left of the vertical, the body will roll back. Let i be the angle GI' makes with the vertical; our object will be to find i.

Let ϕ be the angle GI' makes with the common normal at I', viz. $I'C$, and let $GI' = r$. Let $I'J''$ be any further arc δs over which the body may be made to roll. Let ρ, ρ' be the radii of curvature of the upper and lower bodies at I'. Then we have

$$\frac{dr}{ds} = \sin\phi \quad\dots\dots\dots\dots\dots\dots\dots\dots\dots\dots(1).$$

Also $CI'G + I'GJ'' = I'OJ'' = CJ''G + I'CJ''$　　$\therefore \phi + \dfrac{\delta s \cos \phi}{r} = \phi + d\phi + \dfrac{\delta s}{\rho}$,

$$\therefore \frac{d\phi}{ds} = \frac{\cos \phi}{r} - \frac{1}{\rho} \quad\dotfill(2)*.$$

Lastly, let ψ' be the inclination of the normal CC' to the vertical, then $i = \psi' - \phi$ and $d\psi'/ds = 1/\rho'$. Hence by (2)　　$\dfrac{di}{ds} = \dfrac{1}{\rho} + \dfrac{1}{\rho'} - \dfrac{\cos \phi}{r}$ $\dotfill(3)$.

These three equations supply all the conditions of stability. In the position of equilibrium the centre of gravity is vertically over the point of support. Hence $i = 0$. In any other position the value of i is given by Taylor's series, viz.

$$i = \frac{di}{ds}\delta s + \frac{d^2 i}{ds^2}\frac{\delta s^2}{1 \cdot 2} + \&c.$$

If in this series the first differential coefficient which does not vanish is positive and of an odd order, it is clear that the straight line IG will move to the same side of the vertical as that to which the body is moved. The equilibrium will therefore be unstable for displacements on either side of the position of equilibrium. If the coefficient is negative the equilibrium will be stable. If the term is of an even order, it will not change sign with δs, the equilibrium will therefore be stable for a displacement on one side and unstable for a displacement on the other side.

The first differential coefficient is given by (3). The second may be found by differentiating (3) and substituting for $d\phi/ds$ and dr/ds from (2) and (1). The third differential coefficient may be found by repeating this process. In this way we may find any differential coefficient which may be required.

Firstly. Suppose the body such that di/ds is not zero in the position of equilibrium. The condition of stability is therefore that $\dfrac{1}{\rho} + \dfrac{1}{\rho'} - \dfrac{\cos \phi}{r}$ is negative. This leads to the rule already considered in Art. 244.

Secondly. Suppose the body such that in the position of equilibrium the centre of gravity lies on the circle of stability. We then have $di/ds = 0$. Differentiating (3) and substituting for $(\cos \phi)/r$ its value $1/\rho + 1/\rho'$ we find

$$\frac{d^2 i}{ds^2} = \frac{d}{ds}\left(\frac{1}{\rho} + \frac{1}{\rho'}\right) + \tan \phi \left(\frac{1}{\rho} + \frac{1}{\rho'}\right)\left(\frac{1}{\rho} + \frac{2}{\rho'}\right) \quad\dotfill(4).$$

Unless this vanishes the equilibrium will be stable for displacements on one side and unstable for displacements on the other side of the position of equilibrium.

Thirdly. Suppose the second differential coefficient given by (4) is also zero in the position of equilibrium. We find by differentiating (3) twice and substituting for r as before

$$\frac{d^3 i}{ds^3} = \frac{d^2}{ds^2}\left(\frac{1}{\rho} + \frac{1}{\rho'}\right) + \left(\frac{1}{\rho} + \frac{1}{\rho'}\right)\left\{\left(\frac{1}{\rho} + \frac{2}{\rho'}\right)\frac{1}{\rho} - \tan \phi \frac{d}{ds}\frac{1}{\rho} - 3\tan^2 \phi \left(\frac{1}{\rho} + \frac{1}{\rho'}\right)\left(\frac{1}{\rho} + \frac{2}{\rho'}\right)\right\}.$$

* The equation (2) is useful for other purposes besides that of finding the conditions of stability. For example it may be very conveniently used in the differential calculus to find the conic of closest contact at any point I of a curve. If ϕ be the angle between the central radius and the radius of curvature ρ at any point P of a conic, it may be shown that $\tan \phi = -\frac{1}{3}\dfrac{d\rho}{ds}$, where ϕ is positive when measured behind the normal as P travels along the conic in the direction in which the arc s is measured. Suppose G to be the centre of the conic, then assuming this value of ϕ, the distance r of the centre of the conic from I is given by the equation (2) in the text.

Generally the equation (2) is useful to find the point of contact with its envelope of a straight line IG drawn through each point of a curve making with the normal an angle ϕ which is a given function of s.

The equilibrium is stable or unstable according as this expression is negative o positive.

250. Ex. 1. A body rests in neutral equilibrium to a first approximation on the surface of another, and both are symmetrical about the common normal. Show that the equilibrium cannot be stable unless either the point of contact is the summit of the fixed surface or $\rho' = -2\rho$.

Ex. 2. A body rests in neutral equilibrium to a second approximation on a rough inclined plane. Show that the equilibrium is stable or unstable according as $d^2\rho/ds$ is positive or negative.

Ex. 3. A body rests in equilibrium on the surface of another body fixed in space and the centre of gravity G of the first body is acted on by a central force tending to some point O in GI produced and varying as the distance therefrom. If G' be taken on IG so that $\dfrac{1}{IG'} = \dfrac{1}{IG} + \dfrac{1}{IO}$, the equilibrium is stable or unstable according as G' lies within or without the circle of stability.

251. Rocking Stones in three dimensions. The upper body being in its position of equilibrium, let the common tangent plane at the point of contact O be taken as the plane of xy. Let the equations to the upper and lower bodies be respectively
$$\left.\begin{array}{l} 2z = ax^2 + 2bxy + cy^2 + \&\text{c}. \\ -2z' = a'x^2 + 2b'xy + c'y^2 + \&\text{c}. \end{array}\right\} \quad \ldots\ldots\ldots\ldots\ldots\ldots(1).$$

In the standard case, therefore, the two bodies have their convexities turned towards each other. We shall now suppose the upper body to be displaced from its position of equilibrium by rolling over the lower along the axis of x through a small arc ds. Take $OP = OP' = ds$.

We have first to determine how the upper body must be rotated to bring the tangent plane at P into coincidence with that at P'. Referring to equations (1), we

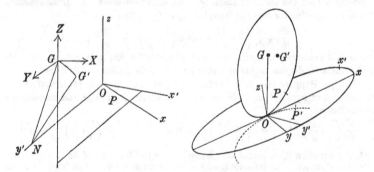

see that the tangents at P and P' to OP and OP' make angles with the plane of xy which are $dz/dx = ads$ and $dz'/dx = -a'ds$. To make these tangents coincide we must rotate the upper body round Oy through an angle $\omega_2 = (a + a') \, ds$. Consider next the tangents at P and P' which are perpendicular to OP and OP'; these make angles with the plane of xy which are $dz/dy = bds$ and $dz'/dy = -b'ds$. To make these tangents coincide we must rotate the upper body round Ox through an angle $\omega_1 = -(b + b') \, ds$. Taking both these rotations either simultaneously or one after the other, the upper body will be rolled along the arc $OP = ds$.

The two rotations ω_1 and ω_2 about the axes of x and y are equivalent to a resultant rotation Ω about some axis Oy'. If the angle $xOy' = i$, we have $\Omega \cos i = \omega_1$ and $\Omega \sin i = \omega_2$. The arc of rolling Ox and the axis of rotation Oy are not necessarily at right angles to each other; either being given, the other can be found by these relations.

252. The body being placed at rest in its new position, the centre of gravity G is no longer in the vertical through the point of contact. The weight will therefore make the body begin to move. *Let us suppose that the body is constrained either to go back to its position of equilibrium by the way it came or to recede further on that course.* The equilibrium will then be stable or unstable according as the moment of the weight about a parallel to Oy' through the new point of contact tends to bring the body back to or further from the position of equilibrium.

It will be found more convenient to refer the displacement of G to the rectangular axes Ox', Oy', Oz instead of the original axes. Let x', y', z be the coordinates of G in the position of equilibrium, let $r = OG$ and let a', β', γ be the direction angles of OG. Then $x' = r \cos a'$, $y' = r \cos \beta'$, $z = r \cos \gamma$.

If we draw GN a perpendicular on Oy', the point G will be displaced by the rotation Ω along a small arc GG' of a circle whose plane is parallel to $x'z$, whose centre is N and radius NG. The displacements of G parallel to x' and z are therefore Ωz and $-\Omega x'$. The resolved forces on G parallel to the axes x', y', z are

$$X = -W \cos a', \quad Y = -W \cos \beta', \quad Z = -W \cos \gamma,$$

where W is the weight of the body. The moment of these about a parallel to Oy' drawn through the new point of contact P is

$$M = (z - \Omega x') X - (x' + \Omega z - ds \sin i) Z$$
$$= \{r\Omega (\cos^2 a' + \cos^2 \gamma') - ds \sin i \cos \gamma\} W.$$

The equilibrium is therefore stable or unstable according as the sign of M is negative or positive.

253. We observe that Ω and i do not depend on the curvatures a, a' or b, b' but on their sums $a + a'$, $b + b'$. *If, then, we replace the rocking body by another having the curvatures of its normal sections equal to the relative curvatures of the given bodies, and make this new body roll on a rough plane inclined to the horizon at an angle γ, the conditions of stability are unaltered.* The equation of this new body is

$$2z = (a + a') x^2 + 2 (b + b') xy + (c + c') y^2 + \&c. \quad\ldots\ldots\ldots\ldots(2).$$

The indicatrix is obtained by rejecting the terms included in the &c., and giving z any constant value. This conic may be called the relative indicatrix of the solids given by (1). *It must be an ellipse* for otherwise rolling would be impossible. The equation of the axis of y' is $\omega_2 x = \omega_1 y$, i.e. $(a + a') x + (b + b') y = 0$, which is the conjugate of the axis of x. It follows that *the axis of rotation Oy' and the tangent Ox to the arc of rolling are conjugate diameters in the relative indicatrix.*

Let ρ, ρ' be the radii of relative curvature of the normal sections drawn through the arc of rolling Ox and the conjugate Oy'; ρ_1, ρ_2 the principal radii of curvature. Since each ρ is proportional to the square of the corresponding diameter of the indicatrix, it follows from a property of conjugates that $\rho\rho' \sin^2 i = \rho_1\rho_2$.

254. *To discuss the sign of the moment M,* we substitute for $\Omega \sin i$ its value $(a + a') ds$, i.e. ds/ρ. The expression then becomes

$$M = \left(r \sin^2 \beta' - \frac{\rho_1\rho_2}{\rho'} \cos \gamma \right) \frac{Wds}{\rho \sin i} \quad\ldots\ldots\ldots\ldots\ldots\ldots(3).$$

The equilibrium is stable or unstable for any given displacement according as the first factor is negative or positive.

If the rocking body rest on the summit of the fixed body, the centre of gravity G lies in the common normal Oz and therefore $\beta' = \frac{1}{2}\pi$ and $\gamma = 0$. We then have

$$M = \left(r - \frac{\rho_1 \rho_2}{\rho'} \right) \frac{W ds}{\rho \sin i} \dots\dots\dots\dots\dots\dots\dots\dots(4).$$

Considering displacements in all directions, we see that *if OG, i.e. r, is less than the least radius of relative curvature of the arc of rolling, the equilibrium is wholly stable, if OG is greater than the greatest radius of relative curvature the equilibrium is wholly unstable.* If OG lies between these limits the equilibrium is stable for some displacements and unstable for others, the separating displacement being that one in which the radius of curvature ρ' of the conjugate arc is equal to $\rho_1 \rho_2 / r$.

Ex.   A solid paraboloid of revolution is bounded by a plane perpendicular to the axis at a distance from the vertex equal to nine-eighth's of the latus rectum. Prove that it will rest in stable equilibrium with one end of the latus rectum of the generating parabola in contact with a horizontal plane.   [Coll. Ex., 1891.]

255.   Lagrange's proof of the principle of virtual work. Let a body ABC be acted on by any commensurable forces P, Q, R &c. at the points A, B, C &c. Let these forces be multiples l, m, n &c. of some force $2K$. At the point A of the body let a small smooth pulley be attached, and opposite to it at some point A' fixed in space let an equal pulley be fixed so that AA' is the direction of the force P. Let a fine string be wound round these two pulleys so as to go round each l times. It is clear that, if the tension of this string were K, the force exerted at A would be equal to the given force P and act in the same direction.   Imagine similar pulleys

to be placed at B, C &c. and opposite to them at B', C' &c. Let the same string go round the pulleys B, B' m times, and round C, C' n times, and so on. Let one extremity of this string be attached to a point O fixed in space. Let the other extremity of the string after passing over a smooth pulley D fixed in space be attached to a weight K. By this arrangement, all the forces P, Q, R &c. of the system have been replaced by the pressures due to the tension K of the string.

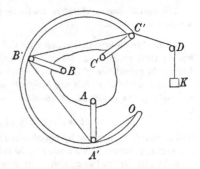

Suppose now the body receives any small displacement so that the pulleys A, B, C &c. are made to approach A', B', C' &c. respectively by small spaces α, β, γ &c. which may be positive or negative. Since the string passes round each of the pulleys A, A' l times, the string is shortened by $2l\alpha$ when these pulleys are brought nearer by a distance α. Similarly the string is shortened by $2m\beta$ when B and B' are brought closer, and so on. As the lengths OA', $A'B'$, &c. are all invariable it is clear that by this displacement the weight K will descend a space s where $s = 2(l\alpha + m\beta + \&c.)$. It is also clear that, since $P = 2lK$, $Q = 2mK$ &c., their work is $2K(l\alpha + m\beta + \&c.)$ i.e. the work of the forces due to the displacement is equal to Ks.

Lagrange reasons thus: if there were any displacement of the system which would permit the weight to descend, the weight K, always tending to descend, would necessarily descend and produce that displacement.   It follows that, if the system is in equilibrium, no possible displacement can permit the weight K to descend. Hence $s = 0$ and the virtual work of all the forces is equal to zero.

Lagrange goes on to remark that, if the quantity $la + m\beta + \&c.$ instead of zero were negative, this condition would appear to be sufficient for equilibrium, for it is impossible that the weight K would *ascend* of itself. But he points out that, if in any displacement the value of $la + \&c.$ is negative, it will become positive by giving the system a displacement in an exactly opposite direction. This displacement would cause the weight K to descend, and thus equilibrium would be destroyed.

The argument concerning the descent of K has been admitted as sound by many eminent mathematicians. Yet it does not appear to be so evident and elementary as to entitle the principle of virtual work (thus proved) to become the basis of a science. It has also been objected that it is not true without further limitations, for if a heavy particle were placed in unstable equilibrium at the highest point of a fixed smooth sphere, a small displacement would enable the particle to descend notwithstanding that it is in equilibrium.

256. Conversely, if the equation $la + \&c. = 0$ holds for all possible infinitely small displacements of the system, the system will be in equilibrium. For the weight remains immoveable in all these displacements so that there is no reason why the forces which act on the system should act so as to move the system in any one direction or its opposite. The system therefore will be in equilibrium.

The mode in which Lagrange proves this converse is certainly open to many objections. For these we refer the reader to De Morgan's criticism in the article *Virtual Velocities* in *Knight's English Cyclopædia*. The writer of that article suggests another mode of arranging Lagrange's proof which obviates some of the objections usually made to it. But this new method is itself not free from objection.

CHAPTER VII.

FORCES IN THREE DIMENSIONS.

257. *To find the resultants of any number of forces acting on a body in three dimensions. Poinsot's method.*

Let the forces be P_1, P_2, &c., and let them act at the points A_1, A_2, &c. Let O be any point arbitrarily chosen. It is proposed to reduce these forces to a single force acting at O and a couple.

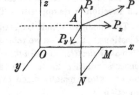

Let the point O be taken as the origin of a system of rectangular coordinates. Let P be any one of the forces, let $x = OM$, $y = MN$, $z = NA$ be the coordinates of its point of application A.

We begin by resolving P into its three axial components P_x, P_y, P_z; we shall then transfer each of these (as in Art. 104) to act at the point O by introducing into the system the appropriate couple. At M apply two opposite forces each equal and parallel to P_z, and at O apply two other opposite forces each also equal and parallel to P_z. Then since P_z may be supposed to act at N, the force P_z is equivalent to a force P_z acting at O, and two couples whose moments are yP_z and $-xP_z$, and whose planes are respectively parallel to yz and xz. The signs + and − are given according as they tend to rotate the body in the positive or negative directions of the coordinate planes in which they act. In the same way, by drawing a perpendicular from A on the plane yz, we can prove that the component P_x may be replaced by an equal force acting at the point O together with two couples zP_x and $-yP_x$ acting in the planes xz, xy respectively. Lastly, the component P_y may be replaced by an equal force at O, and the two couples xP_y and $-zP_y$ acting in the planes xy, yz. Summing up, we see that the force P may be replaced by the three axial components P_x, P_y, P_z

acting at O, and three couples whose moments are $yP_z - zP_y$, $zP_x - xP_z$, $xP_y - yP_x$, and whose planes are yz, zx, xy respectively.

Repeating this for all the given forces, we see that they may be replaced by three forces X, Y, Z acting along the axes of coordinates, and three couples whose moments are L, M, N, and whose axes are the axes of coordinates, where

$$X = \Sigma P_x, \qquad L = \Sigma \, (yP_z - zP_y),$$
$$Y = \Sigma P_y, \qquad M = \Sigma \, (zP_x - xP_z),$$
$$Z = \Sigma P_z, \qquad N = \Sigma \, (xP_y - yP_x).$$

These are called the *six components of the forces.*

The three components X, Y, Z may be compounded into a single force. Let R be its magnitude, and (l, m, n) the direction cosines of its positive direction, then

$$Rl = X, \quad Rm = Y, \quad Rn = Z,$$
$$R^2 = X^2 + Y^2 + Z^2.$$

This force is called by Moigno the *principal force* at the point O.

The three components L, M, N in the same way may be compounded into a single couple whose moment G and the direction cosines (λ, μ, ν) of whose axis are given by

$$G\lambda = L, \quad G\mu = M, \quad G\nu = N,$$
$$G^2 = L^2 + M^2 + N^2.$$

The couple G is called the *principal couple* at the point O. The components L, M, N of the principal couple are also called the moments of the forces about the axes.

258. The base of reference O to which the forces have been transferred, has been taken as the origin of coordinates. But when it is necessary to distinguish between these points we must modify the expressions for the components. Let some point O' whose coordinates are ξ, η, ζ be the base of reference. The expressions for the six components for this new base may be deduced from those for the origin by writing $x - \xi$, $y - \eta$, $z - \zeta$ for x, y, z.

The expressions for the components of the force R do not contain x, y, z, hence *the principal force R is the same in magnitude and direction whatever base is chosen.*

The expressions for the components of the couple G become

$$L' = \Sigma \, \{(y - \eta) \, P_z - (z - \zeta) \, P_y\} = L - \eta Z + \zeta Y,$$
$$M' = \Sigma \, \{(z - \zeta) \, P_x - (x - \xi) \, P_z\} = M - \zeta X + \xi Z,$$
$$N' = \Sigma \, \{(x - \xi) \, P_y - (y - \eta) \, P_x\} = N - \xi Y + \eta X.$$

Thus the magnitude and the axis of the principal couple G are in general different at different bases.

259. Conditions of equilibrium. It has been proved in Art. 105 that the forces on a body can be reduced to a single force R and a single couple G. By the same reasoning as in Art. 109 it is necessary and sufficient for equilibrium that these should separately vanish. We therefore have $R = 0$ and $G = 0$.

If the axes of reference are at right angles, these lead at once to the six conditions

$$X = 0, \quad Y = 0, \quad Z = 0, \quad L = 0, \quad M = 0, \quad N = 0\,;$$

we may, however, put these results into a more convenient form.

In order to make the resultant force R zero, *it is necessary and sufficient that the sum of the resolutes of all the forces along each of any three straight lines (not all parallel to the same plane) should be zero.* To prove this, let OA, OB, OC be parallel to the three straight lines. If the resolute of R along OA is zero, it is evident that either R is zero, or the direction of R is perpendicular to OA. If R is not zero, its direction is perpendicular to each of three straight lines meeting in O, not all in one plane, which is impossible.

In the same way, since couples are resolved according to the same laws as forces, we infer that to make the principal couple G zero, it is necessary and sufficient that the component couple of all the forces about each of any three straight lines intersecting in the base O but not all in one plane, should be zero. It will be presently seen that the moment of the component couple for any axis through O is also the moment of the forces about that axis, Art. 263.

Since a couple may be moved into a parallel plane without altering its effect, it is clear that, *when the force R is zero*, the moments about all parallel straight lines are equal. It is therefore sufficient for equilibrium that the *moment of the forces about each of any three straight lines (whether intersecting or not) should be zero, but all three must not be parallel to the same plane, and no two must be parallel to each other.* The method of finding these moments will be more fully explained a little further on.

260. Components of a force. Usually we suppose a force to be given when we know its magnitude and the equations of its line of action. We see from the results of the proposition in Art.

257 that it will sometimes be more convenient to determine a force P by the values of its six components, viz. P_x, P_y, P_z, and $yP_z - zP_y$, $zP_x - xP_z$, $xP_y - yP_x$. The advantage of this representation is that the resulting effect of any number of forces is found by adding their several corresponding components.

If we wish to represent the line of action of the force apart from the force itself, we may regard the straight line as the seat of some force of given magnitude, and suppose the line itself determined by the six components of this chosen force. Let (l, m, n) be the direction cosines of the straight line, (x, y, z) the coordinates of any point on it. Then, if the force chosen is a unit, the six components or coordinates* of the line are

$$l, m, n, \lambda = yn - zm, \mu = zl - xn, \nu = xm - yl,$$

with the obvious relation

$$l\lambda + m\mu + n\nu = 0 \dots\dots(1).$$

If a force P act along this straight line, its six components or coordinates are $Pl, Pm, Pn; P\lambda, P\mu, P\nu.$

If we compound several forces together, the six components become

$$X = \Sigma Pl,\ Y = \Sigma Pm,\ Z = \Sigma Pn;\ L = \Sigma P\lambda,\ M = \Sigma P\mu,\ N = \Sigma P\nu,$$

but the relation

$$XL + YM + ZN = 0 \dots\dots(2)$$

is not necessarily true.

261. We have seen in Art. 257 that all these forces may be joined together so as to make a single force R and a couple G. This combination of a force and a couple has been called by Plücker a *dyname*. The six quantities X, Y, Z, L, M, N are the components of the dyname. The three former components are multiples of some unit force, the three latter of some unit couple.

It will be shown further on that when the coordinates of the dyname satisfy the condition (2), either the force R or the couple G of the dyname is zero.

262. Ex. 1. The six components of a force are 1, 2, 7; 4, 5, -2. Show that the magnitude of the force is $\sqrt{54}$, and that the equations to its line of action are $$(7y - 2z)/4 = (z - 7x)/5 = (2x - y)/(-2) = 1.$$

Ex. 2. The six components of a dyname are 1, 2, 3; 4, 5, 6. Show that the magnitude of the force is $\sqrt{14}$, and that its direction cosines are proportional to 1, 2, 3. If this force act at the origin the magnitude of the couple is $\sqrt{77}$, and the direction cosines of its axis are proportional to 4, 5, 6.

* The six coordinates of a line are described in Salmon's *Solid Geometry* (fourth edition, Art. 51) from an analytical point of view. See also Cayley, *Quart. Journal*, 1860; *Camb. Trans.* 1867; Plücker, *Phil. Trans.* 1865 and 1866.

263. Moment of a force. It has already been stated that the expressions for L, M, N in Art. 257 are usually called *the moments of the forces about the axes of x, y, z* respectively. These expressions are

$$L = \Sigma\,(yP_z - zP_y), \qquad M = \Sigma\,(zP_x - xP_z), \qquad N = \Sigma\,(xP_y - yP_x).$$

To show how far this definition agrees with that already given in Art. 113, let us examine how the expression for N has been obtained. The force P has been resolved into its components P_x, P_y, P_z; the two former act in a plane perpendicular to the axis of z, hence by the definition given in Art. 113, the expressions yP_x and $-xP_y$ are respectively equal to their moments about that axis. The latter P_z acts parallel to the axis of z, and if the moment of this component is defined to be zero, the expression N will become the moment of the forces about the axis of z. Let Q be the resultant of the two components P_x, P_z, then the moment of Q about the axis of z is equal to the sum of the moments of P_x and P_z, Art. 116.

Since any straight line may be taken as the axis of z, this explanation applies to all straight lines. It appears therefore that the moment of the component couple for any axis is the same as the moment of all the forces about that axis.

We thus arrive at the following definition of the moment of a force about any straight line. Let the straight line be called CD. *Resolve the force P into two components, one parallel and the other perpendicular to the straight line CD. The moment of the former is defined to be zero. The moment of the latter is obtained by multiplying its magnitude by the shortest distance between it and the given straight line CD.*

It is evident that this shortest distance is equal to the shortest distance between the original force P and the straight line CD, each being equal to the distance between CD and the plane of the components. Let r be the length of this shortest distance. Let θ be the angle between the positive directions of the force P and the line CD, then the resolved part of the force P perpendicular to CD is $P \sin \theta$. We therefore find that *the moment of the force P about CD is equal to $Pr \sin \theta$.*

When the moments of several forces round the same straight line CD are to be added together, we must take care that these have their proper signs. Any direction of rotation round CD

having been chosen as the positive direction, the moment of any force is to be taken as positive when the force acts round CD in the positive direction.

264. It follows from Art. 263 that, if two equal forces act along the positive directions of two straight lines AB, CD, the moment of the former about CD is equal to the moment of the latter about AB.

The product $r \sin \theta$ is sometimes called *the moment of either of the straight lines AB, CD about the other*. Let i be the moment of one straight line about the other, and let either line be occupied by a force P. Then the moment of P about the other line is Pi.

265. In some cases it may be necessary to take account of the signs of r and θ. Supposing the positive direction of the common perpendicular to AB and CD to have been already determined, the shortest distance r must be measured in that direction. The angle θ must then be measured in any plane perpendicular to r from the projection of one line to the projection of the other in such a direction that when r and $\sin \theta$ are positive, a positive force acting along either line will tend to produce rotation round the other in the positive direction. See Art. 97.

266. *Geometrical representation of i.* The volume of a tetrahedron is known* to be equal to one-sixth of the continued product of the lengths of two opposite edges, the shortest distance between the edges and the sine of the angle between them. *Let AB, CD be any lengths conveniently situated on the two straight lines. The mutual moment of the two lines is equal to* $\dfrac{6V}{AB \cdot CD}$, *where V is the volume of the tetrahedron whose opposite edges are AB, CD.*

Analytical representation of i. Let (fgh), $(f'g'h')$ be the coordinates of A, C, and (lmn), $(l'm'n')$ the direction cosines of the positive directions of AB, CD. *The mutual moment of AB, CD, is the determinant in the margin.* The order of the terms in the determinant is as follows; if f, g, h precede f', g', h' in the first row, then l, m, n precedes l', m', n' in the order of the rows.

$$\begin{vmatrix} f-f', & g-g', & h-h' \\ l, & m, & n \\ l', & m', & n' \end{vmatrix}$$

To prove this we take C as origin, and let $x=f-f'$, $y=g-g'$, $z=h-h'$. The required moment is then $\lambda l' + \mu m' + \nu n'$, where λ, μ, ν have the meanings given in Art. 260.

---

* To find the volume of a tetrahedron. Pass a plane through CD and the shortest distance EF between CD and the opposite edge. Then since the tetrahedron $ABCD$ is the sum or difference of the tetrahedrons whose vertices are A and B and common base is DEC, its volume is one third the area DEC multiplied by $AB \cdot \sin \theta$, where θ is the angle AB makes with the plane DEC.

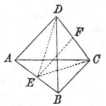

If a straight line AB cut a plane in E and be at right angles to a straight line EF in that plane, its inclination to the plane is the angle it makes with a straight line drawn in the plane perpendicular to EF. Euc. xi, 11. But CD lies in the plane and is perpendicular to EF, hence θ is equal to the angle between the opposite edges AB, CD. The volume is therefore equal to $\tfrac{1}{6} AB \cdot CD \cdot EF \cdot \sin \theta$.

267. Ex. 1. Two straight lines are given by their six coordinates $(lmn\lambda\mu\nu)$, $(l'm'n'\lambda'\mu'\nu')$: show that their mutual moment is $i = l\lambda' + m\mu' + n\nu' + l'\lambda + m'\mu + n'\nu$. This quantity is therefore invariable for the same two lines, to whatever rectangular axes their coordinates are referred. If $i = 0$, the lines intersect.

Other theorems on the moments of lines are given in Scott's *Determinants*.

Ex. 2. If $(xyzu)$, $(x'y'z'u')$ are the tetrahedral coordinates of any two points H, K on the line of action of a force P, show that the moment of the force about the edge AB of the tetrahedron, is $P \cdot \dfrac{6V}{HK \cdot AB} \begin{vmatrix} z, & z' \\ u, & u' \end{vmatrix}$.

If the force, when positive, acts from H towards K and the terms in the determinant are taken in the order shown, this expression gives the moment of the force round AB in the direction from the corner C to the corner D.

Ex. 3. If in a tetrahedron the mutual moments of the opposite edges are equal, prove that the product of their lengths are also equal. If (r, s, t) are the lengths of the lines joining the middle points of opposite edges and (α, β, γ) are the angles at which they intersect, prove also that

$$r^4 - 2r^2s^2\cos^2\gamma + s^4 = s^4 - 2s^2t^2\cos^2\alpha + t^4 = t^4 - 2t^2r^2\cos^2\beta + r^4. \quad \text{[St John's, 1891.]}$$

Ex. 4. Two triangles ABC and $A'B'C'$ are seen in perspective by an eye placed at O; forces P, Q, R act in BC, CA and AB, another set P', Q', R' in $C'B'$, $A'C'$ and $B'A'$ respectively, and the whole system is in equilibrium. Show that

$$\frac{\Delta \cdot P \cdot OA'}{BC \cdot AA'} = \frac{\Delta' \cdot P' \cdot OA}{B'C' \cdot AA'} = \frac{\Delta \cdot Q \cdot OB'}{CA \cdot BB'} = \frac{\Delta' \cdot Q' \cdot OB}{C'A' \cdot BB'} = \frac{\Delta \cdot R \cdot OC'}{AB \cdot CC'} = \frac{\Delta' \cdot R' \cdot OC}{A'B' \cdot CC'},$$

where Δ and Δ' are the volumes of the tetrahedra $OABC$ and $OA'B'C'$ respectively.
[Math. Tripos, 1883.]

The six lines OA, OB, OC, AB, BC, CA form a tetrahedron. If we equate to zero the sum of the moments of the six forces about the edge OA, we find that the first and second of the above given expressions are equal. In the same way taking moments about the edge AB, we find that the second and fourth are equal. It follows by symmetry that all the six expressions are equal. The moments may be found by using the rule given in Art. 266.

268. Problems on Equilibrium. Ex. 1. *A body, free to turn about a straight line as a fixed axis, is acted on by any forces. It is required to find the condition of equilibrium and the pressure on the axis.*

Let the straight line be the axis of z, and let x, y be two perpendicular axes.

The pressures on the elements of length of the axis constitute a system of forces. If the body is free to slide smoothly along the axis, each of these pressures will act perpendicularly to the axis. But as this limitation does not simplify the result, we shall suppose the direction of the pressure to be perfectly general. Taking any arbitrary point B on the axis as a base of reference, each pressure may be transferred to act at B, by introducing a couple whose plane passes through the axis. All the pressures are therefore equivalent to a resultant pressure which acts at B together with a resultant couple whose plane passes through the axis. Let one of the forces of this couple act at B and let the arm be so altered (if necessary) that the other force acts at some other arbitrary point C of the axis. Then compounding the forces which act at B, we see that *the pressures on all the elements of length of the axis are equivalent to two pressures which may be made to act at any two arbitrary points B, C of the axis. We may suppose the body attached to its axis at these two points by smooth hinges.*

Let F_x, F_y, F_z and G_x, G_y, G_z be the resolutes of the pressures at B and C respectively. Let b, c be the ordinates of these points. Let X, Y, Z, L, M, N be the six components of the given forces. Then resolving parallel to the axes and taking moments as in Art. 257,

$$\left.\begin{aligned} F_x + G_x + X &= 0 \\ F_y + G_y + Y &= 0 \\ F_z + G_z + Z &= 0 \end{aligned}\right\}, \qquad \left.\begin{aligned} -F_y b - G_y c + L &= 0 \\ F_x b + G_x c + M &= 0 \\ N &= 0 \end{aligned}\right\}.$$

The last equation determines the condition of equilibrium, and shows that the body will turn about the axis unless the moment of the given forces about it is zero.

We have therefore five equations to determine the six component pressures on the axis. The pressures F_x, F_y, G_x, G_y are obviously determinate, but only the sum of the components F_z, G_z can be found.

The solution of these equations will be simplified by a proper choice of the arbitrary points B and C. The position of the origin is generally determined by the circumstances of the problem. If we place B at the origin we have $b = 0$, and the values of G_y, G_z become evident by inspection.

Suppose for example the body to be a heavy door constrained to turn round an axis inclined at an angle a to the vertical. In this case, since the moment of the forces about the axis must be zero, the centre of gravity of the door must lie in the vertical plane through the axis. Let us take this plane as the plane of xz, the axis of the door being as before the axis of z. Let $\bar{x}$, 0, $\bar{z}$ be the coordinates of the centre of gravity, and let W be the weight of the door. To simplify the moments we resolve W parallel to the axes; we therefore replace W by the two components $W \sin a$ and $-W \cos a$ acting at the centre of gravity parallel to the axes of x and z. We shall choose the arbitrary point B to be at the origin, while the other C is at a distance c from it. Resolving and taking moments as before, we have

$$\left.\begin{aligned} F_x + G_x + W \sin a &= 0 \\ F_y + G_y &= 0 \\ F_z + G_z - W \cos a &= 0 \end{aligned}\right\}, \qquad \left.\begin{aligned} -G_y c &= 0 \\ G_x c + W\bar{z} \sin a + W\bar{x} \cos a &= 0 \end{aligned}\right\}.$$

It follows from these equations that F_y and G_y are both zero, so that the resultant pressures act in the vertical plane through the axis. The values of F_x, G_x and $F_z + G_z$ may be easily found.

Ex. 2. *Three equal spheres, whose centres are A, B, C, are placed on a smooth horizontal plane and fastened together by a string which surrounds them in the plane*

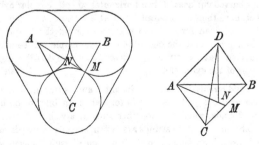

of their centres, and is just not tight. A fourth equal sphere, whose centre is D, is placed on the top of these touching all three. Prove that the tension of the string is $T = W/3\sqrt{6}$.

Let R be the reaction of any one of the lower spheres on the upper, DN a perpendicular from D on the plane ABC, then $3R \cos ADN = W$. Consider next the sphere whose centre is A; the other two of the lower spheres exert no pressure on it. The resolved part of R in the direction NA balances the two tensions of the parts of the string parallel to AB and AC. Hence $R \cos DAN = 2T \cos BAN$. The angle $BAC = 60°$, and

$$\sin ADN = \frac{AN}{AD} = \tfrac{2}{3}\frac{AM}{AD} = \tfrac{2}{3}\frac{2r \sin 60°}{2r}.$$

We now easily find T in terms of W.

Ex. 3.   Four equal spheres rest in contact at the bottom of a smooth spherical bowl, their centres being in a horizontal plane. Show that, if another equal sphere be placed upon them, the lower spheres will separate if the radius of the bowl be greater than $(2\sqrt{13}+1)$ times the radius of a sphere.   [Math. Tripos, 1883.]

Ex. 4.   Six thin uniform rods, of equal length and equal weight W, are connected by smooth hinge joints at their extremities so as to constitute the six edges of a regular tetrahedron; one face of the tetrahedron rests on a smooth horizontal plane. Show that the longitudinal strain of each of the rods of the lowest face is $W/2\sqrt{6}$.   [Coll. Ex.]

Ex. 5.   A heavy uniform ellipsoid is placed on three smooth pegs in the same horizontal plane, so that the pegs are at the extremities of a system of conjugate diameters. Prove that there will be equilibrium, and that the pressures on the pegs are one to another as the areas of the conjugate central sections.   [Coll. Ex.]

Ex. 6.   Four equal heavy rods are jointed to form a square. One side is held horizontal and the opposite one is acted on by a given couple whose axis is vertical. Show that in a position of equilibrium the lower rod makes an angle $2\sin^{-1} G/Wl$ with the upper, G being the couple, and W and l the weight and length of a rod. Find the action at either of the lower hinges.   [Coll. Ex., 1880.]

Ex. 7.   An equilateral triangular lamina, weight W, hangs in a horizontal position with its angles suspended from three points by vertical strings each equal in length to the diameter $2a$ of the circle circumscribing the triangle. Prove that the couple required to keep the lamina at a height $2(1-n)a$ above its initial position is $Wa\sqrt{(1-n^2)}$.   [Coll. Ex., 1886.]

Ex. 8.   A weightless rod, of length $2l$, rests in a given horizontal position with its ends on the curved surfaces of two horizontal smooth circular cylinders, each of radius a, which have their axes parallel and at a distance $2c$. The rod is acted on at its centre by a given force P and a couple. Find the couple when there is equilibrium, and prove that the magnitude of the couple will be least when P acts vertically, provided that $c < l \sin\phi + \tfrac{1}{2} a\sqrt{2} \sec \tfrac{1}{2}\phi$, where ϕ is the angle between the rod and the axes of the cylinders.   [Math. Tripos, 1889.]

Ex. 9.   A solid circular cylinder, of height h and radius a, is enclosed in a rigid hollow cylinder which it just fits, and is formed of an infinite number of parallel equally elastic threads, which will together support a weight W when stretched to a length $2h$. The ends of these strings are fastened firmly to two discs, one of which is then turned through an angle a in its own plane: assuming each thread to form a helix, prove that there is a force exerted in the direction of the axis of the cylinder

equal to $\dfrac{2W}{a^2}\left(\dfrac{a^2}{2}-\dfrac{h}{a^2}\sqrt{h^2+a^2a^2}+\dfrac{h^2}{a^2}\right).$ [Math. Tripos, 1871.]

Ex. 10. Three equal heavy spheres, of weight W and radius a, are suspended from a fixed point by three equal strings each of length l. A very light smooth spherical shell of radius b is placed symmetrically on the top of them, and water is poured very gently into it. Show that the greater the amount of water poured in the closer must the three lower spheres be to one another in order that equilibrium may be possible, and that equilibrium will be impossible if the weight of the water poured in exceed nW, where n is the positive root of the equation

$$n^2\,(l-b)\,(l+2a+b)+(2n+3)\,(a^2-6ab-3b^2)=0,$$

it being assumed that b is so small as to admit of the strings being straight.

[Math. Tripos, 1890.]

269. Ex. 1. *A heavy rod OAB can turn freely about a fixed point O, and rests over the top CAD of a rough wall. If OC be a perpendicular from O on the top of the wall, prove that the angle θ which the rod makes with OC when the equilibrium is limiting is given by $\mu=\tan\beta\sin\theta$, where β is the angle OC makes with the perpendicular OE drawn from O to the vertical face of the wall.*

To assist the description of the figure, let OAB be called the axis of x. Let z be normal to the plane AOC, and let y be perpendicular to x and z. The weight W of the rod acting at G is equivalent to $W\cos\beta$ parallel to z, and $W\sin\beta$ acting parallel to CO. This latter is equivalent to $W\sin\beta\cos\theta$ and $W\sin\beta\sin\theta$ parallel to x and y respectively.

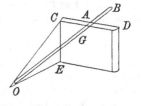

The reaction R at A is perpendicular to both OA and CD, and is therefore parallel to z. The point A of the rod can only move perpendicularly to OA. The friction therefore acts, not along the top of the wall, but opposite to the direction of motion, i.e. parallel to y.

Taking moments about y and z respectively, we have

$$W\cos\beta\,.\,OG=R\,.\,OA,\qquad W\sin\beta\sin\theta\,.\,OG=\mu R\,.\,OA.$$

These give $\mu=\tan\beta\sin\theta$.

Ex. 2. Three equal heavy spheres, each of weight W, are placed on a rough ground just not touching each other. A fourth sphere of weight nW is placed on the top touching all three. Show that there is equilibrium if the coefficient of friction between two spheres is greater than $\tan\frac12\alpha$, and that between a sphere and the ground is greater than $\tan\frac12\alpha\,.\,n/(n+3)$, where α is the inclination to the vertical of the straight line joining the centres of the upper and one lower sphere.

Ex. 3. A pole of uniform section and density rests with one end A on the ground (which is sufficiently rough to prevent any motion of that end) and with the other against a rough vertical wall whose coefficient of friction is μ. If AB be the limiting position of the pole for any position of A, AN the perpendicular from A on the wall, α the angle BAN, and θ the inclination of BN to the vertical, prove that $\tan\alpha\tan\theta$ is constant, and find the whole friction exerted at B. Find also the equation to the locus of B on the wall, N being fixed, and prove that the deviation of B from the vertical through N is greatest when $\alpha=\theta=\tan^{-1}\sqrt{\mu}.$ [Coll. Ex., 1886.]

Ex. 4. A narrow uniform rod of length $2a$ rests in an oblique position with one end on a rough horizontal table and the other against a rough vertical wall, the coefficients of friction at the table and wall being μ_1 and μ_2, and the distance of the foot of the rod from the wall being k; show that the rod is on the point of slipping at the lower end if the vertical plane in which it lies makes an angle θ with the wall given by $k\mu_1 (\mu_2^2 \sin^2 \theta - \cos^2 \theta)^{\frac{1}{2}} = k - 2\mu_1 (4a^2 \sin^2 \theta - k^2)^{\frac{1}{2}}$, and that the inclination of the tangential action at the upper end to the horizon is then $\sec^{-1}(\mu_2 \tan \theta)$.

[Math. Tripos, 1887.]

Ex. 5. A curtain is supported by an anchor ring capable of sliding on a horizontal cylinder by means of a hook fixed at that point of the ring which is lowest when the curtain is hanging. Show (1) that the ring may touch the cylinder at one or two points but not more, (2) that if there be double contact and the weight of the ring can be neglected the ring will not slip along the cylinder however it be pulled unless the coefficient of friction be less than $\dfrac{(2a+b)\cos\theta}{(2a+b)\sin\theta - b}$, in which b is the radius of the generating circle, a that of the circle described by its centre and θ the inclination of the plane of this latter circle to the axis of the cylinder. [Math. T.]

For the sake of the perspective take the axis of the anchor ring as axis of z, and let the plane of the circle whose radius is a be the plane of xy. Let the axis of x pass through the hook. Let B, B' be the two points of contact of the cylinder and ring, B' being nearest the hook. Let $(R, \mu R)$ $(R', \mu R')$ be the reactions at these points, then these four forces lie in the plane xz. Taking moments about an axis through the hook and solving, we find

$$\mu = \frac{(2a+b)\cos\theta - \rho b \cos\theta}{(2a+b)\sin\theta - b + \rho b (1 + \sin\theta)},$$

where ρ is the ratio of R' to R. As long as there is double contact R and R' are both positive. But if μ is greater than the value given in the question, this equation shows that ρ must be negative.

Ex. 6. A solid heavy cone, placed with a generating line in contact with a rough vertical wall, can turn freely about its vertex which is fixed, and is acted on by a couple whose moment is L and whose plane is parallel to the base. Prove that in equilibrium the inclination θ to the vertical of the generating line in contact with the wall is given by $L = \frac{3}{4} Wh \sin \theta \tan \alpha$, where α is the semi-vertical angle of the cone and h its altitude. If the rim only of the cone is rough, prove that the least value of the coefficient of friction is $2 \tan \theta \cdot \operatorname{cosec} 2\alpha$.

The central axis and the invariants.

270. Poinsot's Central Axis. Any base O having been chosen, the forces of a system have been reduced to a force R acting at O and a couple G. We shall now examine whether this representation of the forces can be further simplified by a proper choice of the base.

Let θ be the angle between the direction of the force R and the axis of the couple G. We may resolve G into two couples, one $G \cos \theta$ whose plane is perpendicular to R, and the other $G \sin \theta$ whose plane contains that force. This latter couple together with the force R may be replaced by a single force in its plane equal and parallel to R, but situated at a distance $G \sin \theta / R$ from O.

We have therefore *reduced the system to a force R* (acting in a direction parallel to the principal force at any base) *together with a couple whose plane is perpendicular to the force.* The line of action of this force R is called Poinsot's central axis.

To construct *geometrically* the central axis when the couple G and the force R at any base of reference O are given, we notice that (1) the central axis is parallel to R, (2) it is at a distance $G \sin \theta / R$ from R, (3) the perpendicular from O on the central axis is at right angles both to R and the axis of G, (4) the perpendicular from O must be so drawn that its foot is moved by the couple $G \sin \theta$ in the same direction as that in which R acts.

271. Screws and wrenches. A body is said to be screwed along a straight line when it is rotated round this straight line as an axis through any small angle $d\theta$, and at the same time translated parallel to the axis through a small distance ds. The ratio $ds/d\theta$ is called the pitch of the screw. If the pitch is uniform, it may also be defined as the space described along the axis when the angle of rotation is a radian, i.e. a unit of circular measure. The pitch of a screw is therefore a length. For the sake of brevity the axis of the screw is often called the screw.

The term *wrench* has been applied by Sir R. Ball to denote a force and a couple whose axis coincides with or is parallel to the force. The phrase *wrench on a screw* denotes a force directed along the axis of the screw and a couple in a plane perpendicular to the screw, the moment of the couple being equal to the product of the force and the pitch of the screw. The force is called the intensity of the wrench. When the pitch of the screw is zero the wrench is simply a force. When the pitch is infinite the wrench reduces to a couple. The phrase wrench on a screw is sometimes abbreviated into the single word, wrench.

A wrench is a dyname in which the direction of the force is perpendicular to the plane of the couple.

To determine a screw five quantities are necessary. Four are required to determine the position of the axis, for example the coordinates of the points in which it cuts two of the coordinate planes. One more is necessary to determine the pitch. To determine a wrench on a screw a sixth quantity is required, viz. the magnitude of the force.

272. Screws are distinguished as right or left-handed according to the direction in which the body is rotated for the same translation. Let an observer stand with his back along the axis, so that the translation is called positive when it is in the direction from the feet to the head. The screw is then called right or left-handed according as the rotation appears to be opposite to or the same as that of the hands of a watch ; see Art. 97.

As an example, the common corkscrew is a right-handed screw. As another example, let the reader push his two hands forward horizontally, turning at the same time his right thumb to the right and his left thumb to the left. The motion of the right hand will illustrate a right-handed screw, that of the left a left-handed screw.

In this chapter the figures are drawn in agreement with the system of coordinates usually adopted in solid geometry. The left-handed screw will therefore represent the conventions adopted to distinguish the positive and negative directions of rotation and translation. By interchanging the positions of the axes of x and y the figures may be adapted to the other system.

273. The equivalent wrench. *A system of forces is given by its six components* X, Y, Z, L, M, N *referred to any rectangular axes with the origin* O *as the base of reference. It is required to find analytical expressions for the equivalent wrench.*

It is obvious that the axis of the equivalent wrench is Poinsot's central axis, and that it is parallel to the principal force R at any base of reference.   Hence

(1)   the direction cosines of the central axis are
$$l = X/R, \quad m = Y/R, \quad n = Z/R,$$

(2)   the force or intensity of the wrench is R.

(3)   Let Γ be the required couple of the wrench. Then by Poinsot's theorem all the forces are statically equivalent to R and Γ, so that the moment of all the forces of the system about any straight line is equal to that of R and Γ about the same line. If this straight line be parallel to the central axis, the moment of R is zero and that of the couple is Γ. It follows that the *moment of the forces of a system about all straight lines parallel to the central axis are equal to the moment about the central axis.*

The principal force R at the origin is parallel to the central axis, hence, if θ be the angle the axis of G makes with R,

$$\Gamma = G \cos \theta = Ll + Mm + Nn.$$

$$\therefore \; \Gamma R = LX + MY + NZ.$$

The pitch of the screw on which the wrench acts is therefore

$$p = \frac{\Gamma}{R} = \frac{LX + MY + NZ}{R^2}.$$

(4) Let $(\xi \eta \zeta)$ be the coordinates of any point on the central axis. When this point is chosen as the base, the components L', M', N' of the couples are given in Art. 258 and these components are proportional to the direction cosines of the axis of the principal couple. We have therefore by (1)

$$\frac{L - \eta Z + \zeta Y}{X} = \frac{M - \zeta X + \xi Z}{Y} = \frac{N - \xi Y + \eta X}{Z}.$$

These are therefore the equations to the central axis.

If we multiply the numerator and denominator of each fraction by X, Y, Z respectively and add them together, we see that *each fraction is equal to the expression found above for the pitch p.*

274. *If X, Y, Z are each equal to zero* the principle on which these equations have been obtained becomes nugatory. But in this case the given system is equivalent to a resultant couple. Any straight line parallel to its axis is the central axis.

If the couple $\Gamma = 0$, the given system is equivalent to a single force R. Since the components L', M', N', at any point $(\xi \eta \zeta)$ on this force are zero, we have

$$L - \eta Z + \zeta Y = 0, \quad M - \zeta X + \xi Z = 0, \quad N - \xi Y + \eta X = 0.$$

Any two of these are the equations of the single resultant.

275. We may obtain the equations to the central axis in another way. The moments of the force R and the couple Γ about the axes are L, M, N. Hence the moments of the force R alone are $L - \Gamma l$, $M - \Gamma m$, $N - \Gamma n$, i.e. they are $L - Xp$, $M - Yp$, $N - Zp$. The six components of the force R are therefore X, Y, Z, $L - Xp$, $M - Yp$, $N - Zp$. These are the six coordinates of the central axis.

276. Conversely, *the equivalent wrench being given, we may find the six components of the forces at any base of reference.*

Let Oz be the given axis of the wrench, and let O' be any point at which the components are required. Let $O'O$ be a perpendicular on Oz and let $OO' = r$. Let $O'C$ be parallel to Oz and $O'B$ perpendicular to the plane $O'Oz$.

The force R acting along Oz may be transferred to act along $O'C$ by introducing the couple Rr with $O'B$ for axis. The couple Γ may also be transferred from its axis Oz to $O'C$. Compounding these two couples we have a resultant couple G whose axis $O'A$ lies in the plane $BO'C$ and makes an angle θ with $O'C$, where

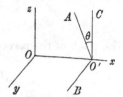

$$G^2 = \Gamma^2 + R^2 r^2, \qquad \tan\theta = Rr/\Gamma.$$

277. From these values of R and G we may draw several obvious conclusions.

(1) We see that G is always numerically greater than Γ, so that the principal couple G is least when the base of reference is on the central axis.

(2) Since OO' may be drawn in any direction from Oz, it follows that the locus of the base at which the principal couple G has a given value is a right circular cylinder whose axis is the central axis.

(3) The locus of the axis, viz. $O'A$, of the principal couple of given magnitude is a system of hyperboloids of revolution.

278. Examples. Ex. 1. The equivalent wrench being given, show that the base on a given straight line at which the principal couple is least is the point at which the straight line is intersected by the shortest distance between itself and the central axis. Find also the base at which the axis of the principal couple makes the least angle with the given straight line.

Ex. 2. The base being the origin of coordinates, show that the plane containing the force R and the axis of G is given by the determinantal equation in the margin. Show also that the minors of the first row, after division by R^2, are the coordinates of the foot of the perpendicular from the origin on the central axis. Thence find the equations to the central axis regarding it as a straight line drawn through this point parallel to R.

$$\begin{vmatrix} \xi & \eta & \zeta \\ X & Y & Z \\ L & M & N \end{vmatrix} = 0.$$

Ex. 3. Twelve equal forces occupy the edges of a cube, the parallel forces acting in the same direction: prove that their central axis is a diagonal. If the forces are replaced by twelve equal couples whose axes occupy the edges, prove that their central axis is parallel to a diagonal.

Ex. 4. Six equal forces act along the edges AB, BC, CA, DA, DB, DC of a regular tetrahedron: show that their central axis is the perpendicular from the corner D of the tetrahedron on the face ABC.

Ex. 5. Six forces act along the edges AB, BC, CA, AD, BD, CD of a tetrahedron, each force being proportional to the length of the edge along which it acts. Show that their central axis is parallel to DG and is at a distance $\frac{2}{3}\Delta\cos\phi/DG$ from it, where Δ is the area of the face ABC, G its centre of gravity, and ϕ the angle DG makes with the face.

Ex. 6. Any number of forces are represented in magnitude, direction, and position by the straight lines, A_1A_1', $A_2A_2'...A_nA_n'$ and G, G' are the centres of gravity of equal particles placed at $A_1...A_n$ and $A_1'...A_n'$ respectively. Prove that the central axis of these forces is parallel to GG'. If these forces intersect any plane drawn perpendicular to GG' in B_1, $B_2,...B_n$ prove that the central axis intersects this plane in the centre of gravity of particles placed at B_1, $B_2,...B_n$ whose weights are proportional to the resolved parts of the forces parallel to GG'.

[Coll. Ex., 1889.]

Ex. 7. A system of forces intersects the plane of xy and a plane $z=h$ in the points $A_1A_2...$, $A_1'A_2'...$ respectively; their magnitudes are $a_1.A_1A_1'$, $a_2.A_2A_2'...$ and the pitch of the equivalent wrench is p. Prove that the central axis intersects these planes in the points H, H' whose coordinates (ξ, η), (ξ', η') are given by

$$\xi' - x' = \xi - x = (y' - y)\,p/h, \qquad \eta' - y' = \eta - y = -(x' - x)\,p/h,$$

where (xy) are the coordinates of the centre of gravity G of masses a_1, $a_2,...$ placed at $A_1A_2...$ and $x'y'$ those of the centre of gravity G' of the same masses placed at $A_1'A_2'....$

Show also that (1) GH is perpendicular to GK' and equal to $GK'.p/h$ where K' is the projection of G' on the plane of xy, and (2) HH' is parallel to GG'.

Ex. 8. Prove that the trilinear coordinates $\alpha\beta\gamma$ of the point in which the central axis of a system of forces cuts the plane of any triangle ABC are given by

$$Z\alpha = M_1 - X_1 p, \qquad Z\beta = M_2 - X_2 p, \qquad Z\gamma = M_3 - X_3 p,$$

where M_1, M_2, M_3 are the moments of the forces about the sides, X_1, X_2, X_3, the resolutes along the sides of the triangle, Z the resolute perpendicular to its plane, and p is the pitch.

Regarding AB as the axis of x and the plane of the triangle as that of xy, the ordinate η, found by putting $\zeta=0$ in the equation of the central axis, Art. 273, is the trilinear coordinate γ.

279. Invariants of a system. It follows from the third result of Art. 273 that, whatever base is chosen and whatever the directions of the rectangular axes may be, the quantity $I = LX + MY + NZ$ is invariable and equal to ΓR. The square of the resultant force, viz. $R^2 = X^2 + Y^2 + Z^2$ is also invariable. *These two quantities, viz. I and R^2, are called the invariants.* When the invariants I and R^2 are known, a third invariant, viz. the pitch $p = I/R^2$, can be immediately deduced.

If the forces of the system are such that the first of these invariants is zero, it follows that either $R = 0$ or $\Gamma = 0$. *The condition that the forces should be equivalent to either a single force or a single couple is therefore $I = 0$.* We may distinguish between these two cases by examining the second invariant. *If the forces are to be equivalent to a single force we must have as a second condition R not equal to zero.*

280. When two systems of forces P_1, P_2 &c. and Q_1, Q_2 &c. are given we form the two expressions

$$\Sigma PQr \sin(P, Q), \qquad \Sigma PQ \cos(P, Q),$$

where r is the shortest distance between the forces P, Q, and (P, Q) is the angle between these forces, the products being taken with their proper signs. Then *each of these expressions is invariable when we change either system into any equivalent system of forces.* This theorem is given by Chasles, *Liouville's Journal*, 1847.

To prove this consider both systems as one, then however the forces may be changed, the invariant I of the united systems remains the same. Hence

$$\Sigma P_1 P_2 r_{12} \sin (P_1, P_2) + \Sigma Q_1 Q_2 r'_{12} \sin (Q_1, Q_2) + \Sigma PQr \sin (P, Q)$$

is invariable. But each of the two first terms is invariable. Hence the last term is also invariable.

In just the same way by considering the invariant R^2 we may show that $\Sigma PQ \cos (P, Q)$ is also invariable.

281. *To find the invariants of a system of forces.* To find the invariants of two forces P_1, P_2 we refer to the figure of Art. 276. Let the line of action of the force P_1 be the axis of z, let the line of action of P_2 be $O'A$, and let the shortest distance OO' between these forces be the axis of x. The components of the forces are

$$X = 0, \qquad Y = P_2 \sin \theta, \qquad Z = P_1 + P_2 \cos \theta,$$
$$L = 0, \qquad M = -P_2 r \cos \theta, \qquad N = P_2 r \sin \theta.$$

Since the invariants are independent of all axes, we have

$$I = LX + MY + NZ = P_1 P_2 r \sin \theta,$$
$$R^2 = P_1^2 + P_2^2 + 2P_1 P_2 \cos \theta.$$

Since $I = P_1 N$, it follows that the *invariant of two forces is equal to either force multiplied by the moment of the other force about the first.*

Let the positive direction of a straight line be determined by the signs of the direction cosines of the line. The positive direction of rotation round that line is then determined by the rule in Art. 272 or Art. 97. The sign of the invariant of two forces is positive or negative according as the sign of either force and that of the moment of the other are like or unlike.

The forces P_1, P_2 being represented by two lengths measured along their respective lines of action, the *invariant I is equal to six times the volume of the tetrahedron having these lengths for opposite edges.* This tetrahedron is sometimes called the tetrahedron constructed on two forces. See Art. 266.

To find the invariant I of any number of forces P_1, P_2 &c. Taking any rectangular axes, the six components are given in Art. 257. It follows that I is a quadratic function of P_1, P_2 &c. of the form

$$I = A_{11} P_1^2 + A_{22} P_2^2 + 2A_{12} P_1 P_2 + \&c.$$

where A_{11} &c. are all independent of the magnitudes of the forces. When all the forces except P_1, P_2 are put zero this expression should reduce to $P_1 P_2 r_{12} \sin (P_1, P_2)$, where (P_1, P_2) expresses the angle between the directions of the forces. Hence $A_{11} = 0$, $A_{22} = 0$; applying the same reasoning to the other forces, we infer that

$$I = \Sigma P_1 P_2 r_{12} \sin (P_1, P_2).$$

It follows that *I is half the sum of each force multiplied by the sum of the moments of all the other forces about it, each moment being taken with its proper sign.*

It also follows that *the invariant of any number of forces is the sum of their invariants taken two and two with their proper signs.*

Any number of systems of forces being given the invariant I of the whole is the sum of the invariants of each separate system plus the invariants of each two systems.

For in this summation any one force is taken in combination with every other force in the partial invariant in which they both occur.

282. *The invariant I of a force R and a couple whose moment is G is RG* cos θ, where θ is the angle the direction of the force makes with the axis of the couple. For by definition $I = R\Gamma = RG \cos \theta$.

The invariant I of two couples G, G', is zero. To prove this we move the couples in their own planes until each has a force acting parallel to the intersection of the planes. The four forces being now parallel, the invariant of every two is zero, and therefore their sum is zero.

The invariant of two wrenches whose forces are P, P', and pitches p, p', is

$$P^2 p + P'^2 p' + PP' \{(p+p') \cos \theta + r \sin \theta\}.$$

This is seen to be true by adding together the six invariants of the forces P, P', and the couples Pp, $P'p'$, taken two and two, Art. 281.

Ex. If the system is equivalent to the forces X, Y, Z, acting along oblique axes and the couples L, M, N, whose axes coincide with the oblique axes, show that the invariant I is

$$I = LX + MY + NZ + (YN + ZM) \cos(y, z) + (ZL + XN) \cos(z, x) + (XM + LY) \cos(x, y).$$

283. Examples. Ex. 1. Forces la, mb, nc act in three non-intersecting edges of a parallelepiped, where a, b, c are the lengths of those edges. Prove that, if the system be reduced to a wrench, the product of the force and couple of that wrench is $(lm + mn + nl) V$, where V is the volume of the parallelepiped. [St John's, 1890.]

Ex. 2. A system of n given forces is combined with another force P, which is given in magnitude and passes through a fixed point; prove that, if the $n+1$ forces have a single resultant, P must lie on a right circular cone, and that, if their least principal moment be constant, it must lie on a cone of the fourth degree. In the second case, prove that if the n forces reduce to a couple, the central axis of the $n+1$ forces lies on a hyperboloid of revolution. [Math. Tripos, 1871.]

Ex. 3. If a system, consisting of two forces whose lines of action are given and a couple whose plane is given, admit of a single resultant, prove that the direction of this resultant lies upon a certain hyperbolic paraboloid. [Math. Tripos.]

Ex. 4. A rigid body is acted upon by three forces $2P \tan A$, $-P \tan B$, $2P \tan C$ along three edges of a cube which do not meet, symmetrically chosen with respect to the axes of coordinates drawn parallel to them through the centre of the cube. Prove that the forces are equivalent to a single force acting along the line whose equations are $2a \cot B - x \cot A = 2y \cot B + a \cot A = -z \cot C$, where $2A$, $2B$, $2C$ are the angles of a triangle whose sides are in arithmetical progression, and $2a$ is the edge of the cube. [Math. Tripos, 1867.]

Ex. 5. If the rectangle under the three pairs of opposite edges of a tetrahedron are equal to each other, show that four equal forces acting along the sides taken in order of the skew quadrilateral formed by leaving out one pair of opposite edges are equivalent to a single resultant force; and that the lines of action of the three single resultants obtained by leaving out different pairs of opposite edges in succession are the three diagonals of the complete quadrilateral in which the faces of the tetrahedron are cut by a certain plane. [Coll. Ex., 1889.]

On Screws and Wrenches.

284. *To find the resultant wrench of two given wrenches, or of two given forces.* Analytical method.

Let P, P' be the forces, p, p', the pitches of the given wrenches. Let θ be the inclination of the two axes and h the shortest distance between them. It is clear that *if the resultant wrench of two given forces is required*, we merely put $p = 0$, $p' = 0$ in the following process.

Let R be the force of the resultant wrench, ϖ its pitch. By equating the invariants of the given wrenches to those of their resultant, we have

$$R^2\varpi = P^2 p + P'^2 p' + PP' \{(p + p') \cos \theta + h \sin \theta\},$$
$$R^2 = P^2 + P'^2 + 2PP' \cos \theta.$$

These equations determine the magnitude of the resultant wrench. We easily deduce

$$R^2 \{\varpi - \tfrac{1}{2}(p + p')\} = \tfrac{1}{2}(P^2 - P'^2)(p - p') + PP'h \sin \theta.$$

285. We have next to find the position in space of the axis of the resultant wrench. Let AA' be the shortest distance between the axes AF, $A'F'$ of the given wrenches, the arrows indicating the positive directions in which the forces P, P' act. Since Poinsot's central axis is parallel to the resultant of the forces P, P', transferred to any base *the central axis must be perpendicular to AA'.* Again since the moment of both the given wrenches about AA' is zero, the moment about the same line of R and the couple Γ (whose axis has been proved perpendicular to AA') is also zero. This requires that *the central axis should intersect the shortest distance AA' in some point O.*

Let AA' be taken as the axis of x, and let the required central axis be the axis of z. Let γ, γ', be the inclinations of $AF, A'F'$ to the central axis, then $\theta = \gamma + \gamma'$. By resolving the forces we have

$$R \sin \gamma = P' \sin \theta, \quad R \cos \gamma = P + P' \cos \theta,$$
$$R \sin \gamma' = P \sin \theta, \quad R \cos \gamma' = P' + P \cos \theta \quad \Big\} \dots\dots(1).$$

Let C be the middle point of AA', $CO = \xi$. Equating the moments about a parallel to Oy drawn through C of the given wrenches and their resultant wrench we have

$$R\xi = \tfrac{1}{2}h(P \cos \gamma - P' \cos \gamma') - Pp \sin \gamma + P'p' \sin \gamma'.$$

Substituting for $\sin \gamma$, $\cos \gamma$, &c. from (1) we have

$$R^2\xi = \tfrac{1}{2}h\,(P^2 - P'^2) - PP' \sin \theta\,(p - p').$$

This equation determines the distance ξ of the central axis of the two wrenches from the middle point of the shortest distance measured positively towards P. A formula equivalent to this was given in the Math. Tripos, 1887.

Ex. Prove that the central axis of two given forces P, P' divides their shortest AA' distance in the ratio $P'(P' + P\cos\theta) : P(P + P'\cos\theta)$ which is independent of the length of AA', the angle between the forces being θ.

286. *To find the resultant wrench of two wrenches whose axes intersect in some point A.* The magnitudes of Γ and R are found by the same invariants as in the last proposition, but the determination of the position in space of the resultant axis is much simplified.

Let the resultant R of the forces P, P', act at A in the direction AB and make angles γ, γ' with AF, AF'. Then $R\sin\gamma = P'\sin\theta$, $R\sin\gamma' = P\sin\theta$. Following the rule given in Art. 270 to construct the central axis we find the component of the couples about a straight line AD drawn perpendicular to R in the plane of the forces. This component is

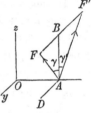

$$Pp \sin \gamma - P'p' \sin \gamma' = PP' \sin \theta\,(p - p')/R.$$

We now *measure a distance AO in a direction normal to the plane of the forces equal to $PP' \sin\theta\,(p - p')/R^2$, and draw a parallel Oz to the direction of R. Then Oz is the central axis.*

To determine on which side of the plane of the forces AO should be drawn, we notice that the couple $Pp\sin\gamma$ should turn AO round A towards the direction of R.

287. The Cylindroid. This surface has been used by Sir R. Ball for the purpose of resolving and compounding wrenches. Following his line of argument we shall first examine a special case, and thence deduce the general solution.

To find the resultant of two wrenches of given intensities on screws of given pitches which intersect at right angles. Let the axes of these screws be the axes of x and y. Let X, Y be their forces; p, p' their pitches. Let R be the resultant of the forces X, Y, and let OA be its line of action. Let G be the resultant of the couples Xp, Yp' and let OB be its axis. Let the angle $AOB = \phi$. By resolving G into $G\cos\phi$ about OA and $G\sin\phi$

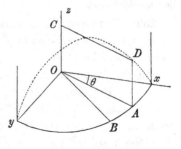

about a perpendicular to OA, it is clear (as in Art. 270) that G and R are together equivalent to a wrench having for its axis a straight line CD parallel to OA such that $OC = (G \sin \phi)/R$. The force along the axis is equal to R and the couple round it is equal to $G \cos \phi$.

Since $G \cos \phi$ and $G \sin \phi$ are the moments about OA and a perpendicular to OA, we see that, if θ be the angle xOA,

$$G \cos \phi = Xp \cos \theta + Yp' \sin \theta = R (p \cos^2 \theta + p' \sin^2 \theta)$$
$$G \sin \phi = - Xp \sin \theta + Yp' \cos \theta = R (p' - p) \sin \theta \cos \theta.$$

Let ρ be the pitch of the resultant wrench and $z = OC$, then

$$\left. \begin{array}{l} \rho = p \cos^2 \theta + p' \sin^2 \theta \\ z = (p' - p) \sin \theta \cos \theta \end{array} \right\} \quad \ldots\ldots\ldots\ldots\ldots (1).$$

Also $X = R \cos \theta$, $Y = R \sin \theta$.

If the wrenches on the axes Ox, Oy, have given pitches but varying forces, the locus of the axis CD of the resultant wrench will be found by writing $\tan \theta = y/x$ and eliminating θ from the second of equations (1). We thus find

$$z (x^2 + y^2) - (p' - p) xy = 0 \ldots\ldots\ldots\ldots\ldots (2).$$

This surface is called the cylindroid.

Describe a cylinder whose axis is the axis of z; as CD travels round Oz beginning at Ox and ending at Oy, thus generating one quarter of the cylindroid, its intersection with the cylinder traces out a curve which is represented in the figure by the dotted line. In the next quarter of the surface, the dotted curve (not drawn) is below the plane of xy, in the third quarter above and so on.

288. Each generating line of the cylindroid, such as CD, is the axis of a screw whose pitch is $p \cos^2 \theta + p' \sin^2 \theta$. Let us then describe the cylinder whose base is the conic $px^2 + p'y^2 = H$, where H is any constant. Let the generating line CD intersect the surface of the cylinder in D. Then the pitch of the screw whose axis is CD is obviously H/CD^2. The base of this cylinder has been called by Sir R. Ball the *pitch conic*.

289. *The forces of any number of wrenches on a given cylindroid being given, it is required to find the resultant wrench and the conditions of equilibrium.*

Let P_1, P_2 &c. be the forces, θ_1, θ_2 &c. their inclinations to the axis of x. Referring to the figure of Art. 287, let CD be the axis of a wrench whose force is P and whose pitch is the pitch appropriate to the axis CD. If θ be the inclination of CD to the axis of x, the resolved parts of P along the axes of x, y and z are

$P \cos \theta$, $P \sin \theta$ and zero respectively. The process of resolving the wrench into its components on the axes being the exact reverse of the process in Art. 287 of compounding the wrenches on the axes, it is clear that the moments of the force P about the axes are $P \cos \theta \cdot p$, $P \sin \theta \cdot p'$ and zero.

Taking all the wrenches, the six components are

$$X = \Sigma P \cos \theta, \qquad Y = \Sigma P \sin \theta, \qquad Z = 0,$$
$$L = \Sigma P \cos \theta \cdot p = Xp, \quad M = \Sigma P \sin \theta \cdot p' = Yp', \quad N = 0.$$

These constitute two wrenches on the axes of x and y, with the same two pitches as before.

By the definition of a cylindroid *the axis of the resultant wrench lies on the same cylindroid.* The pitch ρ and the altitude z of the resultant wrench are given by equations (1) of Art. 287.

290. The necessary and sufficient conditions of equilibrium are $\Sigma P \cos \theta = 0$, $\Sigma P \sin \theta = 0$, for when these vanish all the six conditions of equilibrium are satisfied. It immediately follows that *if the forces of wrenches on the same cylindroid when transferred to act at any one point are in equilibrium, then the wrenches themselves will be in equilibrium.*

For example, the wrenches on any three screws in the same cylindroid are in equilibrium if the force of each is proportional to the sine of the angle between the other two.

To find, also, the resultant wrench of two given wrenches in the same cylindroid we first find the resultant of their forces. The axis of the required wrench is parallel to this resultant and has the pitch appropriate to that axis.

291. We may use this theorem to find the resultant wrench of any two wrenches if we show that a unique cylindroid can be described so as to contain any two given screws.

To prove this, let CD, $C'D'$ be the axes of the two given screws, and let CC' be the shortest distance between them, then CC' must be the z-axis of the cylindroid. Let $CC'=h$, let a be the inclination of the axes CD, $C'D'$ to each other, and ρ, ρ' the pitches of the screws. These four quantities being given, we have to prove that one set of real values can be found for p, p', (z, θ), (z', θ'). Taking the values given for ρ, z, ρ', z' in equations (1) of Art. 287 and joining to them the two equations $z - z' = h$, $\theta - \theta' = a$, we can solve the six resulting equations. The result is that we find unique values for p, p', &c.

292. Work of a wrench. *To find the work done by a wrench on a given screw when the body receives a virtual displacement on any other given screw.*

Let us first find the work done when a given *couple* is moved in its own plane from one position to another. This displacement may be constructed by first translating the couple parallel to itself until one extremity A of its arm AB assumes its new position and then rotating the translated couple about A until the other extremity B assumes its proper position. The work done by the two equal forces during the translation is clearly zero. The work done by the force at A during the rotation is also zero. It remains to find the work done by the force at B.

Let F be the force, a the length of the arm AB, $d\phi$ the angle of rotation. The work done by the force at B is evidently $Fad\phi$. If the angle of displacement is finite, the work done is found by integrating $Fad\phi$. Thus the *work done by a couple of given moment is the product of the moment by the angle of rotation in its own plane.* See Art. 203.

Next let a couple be rotated about an axis in its own plane through any small angle $d\phi$. It is clear that the extremities A, B of the arm begin to move perpendicular to the plane of the forces. The virtual work done by each force is therefore zero.

293. Let us apply these two results to find the work done by a wrench twisted about any screw.

Let p, p' be the pitches of the screw and wrench respectively.
Let θ be the angle between their re-
spective axes and let h be the shortest
distance between them. We suppose
that in the standard case, when θ and
h are positive, the positive direction of
each axis is such that a force acting
along it would produce rotation about
the other axis in the positive direction;
see Art. 265. Let R be the force of the wrench.

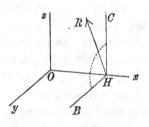

Take the axis of the screw as the axis of z and the shortest distance OH as the axis of x. Let HC and HB be drawn parallel to the axes of z and y respectively. The force R may be resolved into $R\cos\theta$, $R\sin\theta$ along HC and HB. When the body is translated a space $pd\phi$ parallel to the axis of z and rotated an angle $d\phi$ about it, the work of the former force is $R\cos\theta \cdot pd\phi$; the work of the latter is $R\sin\theta \cdot hd\phi$.

The couple Rp' of the wrench may be resolved into two

couples $Rp' \cos \theta$ and $Rp' \sin \theta$ whose axes are HC and HB. The work of the former is $Rp' \cos \theta d\phi$, the work of the latter is zero. *The whole work done is therefore*

$$dW = Rd\phi \{(p + p') \cos \theta + h \sin \theta\}.$$

We notice that this is a symmetrical function of p and p', so that if the two screws are interchanged the work is unaltered.

294. Reciprocal screws.* Two screws are said to be reciprocal when a wrench acting on either does no work as the body is twisted about the other. The analytical condition that two screws are reciprocal is therefore

$$(p + p') \cos \theta + h \sin \theta = 0.$$

Thus, two intersecting screws are reciprocal when either they are at right angles or their pitches are equal and opposite.

It follows from the principle of virtual work that a body free to move only on a screw α is in equilibrium if acted on by a wrench on any screw reciprocal to α.

295. *If a screw σ is reciprocal to each of two given screws, say α and β, it is also reciprocal to every screw on the cylindroid containing α and β.* For a wrench on any third screw γ on this cylindroid may be replaced by two wrenches on the screws α and β, if the forces on α and β are the components of the force on γ (Art. 289). Since the virtual work of each of these when twisted along σ is zero, the screws γ and σ are reciprocal. We may say for brevity that the screw σ is reciprocal to the cylindroid.

296. *A screw σ if reciprocal to a cylindroid must intersect one of the generators at right angles.* The cylindroid, being a surface of the third order, will be cut by the screw σ in three points, and one screw of the cylindroid passes through each of these points. Each of these three screws intersects the screw σ and is reciprocal to it. It follows by Art. 294 that each of these is either perpendicular to σ or has a pitch equal and opposite to that of σ. But since the pitch ρ of a screw on the cylindroid is $p \cos^2 \theta + p' \sin^2 \theta$ there are only two different screws on the same cylindroid of the same pitch, viz. those given by supplementary values of θ. Hence the screw σ must intersect one of the three screws at right angles. Also, as it cannot be perpendicular to more than one screw on the cylindroid (unless it is the nodal line or z axis), the pitches of the two remaining screws must be each equal and opposite to that of σ.

297. Ex. 1. Show that the locus of a screw reciprocal to four screws (no three of which are on the same cylindroid) is a cylindroid.

Since a screw is determined by five quantities it is clear that, when the four conditions of reciprocity are fulfilled, the screw must *in general* be confined to a certain ruled surface. If this surface be not a cylindroid, pass a cylindroid

---

* The theory of reciprocal screws is due to Sir R. Ball and the substance of Arts. 294 to 297 is taken from his book on Screws. To this work the reader is referred for further development.

through any two of its generators, then any screw on this cylindroid will also be reciprocal to the four given screws. The locus therefore would be, not a single ruled surface, but a system of cylindroids.

Ex. 2. Prove that there is in general but one screw reciprocal to five given screws. [As there are five conditions to be satisfied the number of screws is finite. But if there were as many as two there would be a cylindroidal locus of screws.]

Ex. 3. Prove that any two reciprocal screws on the same cylindroid are parallel to conjugate diameters of the pitch conic.

Let ρ, ρ' be the pitches, z, z' the altitudes. Let $z > z'$ and $\theta > \theta'$; Art. 293. It will be seen that a force acting along the positive direction of the axis of either screw would tend to produce rotation round the axis of the other in the *negative* direction. We therefore put $h = z - z'$, $\phi = -(\theta - \theta')$. The condition that the screws are reciprocal is $(\rho + \rho') \cos \phi + h \sin \phi = 0$, Art. 294. Substituting for ρ, ρ', z, z' their values given in Art. 287, this reduces to $p \cos \theta \cos \theta' + p' \sin \theta \sin \theta' = 0$. This is the condition that the axes of the screws are parallel to conjugate diameters of the pitch conic, Art. 288.

On Conjugate Forces.

298. The nul plane. *The locus of all the straight lines, drawn through a given point O, and such that the moment of the system about each vanishes is a plane.*

This plane is called the *nul plane* of O and the point O is called the *nul point* of the plane. Any line about which the moment of the forces is zero is called a *nul line*.

To prove this proposition let us represent the system by a couple G and a force R at O as base. It is at once evident that the moment about a straight line through O cannot be zero unless it lies in the plane of the couple. *The nul plane may therefore also be defined as the plane of the principal couple at O.*

The names *nul-point* and *nul-plane* are due to Moebius, *Lehrbuch der Statik*, 1837. Instead of these the terms *pole* and *polar plane* have been used by Cremona, *Reciprocal Figures*, 1872, translated into French, 1885, into English, 1890. The term *focus* has also been used by Chasles, *Comptes Rendus*, 1843.

299. If any straight line in the nul plane of O and not passing through O were a nul line, the moment of R about it would be zero. This requires that R should either be zero or lie in the nul plane. In the former case the system of forces is equivalent to a single couple, and the nul plane is parallel to the plane of the couple. In the latter, the system is equivalent to a single force, and the nul plane passes through its line of action. In both cases the invariant I of the system is zero.

300. *If the nul plane of a point A passes through another point B, the nul plane of B passes through the point A.*

It follows from the definition of the nul plane of the point A that the straight line AB is a nul line. Hence also the line AB must lie in the nul plane of B.

301. *To find the equation to the nul plane of a given point $(\xi\eta\zeta)$ referred to any system of rectangular axes.*

It is clear that the direction cosines of the plane are proportional to the moments of the forces about axes meeting at the nul point. Hence by Art. 258 the required equation is

$$(L-\eta Z+\zeta Y)x+(M-\zeta X+\xi Z)y+(N-\xi Y+\eta X)z=L\xi+M\eta+N\zeta.$$

Any straight line being given by its equations $(x-f)/l=(y-g)/m=(z-h)/n$, prove that it will be a nul line if
$$\begin{vmatrix} f & g & h \\ X & Y & Z \\ l & m & n \end{vmatrix}=Ll+Mm+Nn.$$

302. *To find the nul point of a given plane* we choose two points conveniently situated on it. The nul planes of these points intersect the given plane in the required nul point. Art. 300.

Ex. 1. If the system be referred to the central axis as the axis of z, prove that the coordinates of the nul point of the plane $z=Ax+By+C$ are $\xi=-pB$, $\eta=pA$, $\zeta=C$, where p is the pitch of the equivalent wrench.

Ex. 2. A plane intersects the central axis in C and makes an angle ϕ with that axis. Show by reasoning similar to that of Art. 270, that the nul point O lies in a straight line CO drawn perpendicular to the central axis so that $CO=\cot\phi\,.\,\Gamma/R$.

Ex. 3. The moments of the forces about the sides of a triangle ABC are respectively M_1, M_2, M_3, and Z is the resolved force perpendicular to the plane of the triangle. Prove (1) that the trilinear coordinates of the nul point O of the plane referred to the triangle ABC are M_1/Z, M_2/Z, M_3/Z; (2) that the nul planes of the three corners A, B, C intersect the plane of the triangle in AO, BO, CO respectively.

303. Conjugate forces. Let O be any point on a given straight line OA. Let the system be reduced to a couple G and a force R at O as base. Pass a plane through R and the given straight line OA, and let it cut the plane BOC of the couple in OB.

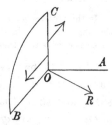

Let us resolve the force R by *oblique resolution* into two forces, one of which F acts along OA and the other F' acts along OB. This force F' may be compounded with the forces of the couple into a single force which also acts in the plane of the couple. Its line of action is parallel to OB and distant G/F' from it. It follows that *all the forces of the system are equivalent to some force F acting along any assumed straight line OA together with a second force F' which acts in the nul plane of the point O.* The forces are given by $F \sin AOB = R \sin ROB$, $F' \sin AOB = R \sin ROA$.

The forces F, F' are called *conjugate forces*, and their lines of action *conjugate lines*.

304. Since O is any point on the straight line OA, it follows that *when O travels along a straight line, the nul plane of O always passes through the conjugate and turns round it as an axis.*

305. *Vanishing of the Invariant I.* When the force R is zero or lies in the nul plane BOC, the system reduces to either a single couple or a single force. In both these cases every point in the plane BOC is a nul point.

If the system is equivalent to a single couple $R=0$, and if the assumed line OA is inclined to the plane of the couple the force F along it is zero; the conjugate is at infinity and its force also is zero. If OA is in the plane of the couple, the force along it forms one force of the couple while the conjugate is the other force, the distance between the conjugates, i.e. the arm of the couple, being arbitrary.

If the system is equivalent to a single resultant, OR lies in the plane BOC. If the assumed line OA does not intersect the single force, the force F along OA is zero, the conjugate being the single resultant. If OA intersects the single resultant, the conjugate is any line in their plane passing through that intersection, the conjugate forces being found by resolving the single resultant in their directions.

Conversely, since $I = FF'r \sin \theta$, (Art. 281) we see that *when the invariant is zero either one conjugate force is zero, or the two conjugates lie in one plane.*

306. *To find the conjugate of a nul line.* In this case OA lies in the nul plane of O, and if R is not zero and does not also lie in that plane the straight lines OA, OB, are opposite to each other, Art. 303. The components of R, viz. F and F', are therefore both infinite so that the two forces F, F' act in opposite directions along the same straight line OA. Such lines may therefore be called *self-conjugate*. They have also been called *double lines* by Cremona.

In the limiting case when the invariant I is zero, any line lying in the plane of the single couple or intersecting the single resultant is a line of nul moment. We have seen above that their conjugates are indeterminate.

307. It has been proved that the conjugate of every line passing through a given point O lies in the nul plane of O, we shall now show that *the conjugate of every straight line in that plane passes through the nul point.*

It is evident that if one conjugate intersect a line of nul

moment, the other conjugate must either intersect that line or its force must be zero. Now the nul lines of the plane BOC radiate from O and are intersected by any chosen line DE in that plane. It follows that the conjugate of DE must also intersect them or its force must be zero. If I is finite the conjugate force cannot also lie in that plane or be zero, it must therefore pass through the nul point O. If $I = 0$ every point in the plane is a nul point and the theorem is again true.

308. *To find the equation of the conjugate of the given line*

$$(x - f)/l = (y - g)/m = (z - h)/n \dots\dots\dots\dots(1).$$

It follows from Art. 304, that if any two points O, O' are chosen on the given line OA, their nul planes intersect on the conjugate. The nul planes of the point (fgh) and of another point at infinity whose coordinates are proportional to l, m, n are (Art. 301) respectively

$$(L - gZ + hY)x + (M - hX + fZ)y + (N - fY + gX)z = Lf + Mg + Nh$$
$$(-mZ + nY)x + (-nX + lZ)y + (-lY + mX)z = Ll + Mm + Nn.$$

These are the equations to the conjugate. They also take the form

$$\begin{vmatrix} x, & y, & z \\ X, & Y, & Z \\ f, & g, & h \end{vmatrix} = L(f - x) + M(g - y) + N(h - z), \qquad \begin{vmatrix} x, & y, & z \\ X, & Y, & Z \\ l, & m, & n \end{vmatrix} = Ll + Mm + Nn.$$

The line of action of the force F being given as above by the equations (1), an analytical expression for the magnitude of F can be found which may be used when the position and magnitude of the conjugate force F' are not required. If we reverse the force F and join it to the given system, the compound system will be equivalent to a single force. The invariant of the compound system is therefore equal to zero. If l, m, n are the actual direction cosines of the given line of action of the force F, the components of the compound system are

$$X' = X - Fl, \qquad\qquad L' = L + Fmh - Fng,$$
$$Y' = Y - Fm, \qquad\qquad M' = M + Fnf - Flh,$$
$$Z' = Z - Fn, \qquad\qquad N' = N + Flg - Fmf.$$

Equating the invariant $L'X' + M'Y' + N'Z'$ to zero, we find

$$\frac{LX + MY + NZ}{F} = Ll + Mm + Nn - \begin{vmatrix} f, & g, & h \\ X, & Y, & Z \\ l, & m, & n \end{vmatrix}.$$

In this manner a unique value of F has been found. The value of F can be infinite when the right-hand side is zero; this occurs when the given line is a nul line, Art. 301.

The value of F being known, all the six components of the compound system are known. The magnitude and line of action of the single resultant F' may then be found by equations (4) of Art. 273, whence $F'^2 = X'^2 + Y'^2 + Z'^2$ and $\Gamma = 0$.

309. *To determine the arrangement of the conjugate forces about the central axis.*

We know by Art. 285 that the central axis intersects at right angles the shortest distance between any two conjugates. Let Oz be the central axis; R, Γ, the given force and couple. Let F, F', be two con- jugate forces acting along AF, $A'F'$; AA' being the shortest distance be- tween them. Let $OA = a$, $OA' = a'$ measured positively from O in oppo- site directions, $h = a + a'$.

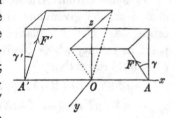

The force R may be replaced by two parallel forces acting at A, A', respectively equal to Ra'/h and Ra/h, Art. 79. The couple Γ is equivalent to two forces acting at the same points parallel to the axis of y equal to $\pm \Gamma/h$. Since the forces acting at A, A' have F, F' for their resultants, we find

$$\begin{array}{ll} \Gamma = Ra' \tan \gamma, & F^2 h^2 = \Gamma^2 + R^2 a'^2 \\ \Gamma = Ra \tan \gamma', & F'^2 h^2 = \Gamma^2 + R^2 a^2 \end{array} \Bigg\} \dots\dots (1).$$

When any arbitrary line AF is chosen as the seat of one force, a and γ are given; these equations then determine F, F', γ', a'. We notice also that since the resolved parts of F, F' in the plane xy are equivalent to the couple Γ, $F \sin \gamma = F' \sin \gamma' = \Gamma/h$.

310. If the figure is turned round Oz as an axis of revolution, the conjugates AF, $A'F'$ describe co-axial hyperboloids of revolution whose real axes a, a' are connected by the equations (1). The imaginary axes are $a \cot \gamma$ and $a' \cot \gamma'$; it is easily seen from (1) that each of these is equal to aa'/p where $p = \Gamma/R$ is the pitch of the wrench.

311. It may be a simpler classification to arrange the conjugate forces in a series of planes rather than in hyperboloids. *If the force F' is turned round A so as to describe a plane normal to OA,* the angle γ varies while a is constant. The formulæ (1) then show that γ' is constant, so that *the conjugate F' moves parallel to itself and generates a second plane which passes through OA.* The two planes intersect in a nul line, whose locus when a varies is the paraboloid $pz = -xy$ where p is the pitch of the wrench.

Ex. Any two systems of forces being given show that they will have one common system of conjugate lines real or imaginary. If $OO' = 2c$ is the shortest distance between the axes of the equivalent wrenches, C the middle point of OO', prove that the distances of the common conjugates from C are given by the quadratic $x^2 + (p - p') \cot \theta x + pp' - c^2 - (p + p') c \cot \theta = 0$ where p, p' are the pitches and θ the angle between the axes.

312. Ex. 1. If two straight lines intersect in a point O, their conjugates also intersect, and lie in the nul plane of O. Art. 303.

Ex. 2. A transversal intersects a force and its conjugate. Prove that each intersection is the nul point of the plane which contains the transversal and the other force.

For every straight line drawn through one intersection to cut the other force is a nul line, see also Art. 303.

Ex. 3. The locus of a straight line drawn through a given point O so that the moments about it of two conjugate forces F, F' have a given ratio μ is a plane, which becomes the nul plane of O when $\mu = -1$. Whatever the forces and μ may be, this plane passes through the intersection of the two planes drawn from O to contain the forces, and makes angles ϕ, ϕ' with these two planes such that the given ratio μ is equal to $Fp \sin \phi : F'p' \sin \phi'$. Here p and p' are the perpendicular distances of O from the given straight lines.

313. Ex. 1. Two arbitrary points A, B are taken on a nul line. Prove that the system can be reduced to two conjugate forces acting at A and B, the force at A making a given angle ϕ with AB. Prove also that if ϕ is varied, the locus of the force at each point is the nul plane of the other point.

If ϕ, ϕ' are the angles the conjugate forces make with AB, prove that $G \cot \phi' \pm G' \cot \phi = aX$, where G, G', are the principal couples at A, B, X the force along AB and $a = AB$.

To prove this take A as base (Art. 257) and change the couple G into another whose forces pass through A and B.

Ex. 2. Two planes being given which intersect in a nul line, show that the system can be reduced to two conjugates, one in each plane. [Take A, B of Ex. 1 at the nul points of the planes.]

Ex. 3. If AM, BN are two nul lines, show that the system can be reduced to two finite conjugate forces intersecting both AM, BN.

Let A be any point on AM, the nul plane of A will pass through AM and cut BN in some point B. The rest follows from Ex. 1.

314. The *characteristic of a plane* is the conjugate of the normal at the nul point, Chasles, *Comptes Rendus*, 1843.

Ex. 1. Any two conjugates intersect a plane in M and M': show that MM' passes through the nul point of that plane. Show also that the projections of these conjugates on the plane intersect in the characteristic. [Chasles' theorem.]

Ex. 2. The locus of the axes of the principal couples at all bases situated on a given straight line is a hyperbolic paraboloid. This paraboloid is a plane when the straight line can be a characteristic, and in this case the envelope of the axes of the principal couples is a parabola whose focus is the pole of the plane. [Chasles.]

Let AB be the straight line, CD its conjugate. The axis of the principal couple at any point O on AB is perpendicular to the plane OCD, Art. 303. If the straight line AB were turned round CD as an axis of rotation through any small angle $d\theta$, each point O on AB would move a small space perpendicular to the plane OCD, i.e. it would move a small space along the axis of the principal couple. Hence these axes all intersect two straight lines, viz. AB and its consecutive position, and are all parallel to a plane which is perpendicular to CD. The locus is therefore a hyperbolic paraboloid.

Theorems on forces.

315. Three forces. *If three forces are in equilibrium, they must lie in one plane.*

Let A and B be any two points on two of the forces. Since the moment about the straight line AB is zero, this straight line must intersect the third force in some point C. Let A be fixed and let B move along the second line; the straight line AB will describe a plane, and the second and third forces must lie in this plane. If we fix C and let B move as before, we see that the first force must also lie in the same plane.

Ex. 1. The forces of a system can be reduced to three forces F_1, F_2, F_3 which act along the sides of an arbitrary triangle ABC together with three other forces Z_1, Z_2, Z_3 which act at the corners A, B, C at right angles to the plane of the triangle.

Resolve each force P of the system into two, one in the plane ABC and the other perpendicular to that plane. The former can be replaced by three forces acting along the sides (Art. 120, Ex. 2), and the latter by three parallel forces at the corners (Art. 86, Ex. 1). If P is parallel to the plane ABC we can transfer it to act in the plane by introducing a couple. Turning the couple round in its own plane we can include its forces among those normal to ABC.

Ex. 2. The forces of a system can be reduced to three forces which act at the corners of an arbitrary triangle and satisfy three other conditions.

Replace F_1 by F_1+u at B and $-u$ at C; F_2 by F_2+v at C and $-v$ at A; F_3 by F_3+w at A and $-w$ at B. Compounding the forces at the corners, the arbitrary quantities u, v, w may be used to satisfy three conditions.

Ex. 3. A system of forces is reduced to three acting at fixed points A, B, C. If the force at A is fixed in direction, prove that each of the other two lies in a fixed plane. Show also that these planes intersect along the side BC.

[Coll. Ex., 1891.]

316. Four forces. *If four non-intersecting forces are in equilibrium, they must be generators of the same system of a hyperboloid.* Mœbius, *Lehrbuch der Statik.*

If a straight line move so as always to intersect three given straight lines, called directors, the locus is known to be a hyperboloid and the different positions of the moving straight line form one system of generators. An infinite number of transversals can be drawn to cut three of the forces, but each must intersect the fourth force also, for otherwise the moment of the four forces about that transversal is not zero. Taking any three of these transversals as directors, the four forces lie on the corresponding hyperboloid.

The following theorems will serve as examples, as the proofs are only briefly given.

Ex. 1. If n forces act along generators of the same system and have a single resultant, prove by drawing transversals that the resultant acts along another generator of the same system.

Ex. 2. When two of the forces P, P', act along generators of one system and two Q, Q', along generators of another system, they form a skew quadrilateral. The properties of such a combination of forces have been already considered in Art. 103. Their invariants are given in Arts. 317 and 323.

Prove, by drawing transversals through the intersection of P and Q', that the forces cannot be in equilibrium except when they lie in one plane.

Ex. 3. When three of the forces P_1, P_2, P_3, act along generators of one system and the fourth Q along a generator of the other system, prove that they cannot be in equilibrium except when all the forces lie in a plane. For if every transversal of P_1, P_2, P_3 could intersect Q, this last would intersect all the generators of its own system.

Ex. 4. Four forces act along generators of the same system of a hyperboloid. Their magnitudes are such that if transferred parallel to themselves to act at a point they would be in equilibrium. Prove that they are in equilibrium when acting along the generators.

Let Q be any generator of the other system, which therefore intersects the four forces. Transfer the forces to act at any point of Q, then the transferred forces are in equilibrium and the axes of the four couples thus introduced are perpendicular to Q. The four forces are therefore equivalent to a resultant couple such that either its moment is zero or its axis is perpendicular to *every* position of Q. The latter supposition is impossible. Plücker and Darboux.

Ex. 5. If four forces P_1, P_2, P_3, P_4 are in equilibrium, prove that the invariant of any two is equal to that of the remaining two (this theorem is due to Chasles). Also the invariant of any three of the forces is zero.

Reversing the directions of P_3, P_4, the forces P_1, P_2 become equivalent to P_3, P_4. Their invariants are therefore equal.

Ex. 6. Four forces acting along the straight lines a, b, c, d are in equilibrium. If the symbol ab represent the product of the shortest distance between a, b into the sine of the angle between them, show that the forces acting along these lines are proportional to $(bc \cdot cd \cdot db)^{\frac{1}{2}}$, $(cd \cdot da \cdot ac)^{\frac{1}{2}}$, $(da \cdot ab \cdot bd)^{\frac{1}{2}}$, $(ab \cdot bc \cdot ca)^{\frac{1}{2}}$.

[Cayley, *Comptes Rendus*, 1865.]

We have by Chasles' theorem $P_1P_2 \cdot ab = P_3P_4 \cdot cd$ and $P_1P_3 \cdot ac = P_2P_4 \cdot bd$. Multiplying these together we have the ratio of $P_1^2 : P_4^2$.

317. *Analytical discussion of the hyperboloid.* Refer the system to the axes of the hyperboloid as coordinate axes, and let $a, b, c\sqrt{-1}$, be these axes. Let any generator be

$$\frac{x - a\cos\theta}{a\sin\theta} = \frac{y - b\sin\theta}{-b\cos\theta} = \frac{z}{\pm c},$$

where θ is the eccentric angle of the intersection with the plane of xy, and the generator belongs to one system or the other according to the sign of c. Let P be the force along this generator, X, Y, Z, L, M, N its six components. We see that

$$X = \pm\frac{a}{c}Z\sin\theta, \quad Y = \mp\frac{b}{c}Z\cos\theta, \quad L = bZ\sin\theta, \quad M = -aZ\cos\theta, \quad N = \mp\frac{ab}{c}Z$$

where all the upper signs are to be taken together.

Ex. 1. If four forces act along generators of the same system prove that the six equations of equilibrium reduce to the three $\Sigma Z\sin\theta = 0$, $\Sigma Z\cos\theta = 0$, $\Sigma Z = 0$. This gives an analytical proof of the theorem in Art. 316, Ex. 4.

Ex. 2. Prove that the invariant I of two forces which act along generators of the same system is $I = \mp\dfrac{2ab}{c}Z_1 Z_2 \text{versin}(\theta_1 - \theta_2)$. If the forces act along generators of different systems, their invariant is zero because the generators intersect. If forces act along several generators, the invariant is the sum of the invariants taken two and two, Art. 281.

Ex. 3. When four generators of the same system are given, the ratios of the equilibrium forces are given by

$$\frac{Z_1^2}{\text{vers}(\theta_2 - \theta_3)\,\text{vers}(\theta_3 - \theta_4)\,\text{vers}(\theta_4 - \theta_2)} = \frac{Z_2^2}{\text{vers}(\theta_3 - \theta_4)\,\text{vers}(\theta_4 - \theta_1)\,\text{vers}(\theta_1 - \theta_3)} = \&c.$$

These may be obtained by equating the invariants two and two, as in the proof of Cayley's theorem, Art. 316.

Ex. 4. Four forces in equilibrium act along four generators of a hyperboloid and intersect the plane of the real axes in A_1, A_2, A_3, A_4. Show that the resolved parts of the forces parallel to the imaginary axis are proportional to the areas of the triangles $A_2 A_3 A_4$, $A_3 A_4 A_1$ &c., the forces at adjacent corners of the quadrilateral $A_1 A_2 A_3 A_4$ having opposite signs.

Ex. 5. Forces act along generators of the same kind, say c positive. Prove that the pitch p of the equivalent screw lies between $-ab/c$ and the greater of the quantities bc/a and ca/b. For $p = \dfrac{I}{R^2} = \dfrac{\Sigma L\,.\,\Sigma X + \&c.}{(\Sigma X)^2 + \&c.} = abc\,\dfrac{\eta^2 + \xi^2 - 1}{a^2\eta^2 + b^2\xi^2 + c^2}$ where ξ, η, have been written for $\Sigma Z\cos\theta/\Sigma Z$ and $\Sigma Z\sin\theta/\Sigma Z$. We see at once that $p + ab/c$ is positive and $p - bc/a$ negative if $b > a$.

Ex. 6. Forces act along generators of the same system and the pitch p of the equivalent wrench is given. Prove that the central axis is that generator of the concyclic hyperboloid

$$\left(\frac{bc}{a} - p\right)x^2 + \left(\frac{ca}{b} - p\right)y^2 - \left(\frac{ab}{c} + p\right)z^2 = \left(\frac{bc}{a} - p\right)\left(\frac{ca}{b} - p\right)\left(\frac{ab}{c} + p\right)$$

which intersects the plane of xy in the point

$$x = \frac{ac - bp}{c}\,\frac{\Sigma Z\cos\theta}{\Sigma Z}, \quad y = \frac{bc - ap}{c}\,\frac{\Sigma Z\sin\theta}{\Sigma Z}.$$

Ex. 7. Forces act along generators of the same system and admit of a single resultant, which intersects the plane of *xy* in *D*. Prove that *OD* and the projection of the resultant force are parallel to conjugate diameters.

Ex. 8. Forces act upon a rigid body along generators of the same system of a hyperboloid. Prove that the necessary and sufficient condition of their being reducible to a single resultant is that their central axis should be parallel to one of the generating lines of the asymptotic cone.   [Math. Tripos, 1877.]

Ex. 9. A system of forces have their directions along any non-intersecting generators of a hyperboloid of one sheet; show that the resultant couple at the centre of the hyperboloid lies in the diametral plane of the resultant force, and the least principal moment is $\dfrac{abcR}{a^2+b^2-c^2-D_1{}^2-D_2{}^2}$; D_1 and D_2 being the semi-axes of the section of the hyperboloid by the plane of the couple, and a, b, c the semi-axes of the surface, and R the resultant force. Explain the difficulty in the geometrical interpretation of these results for a single force.   [Math. Tripos, 1880.]

318. *Relation of four forces to a tetrahedron.* Ex. 1. Forces act at the centres of the circles circumscribing the faces of a tetrahedron perpendicular to those faces and proportional to their areas. Prove that they are in equilibrium if they act either all inwards or all outwards.

Ex. 2. Forces act at the corners of a tetrahedron perpendicularly to the opposite faces and proportional to their areas. Prove that they are in equilibrium if they act either all inwards or all outwards.   [Math. Tripos, 1881.]

Let *ABCD* be the tetrahedron, *AK*, *BL* &c. the perpendiculars. Since the product of each perpendicular into the area of the corresponding face is equal to three times the volume of the tetrahedron, the forces are inversely proportional to the perpendiculars along which they act. Let the forces be μ/AK, μ/BL &c.

Let us resolve the force μ/AK into three components which act along the edges *AB*, *AC*, *AD*. The component *F* which acts along *AB* is found by equating the resolutes perpendicular to the plane *ACD*. This gives $F\dfrac{BL}{AB} = \dfrac{\mu}{AK}\cos\theta$, where θ is the angle between the perpendiculars *AK* and *BL*. In the same way we resolve the force μ/BL into components along the edges. The component *F′* which acts along *BA* is found from $F'\cdot\dfrac{AK}{AB} = \dfrac{\mu}{BL}\cos\theta$. Hence *F* and *F′* are equal and opposite forces. In the same way it may be shown that the forces along all the other edges are equal and opposite. The system is therefore in equilibrium.

Ex. 3. Forces act at the centres of gravity of the four faces of a tetrahedron perpendicularly to those faces and proportional to them in magnitude, all inwards or all outwards. Prove that they are in equilibrium.

Joining the centres of gravity we construct an inscribed tetrahedron, the faces of which are parallel to those of the former and proportional to them in area. The given forces act at the corners of this new tetrahedron and are therefore in equilibrium by Ex. 2.

Ex. 4. Forces act at the centres of gravity of the faces of a closed polyhedron in directions perpendicular to the faces and proportional to their areas in magnitude. Prove that they are in equilibrium.

Divide each face into triangles by drawing a sufficient number of diagonals. By joining any internal point *P* to the several corners we divide the polyhedron into tetrahedra. Forces acting at the centres of gravity of the faces of each tetrahedron are in equilibrium by Ex. 3. Removing the equal and opposite forces which act at the centre of gravity of each internal face, the forces which act at the external faces must be in equilibrium.

Ex. 5. Forces act at the middle points of the edges of a closed polyhedron, in directions bisecting the angles between the adjacent faces, and having magnitudes proportional to the product of the length of the edge by the cosine of half the angle between the faces. Prove that they are in equilibrium.

Let forces act at the middle points of the sides of each face in the plane of the face perpendicularly to and proportional to the sides. These are in equilibrium by Art. 37. Compounding the forces at each edge the theorem follows.

319. *Normal forces on surfaces.* Ex. 1. Forces act normally at every element of a closed surface. Prove that they are in equilibrium if each force is either (1) proportional to the area of the element, or (2) proportional to the product of the area by $\frac{1}{\rho} + \frac{1}{\rho'}$ where ρ, ρ' are the principal radii of curvature.

Since the surface may be regarded as the limiting case of a polyhedron, the first theorem follows from Ex. 4.

By drawing the lines of curvature the surface may be divided into rectangular elements which may be regarded as the faces of a polyhedron. The second theorem then follows from Ex. 5. Let $ABCD$ be any element, the external angle between the faces which meet in BC is AB/ρ. The force across this edge is therefore $\frac{1}{2}BC \cdot AB/\rho$ and ultimately acts perpendicularly to the element.

M. Joubert deduces the second of these theorems from the first. He also deduces from the second that normal forces proportional to the quotient of each elementary area by $\rho\rho'$ are in equilibrium. *Liouville's J.* vol. XIII., 1848.

Ex. 2. One-eighth of an ellipsoid is cut off by the principal planes, and along the normal at any point a force acts proportional to the element of surface at that point. Show that all these forces are equivalent to a single force acting along the line $a\,(x - 4a/3\pi) = b\,(y - 4b/3\pi) = c\,(z - 4c/3\pi)$, where $2a$, $2b$, $2c$ are the principal axes of the ellipsoid. [June Exam.]

320. Five forces. *If five finite non-intersecting forces are in equilibrium, they must intersect two straight lines which may be real or imaginary.* Mœbius.

First, we shall prove that any four straight lines a, b, c, d can be cut by two transversals. For, describing the hyperboloid which has a, b, c for directors we notice that the line d cuts this hyperboloid in two points real or imaginary. One generator of the system opposite to a, b, c passes through each of these points and therefore intersects the straight lines a, b, c as well as d. Assuming this lemma we draw the two transversals of any four of the forces. Each of these must intersect the fifth force, for otherwise the moments about them would not be zero. These two transversals may be called *the directors* of the five forces.

321. *Let the shortest distance between two straight lines be taken as axis of z. Let any five forces intersect these straight lines at distances $(r_1 r_1')$ $(r_2 r_2')$ &c. from that axis, and let Z_1, Z_2 &c. be the z resolutes of these forces respectively. Prove that the conditions of equilibrium are $\Sigma Z = 0$, $\Sigma Zr = 0$, $\Sigma Zr' = 0$, $\Sigma Zrr' = 0$.*

Let the origin bisect the shortest distance between the two directors of the forces, and let this shortest distance be $2c$. Let 2θ be the angle between the directors, and let the axes of x and y be its bisectors. The equation to any force may then be written

$$(x - r\cos\theta)/(r - r')\cos\theta = (y - r\sin\theta)/(r + r')\sin\theta = (z - c)/2c.$$

Writing $\quad 1/\mu^2 = (r - r')^2 \cos^2\theta + (r + r')^2 \sin^2\theta + 4c^2,$

and representing the forces by $P_1 \ldots P_5$, the equations of equilibrium formed by resolving along the axes are

$$\Sigma P\mu(r - r')\cos\theta = 0, \quad \Sigma P\mu(r + r')\sin\theta = 0, \quad 2\Sigma P\mu c = 0.$$

The equations of moments are

$$\Sigma(yZ - zY) = \quad \Sigma P\mu(r - r')c\sin\theta = 0,$$
$$\Sigma(zX - xZ) = -\Sigma P\mu(r + r')c\cos\theta = 0,$$
$$\Sigma(xY - yX) = \quad 2\Sigma P\mu rr'\sin\theta\cos\theta = 0.$$

When c and $\sin 2\theta$ are not zero, these six equations reduce to the four given above. These four equations determine the ratios of the five forces $P_1 \ldots P_5$ when the intersections of their lines of action with the directors are known.

322. Let the two directors be moved so that either their mutual inclination 2θ or their distance apart $2c$ is altered, but let them continue to intersect the axis of z at right angles. It follows from these results that equilibrium will continue to exist provided (1) the forces always intersect the directors at the same distances from the axis of z, and (2) the z component of each is unchanged.

When five forces in equilibrium are given in one plane, which besides the three conditions of equilibrium also satisfy the condition $\Sigma Zrr' = 0$, we may by this theorem construct five forces in space which are also in equilibrium.

323. Ex. 1. Any number of forces intersect two directors in the points $ABC\ldots$, $A'B'C'\ldots$, prove that the invariant $I = \sin 2\theta \Sigma Z_1 Z_2 . AB . A'B'/2c$.

Ex. 2. Four forces act along the sides of a skew quadrilateral taken in order and their magnitudes are respectively α, β, γ, δ times the sides along which they act, as in Art. 103, Ex. 5. Prove that the invariant $I = 2c \sin 2\theta (\alpha\gamma - \beta\delta) DD'$ where D, D' are the lengths of the diagonals, $2c$ their shortest distance and 2θ the angle between them.

Ex. 3. Any number of forces intersect two directors and a plane is drawn through each parallel to the other. Find the coordinates of the points in which the central axis intersects these planes. The result is given in Art. 278, Ex. 7.

Ex. 4. Five forces in equilibrium intersect their two directors in the points $ABCDE$ and $A'B'C'D'E'$, and their magnitudes are $a \cdot AA'$, $\beta \cdot BB'$, &c. Prove (1) that the sum of the coefficients a, β, &c. is zero and (2) that

$$\frac{1}{a} \begin{vmatrix} CD \cdot BE, \ DB \cdot CE \\ C'D' \cdot B'E', \ D'B' \cdot C'E' \end{vmatrix} = \frac{1}{\beta} \begin{vmatrix} DE \cdot CA, \ EC \cdot DA \\ D'E' \cdot C'A', \ E'C' \cdot D'A' \end{vmatrix} = \&c. \qquad \text{[Coll. Ex., 1892.]}$$

Ex. 5. Show that the force along AA' is zero when the other four lines cut the two directors in the same anharmonic ratio. This is also a known property of any four generators of a hyperboloid intersected by two fixed lines.

Ex. 6. Show that, if the algebraic sums of the moments of a system of forces about (1) three, (2) four, (3) five straight lines are zero, the central axis of the system (1) lies along one of the generators of a system of concyclic hyperboloids, (2) intersects a fixed straight line at right angles, (3) is fixed. [Math. Tripos, 1888.]

Replace the system by two conjugate forces, one of which cuts the three given straight lines. Then the other force also cuts the same three lines. They are therefore rectilinear generators of a fixed hyperboloid. The first result follows at once by Art. 317, Ex. 6.

Choose one of the conjugates to cut the four given straight lines as in Art. 320. The other also cuts the same four lines. Both these forces are therefore fixed in position. By Art. 285 the central axis cuts the shortest distance between these at right angles.

If the moments about five straight lines are zero, we can by taking two sets of four forces obtain two straight lines each of which is cut at right angles by the central axis. The central axis is therefore fixed.

324. Six forces*. Analytical view. *Forces acting along six straight lines are in equilibrium. Show that, five of these lines and a point on the sixth being given, the sixth line must lie on a certain plane.*

Let a force P be given by its six components Pl, Pm, Pn; $P\lambda$, $P\mu$, $P\nu$, Art. 260. If (fgh) be any point on its line of action, then $\qquad \lambda = gn - hm, \qquad \mu = hl - fn, \qquad \nu = fm - gl.$

Let us suppose that each of the six forces $P_1 \ldots P_6$ is given in this

---

* The theorem that the locus of the sixth force is a plane is due to Mœbius, *Lehrbuch der Statik*, 1837. But he omitted to give a construction for the plane. This defect was supplied by Sylvester "*sur l'involution des lignes droites dans l'espace considérées comme des axes de rotation.*" *Comptes Rendus*, 1861. He gives several theorems on the relative positions of the fifth and sixth lines. The terms "involution" and "polar plane" are due to him. In a second paper in the same volume he states as the criterion for the involution of six lines the determinant given in Art. 327, the moments (12) &c. being replaced by secondary determinants when the equations of the straight lines are given in their most general form. He mentions that Cayley had found a determinant which is the square root of that given by himself and which would do as well to define involution. A proof of this is given by Spottiswoode, *Comptes Rendus*, 1868. See also Scott's *Theory of Determinants*. Analytical and statical investigations connected with involution are given by Cayley, "*On the six coordinates of a line,*" *Cambridge Transactions*, 1867. The extension of the determinant of Art. 327 to six wrenches is given by Sir R. Ball, *Theory of Screws*, 1876.

way, so that $(l_1, m_1, n_1, \lambda_1, \mu_1, \nu_1)$ $(l_2,$ &c.$)$ &c. may be regarded as the coordinates of their several lines of action.

Since the six forces are in equilibrium, they must satisfy the six necessary and sufficient equations given in Art. 259. We have therefore

$$\Sigma Pl = 0, \quad \Sigma Pm = 0, \quad \Sigma Pn = 0; \quad \Sigma P\lambda = 0, \quad \Sigma P\mu = 0, \quad \Sigma P\nu = 0.$$

These six equations will in general require that each of the forces $P_1...P_6$ should be zero. But if we eliminate the ratios of these forces we obtain a determinantal equation which is the condition that the forces should be finite. This determinant has for its six rows the six coordinates of the six given straight lines, viz.

$$\begin{vmatrix} l_1, m_1, n_1, g_1 n_1 - h_1 m_1, h_1 l_1 - f_1 n_1, f_1 m_1 - g_1 l_1 \\ l_2, \&c. \end{vmatrix} = 0.$$

Let us suppose that five of the lines are given and that the sixth is to pass through a given point (f_6, g_6, h_6). Let (x, y, z) be the current coordinates of the sixth line, then writing for $(l_6 \ m_6 \ n_6)$ in the last row their ratios $x - f_6, y - g_6, z - h_6$ this determinantal equation becomes the equation to the locus of the sixth line. It is clearly of the first degree and this proves that the locus of the sixth line is a plane.

325. When six lines are so placed that forces can be found to act along them and be in equilibrium, *the six lines are said to be in involution*. The plane which is the locus of the sixth line when a point O in the line is given is called *the polar plane of O with regard to the five given lines*.

When five lines are so placed that forces can be found to act along them and be in equilibrium, they are in involution with every line taken as a sixth and the force along that sixth is zero. This is briefly expressed by saying that the five lines are in involution.

When lines are in involution any force acting along one of them can be replaced by finite components acting along the remaining lines, provided these remaining lines alone are not in involution.

326. *If the six straight lines are the seats of six wrenches of given pitches, instead of six forces, we may by an extension* *of this determinant form the condition that these wrenches may be in equilibrium.*

Let P be the force of any wrench, p the pitch of its screw. Let $(l, m, n, \lambda, \mu, \nu)$ be the six coordinates of its axis. Then, resolving parallel to the axes of coordinates and taking moments as before, we have

$$\Sigma Pl = 0, \qquad \Sigma Pm = 0, \qquad \Sigma Pn = 0.$$
$$\Sigma P(\lambda + pl) = 0, \quad \Sigma P(\mu + pm) = 0, \quad \Sigma P(\nu + pn) = 0.$$

Eliminating the forces, we have the following six-rowed determinantal equation in which the first line only is written down.

$$\begin{vmatrix} l_1, & m_1, & n_1, & \lambda_1 + p_1 l_1, & \mu_1 + p_1 m_1, & \nu_1 + p_1 n_1 \\ \cdot & \cdot & \cdot & \cdot & \cdot & \cdot \end{vmatrix} = 0.$$

The other lines are repetitions of the first with different suffixes. This determinant has been called the *sexiant* by Ball.

By giving to the pitches $p_1...p_6$ of these screws values either zero or infinity we can express the condition that m forces and n couples $(m + n = 6)$ connected with six given straight lines should be in equilibrium.

327. If we take moments in turn for the six forces $P_1...P_6$ about their lines of action, we obtain six equations of the form

$$P_1 \cdot 0 + P_2 (12) + P_3 (13) + P_4 (14) + P_5 (15) + P_6 (16) = 0,$$

where (12) represents the mutual moment of the lines of action of P_1, P_2 (Art. 264). Eliminating the six forces, we obtain a determinant of six rows equated to zero. This is the necessary condition that the six lines should be in involution.

Taking any five of these equations, we can find the ratios of the six forces. Thus, if I_{12} represent the minor of the constituent in the first row and second column, we have

$$P_1 / I_{11} = P_2 / I_{12} = P_3 / I_{13} = \&c.$$

Since by Salmon's higher algebra $I_{11} I_{22} = I^2_{12}$, we may deduce the more symmetrical ratios

$$P_1^2 / I_{11} = P_2^2 / I_{22} = P_3^2 / I_{33} = \&c.$$

This symmetrical form for the ratios of the forces is given by Spottiswoode in the *Comptes Rendus* for 1868.

328. *We have thus two determinants to define involution. One expresses the condition in terms of the coordinates of the six lines, the other in terms of their mutual moments.* These are not independent, for one determinant is the square of the other. This may be shown by squaring the first and remembering the expression for the mutual moment of two lines given in Ex. 1 of Art. 267.

329. *Let A, B, C, D, E, F be six lines not in involution, then any given force R may be replaced by six components acting along these six lines.*

Let $l'm'n'\lambda'\mu'\nu'$ be the six coordinates of the line of action of R. If $P_1...P_6$ are the six equivalent forces on the given lines, we have by Art. 324 $\Sigma Pl = Rl'$, &c., $\Sigma P\lambda = R\lambda'$, &c. These six equations will determine real values for $P_1...P_6$. *They will be finite* if the determinant of Art. 324 is not zero, i.e. if the given lines are not in involution.

We notice that the value of P_1 is zero if the determinant formed by replacing l_1, m_1, &c. in the first row by $l'm'$ &c. is zero, i.e. if the line of action of R is in involution with $BCDEF$.

Ex. Show that in general there is only one way of reducing a system of forces to six forces which act along six given straight lines. If the lines of action of five of the forces be given and the magnitude and point of application of the sixth, prove that the line of action of the sixth will lie on a certain right circular cone.

[Coll. Exam., 1887.]

330. *If the moments of a system of forces about six straight lines not in involution are zero, the forces are in equilibrium.*

If they are not in equilibrium let (Γ, R) be their equivalent wrench. Let the axis of this wrench be taken as the axis of z, and let the six lines make angles $(\theta_1, \phi_1, \psi_1)$, $(\theta_2, \phi_2, \psi_2)$, &c. with the axes of z, x, y. Let (r_1, r_1', r_1''), (r_2, r_2', r_3'') &c. be the shortest distances between the six lines and the axes of z, x, y.

Since each of the six lines must be a nul line with regard to the wrench, we have for each $\Gamma \cos\theta + Rr \sin\theta = 0$. We shall now prove that, if these six equations can be satisfied by values of Γ and R other than zero, the six lines are in involution.

If forces $P_1...P_6$ can be found acting along these six lines in equilibrium, they must satisfy the six necessary and sufficient equations of equilibrium. These are

$$\Sigma P \cos\theta = 0, \qquad \Sigma P \cos\phi = 0, \qquad \Sigma P \cos\psi = 0,$$
$$\Sigma Pr \sin\theta = 0, \qquad \Sigma Pr' \sin\phi = 0, \qquad \Sigma Pr'' \sin\psi = 0.$$

These six equations in general require that each of the forces $P_1...P_6$ should be zero. But when the six conditions given above are satisfied the two equations $\Sigma P \cos \theta = 0$ and $\Sigma Pr \sin \theta = 0$ follow one from the other. There are therefore only five necessary and sufficient equations connecting the six forces. The ratios of the forces can be found. Hence the lines must be in involution.

If the lines are not in involution, they cannot all six be nul lines of a wrench, i.e. Γ and R must both be zero. It follows that *six equations of moments about six straight lines are insufficient to express the conditions of equilibrium of a system if those six lines are in involution.*

331. *If a system of forces is such that its moment about each of m lines is zero, and its resolute along each of n lines is also zero, where $m + n = 6$, the system is in equilibrium, provided the six lines are such that forces acting along the m lines and couples having their axes placed along the n lines cannot be in equilibrium. The forces and couples are not to be all zero.*

For the sake of brevity, let us suppose that the moments of the system about each of the four lines 1, 2, 3, 4 is zero, and that the resolute along each of the lines 5 and 6 is zero. If the system is not in equilibrium, let (Γ, R) be the equivalent wrench. Let the axes of coordinates and the notation be the same as in Art. 330. We thus have given the four equations

$$\Gamma \cos \theta_1 + Rr_1 \sin \theta_1 = 0, \quad \Gamma \cos \theta_2 + Rr_2 \sin \theta_2 = 0, \quad \&c. = 0,$$

and the two resolutions $R \cos \theta_5 = 0.$ $R \cos \theta_6 = 0.$

These six equations may be called the equations (A).

Let four forces $P_1 \ldots P_4$ act along the four lines $1, \ldots 4$ and let two couples M_5, M_6 have their axes placed along the lines 5, 6. If these can be in equilibrium, they must satisfy the equations

$$P_1 \cos \theta_1 + \ldots + P_4 \cos \theta_4 = 0,$$

$$P_1 r_1 \sin \theta_1 + \ldots + P_4 r_4 \sin \theta_4 + M_5 \cos \theta_5 + M_6 \cos \theta_6 = 0,$$

with four other similar equations obtained by writing ϕ and ψ for θ. These six equations may be called the equations (B).

The equations (B) in general require that the four forces $P_1 \ldots P_4$ and the two couples M_5, M_6 should be zero. But if the equations (A) can be satisfied by values of Γ and R which are not both zero, the six equations (B) are not independent. If we multiply the first by Γ and the second by R and add the products together the sum is evidently an identity by virtue of equations (A). The equations (B) are therefore equivalent to not more than five equations, and thus forces $P_1...P_4$ and couples M_5, M_6, not all zero, may be found to satisfy them.

It follows that, if the six lines are such that the forces $P_1...P_4$ and the couples M_5, M_6 cannot be in equilibrium, the values of Γ and R given by equations (A) must be zero, i.e. the given system is in equilibrium.

332. If four of the six given lines are occupied by the axes of couples, the remaining two having only zero couples or zero forces, it is possible to so choose the four couples that equilibrium shall exist, Art. 99. It follows that *m equations of moments and n equations of resolution are insufficient to express the conditions of equilibrium if m is less than three.*

333. We may also deduce the theorem of Art. 331 from that of Art. 330 by placing some of the lines at infinity.

The expression for the moment of a system of forces about a straight line, drawn in the plane of xz parallel to x and at a distance l from it, is by Art. 258, $L'=L+lY$. If l be very great the condition $L'=0$ leads to $Y=0$. It follows that to equate to zero the resolved part of the forces along y is the same thing as to equate to zero their moment about a straight line perpendicular to y but very distant from it. Now a zero force along such a line at infinity is equivalent to a couple round the axis of y. Since the axis of y is any straight line, it follows that, if a system be such that its moments about m lines are each zero and its resolutes along n lines are also each zero, where $m+n=6$, then the system will be in equilibrium provided the six lines are such that m forces along the m lines and n couples round the n lines cannot be found which are in equilibrium.

334. Geometrical view. *Six forces are in equilibrium. When the lines of action of five are given, the possible positions of the sixth are the nul lines of two determinate forces acting along the two transversals of any four of the five.* From this we can deduce another proof of Mœbius' theorem.

Let us represent the lines of action of the forces $P_1 ... P_6$ by the numbers $1 ... 6$ and the mutual moments of the lines by the symbols (12), (34), &c. Art. 264.

Let a, b be the two transversals which intersect the four straight lines 1, 2, 3, 4 (Art. 320). Since the six forces $P_1...P_6$ are in equilibrium, the moment of P_5 and P_6 about each of these transversals is zero. Hence

$$P_5(5a) + P_6(6a) = 0, \qquad P_5(5b) + P_6(6b) = 0......(1).$$

Eliminating the ratio P_5/P_6, we have

$$(5b)(6a) - (5a)(6b) = 0 .... (2).$$

Thus the sixth line is so situated that the sum of the moments about it of two forces proportional to $(5b)$ and $(-5a)$ acting along a and b is zero. Let us call these forces P_a and P_b; hence

$$P_a(6a) + P_b(6b) = 0 \quad \dots\dots\dots\dots\dots\dots \quad (3).$$

We notice that the positions of the transversals a and b depend on the positions of the lines 1, 2, 3, 4, and are independent of the magnitudes of the corresponding forces. The ratio of the forces applied to these transversals depends on the position of the line 5 relatively to a and b. The transversals a, b and the lines 5, 6 are so related that a, b are nul lines of the forces P_5, P_6 and 5, 6 are nul lines of P_a, P_b.

It follows from this reasoning that when the forces $P_1 \dots P_6$ are varied, so that equilibrium always exists, the sixth line is always a nul line of P_a, P_b. Hence if any point O in the line of action of P_6 is given, that force must lie in the nul plane of O taken with regard to these two forces.

335. Any conjugate forces equivalent to P_a, P_b may also be used. Assuming, for example, any two points A and B, their nul planes with regard to these two forces will intersect in some straight line CD which is the conjugate of AB, Art. 308. *Any straight line intersecting AB and CD will be a nul line and is a possible position of the sixth force.*

336. The sixth line will remain in involution with the five given straight lines 1...5 as it revolves round O in the polar plane of O. The ratios of the forces $P_1...P_6$ will however change.

Let the straight line joining O to the intersection of its polar plane with the transversal a be taken as the sixth line. Then since the sixth line is a nul line of the forces which act along the transversals, it will also intersect the transversal b. Thus the *polar plane of O intersects the transversals a and b in two points which lie in the same straight line with O.*

The position in space of this straight line may be constructed when the four straight lines 1, 2, 3, 4 and the point O are known. Let it be called the line c of the point O with regard to the four lines 1, 2, 3, 4. To construct this line, we first find the two transversals a and b, we then pass a plane through O and each of these transversals. The intersection of these planes is the line c.

If we had begun by finding the two transversals a', b' of some other four of the five given lines say 1, 2, 3, 5, we must have arrived at the same plane as the polar plane of O. Thus by combining the forces in sets of four, we may arrive at five such lines as c. All these lie in the polar plane of O, and any two will determine that plane.

When the four lines 1, 2, 3, 4 and the point O are given, the fifth line being arbitrary, the polar plane of O passes through the fixed straight line c.

337. Since the forces $P_1...P_6$ are in equilibrium the moment of P_5 and P_6 about each of the transversals a, b is zero. Hence as in Art. 334

$$P_5(5a) + P_6(6a) = 0, \quad P_5(5b) + P_6(6b) = 0 \dots\dots\dots\dots\dots(1).$$

When the sixth line is in the position c, the moment of the sixth force about each of the transversals a and b is zero. When the sixth line has revolved in the polar plane of O from this position through an angle θ, the moment of the sixth force may be found by resolving P_6 into two forces, one along the line c and the other along a line d drawn perpendicular to c in the polar plane of O. The moment of the first is zero, that of the second is $(6a) = P_6 \sin\theta . (da)$ or $(6b) = P_6 \sin\theta . (db)$. It follows from either of the equations (1) that the *ratio* $P_5 : P_6$ *is proportional to* $\sin\theta$ *and is therefore greatest when the sixth line is perpendicular to* c.

We have assumed that the moments $(5a)$ and $(5b)$ are not both zero, i.e. that the five given straight lines are not so placed that they all intersect the same two straight lines; see Art. 320. When this happens the lines 1, 2, 3, 4, 5 alone are in involution. The equations (1) then show that the force P_6 is zero when its line of action does not intersect the same directors.

338. Ex. 1.   If A, B, C, D, E, F be six lines in involution, the polar plane of O with regard to A, B, C, D, E is the same as the polar plane of O with regard to A, B, C, D, F, the forces along E, F not being zero.

For let M be *any* straight line through O, then a force acting along M can be replaced by five forces along A, B, C, D, E. But the force along E can be replaced by forces along A, B, C, D, F, hence the force along M is equivalent to forces along A, B, C, D, F, i.e. M lies in the second polar plane. The two polar planes therefore coincide.

Ex. 2.   Supposing two transversals, say a and b, to be known, we may take with regard to these the convenient system of coordinates used in Art. 321. Let $2c$ be the shortest distance between the transversals, 2θ the angle between their directions. Let $(1 + \mu)/(1 - \mu)$ be equal to the known ratio $(5a) : (5b)$, i.e. to the ratio of the moments of the fifth force about the transversals a and b (Art. 334). Show that the polar plane of O is

$$x \sin\theta\,(h + \mu c) + y \cos\theta\,(\mu h + c) - z\,(f \sin\theta + \mu g \cos\theta) = c\,(\mu f \sin\theta + g \cos\theta).$$

This is obtained by substituting in (2) of Art. 334 the Cartesian expression for a moment given in Art. 266.

Tetrahedral Coordinates.

339.   *Show that the forces of any system can be reduced to six forces which act along the edges of any tetrahedron of finite volume.*

Let $ABCD$ be the tetrahedron, let any one force of the system intersect the face opposite D in the point D'. Resolve the force into oblique components, one along DD' and the other in the plane ABC. The former can be transferred to D and then resolved along the edges which meet at D. The second can by Art. 120 be resolved into components which act along the sides of ABC.

We shall suppose that the positive directions of the edges are AB, BC, CA, AD, BD, CD; the order of the letters being such that a positive force acting along any edge tends to produce rotation about the opposite edge in the same standard direction. See Art. 97. We shall represent the forces which act along these sides by the symbols $F_{12}, F_{23}, F_{31}, F_{14}, F_{24}, F_{34}$. The directions of the forces, when positive, are indicated by the order of the suffixes. When we wish to measure the forces in the opposite directions, the suffixes are to be reversed, so that $F_{12} = -F_{21}$.

The ratios of the forces F_{12} &c. to the edges along which they act will be represented by f_{12} &c. The volume of the tetrahedron is V.

Ex. 1. Show that the six straight lines forming the edges of a tetrahedron are not in involution. For, if forces acting along these could be in equilibrium we see, by taking moments about the edges, that each would be zero.

Ex. 2. A force P acts along the straight line joining the points H, K, whose tetrahedral coordinates are (x, y, z, u) (x', y', z', u') in the direction H to K. If this force is obliquely resolved into six components along the edges of the tetrahedron $ABCD$, show that the component F_{12} acting in the direction AB is $P\,\dfrac{AB}{HK}\cdot\begin{vmatrix} x, y \\ x', y' \end{vmatrix}$ where the terms in the leading diagonal follow the order indicated by the directions HK, AB, of the forces.

To prove this we equate the moments of F_{12} and P about the edge CD. The result follows from the expression for the moment given in Art. 267, Ex. 2.

Ex. 3. Two unit forces act along the straight lines HK, LM in the directions H to K and L to M. If the tetrahedral coordinates of H, K, L, M are respectively (x, y, z, u), $(x'$ &c.$)$, $(a, \beta, \gamma, \delta)$, $(a'$, &c.$)$, prove that the moment of either about the other in the standard direction is $\dfrac{6V\Delta}{HK\cdot MN}$ where Δ is the determinant in the margin. $\begin{vmatrix} x, & y, & z, & u \\ x', & y', & z', & u' \\ a, & \beta, & \gamma, & \delta \\ a', & \beta', & \gamma', & \delta' \end{vmatrix}$ The order of the rows is determined by the directions HK, LM in which the forces act; the order of the columns by the positive directions of the edges. This follows from Art. 266. Notice also that this expression is the invariant I of the two unit forces.

Ex. 4. The nul plane of the point whose tetrahedral coordinates are $(a, \beta, \gamma, \delta)$ with regard to the six forces F_{12} &c. is

$$f_{12}\begin{vmatrix} z, u \\ \gamma, \delta \end{vmatrix} + f_{21}\begin{vmatrix} x, u \\ a, \delta \end{vmatrix} + f_{31}\begin{vmatrix} y, u \\ \beta, \delta \end{vmatrix} + f_{14}\begin{vmatrix} y, z \\ \beta, \gamma \end{vmatrix} + f_{24}\begin{vmatrix} z, x \\ \gamma, a \end{vmatrix} + f_{34}\begin{vmatrix} x, y \\ a, \beta \end{vmatrix} = 0.$$

The nul plane of the corner D is $f_{23}x + f_{31}y + f_{12}z = 0$. The areal coordinates of the nul point of the face ABC are proportional to f_{14}, f_{24}, f_{34}.

Ex. 5. Prove that the invariant I of the six forces is

$$I = 6V(f_{12}f_{34} + f_{23}f_{14} + f_{31}f_{24}).$$

Ex. 6. If the six forces have a single resultant prove that it intersects each face in its nul point. Thence find its equation by using Ex. 4.

Ex. 7. Prove that the central axis of the six forces intersects the face ABC in a point whose areal coordinates are proportional to $f_{14} - paX_{23}/6V$, $f_{24} - pbX_{13}/6V$, $f_{34} - pcX_{12}/6V$, where p is the pitch, and X_{23}, X_{13}, X_{12} are the resolutes along the sides a, b, c of the face.

CHAPTER VIII.

Analytical view of reciprocal figures.

340. Two plane rectilineal figures are said to be reciprocal*, when (1) they consist of an equal number of straight lines or edges such that corresponding edges are parallel, (2) the edges which terminate in a point or corner of either figure correspond to lines which form a closed polygon or face in the other figure.

If either figure is turned round through a right angle the corresponding lines become perpendicular to each other but the figures are still called reciprocal.

Any figure being given, it cannot have a reciprocal unless (1) every corner has at least three edges meeting at it, (2) the figure can be resolved into faces such that each edge forms a base for two faces and two only.

The edges meeting at a corner in one figure correspond to the edges which form a closed polygon in the other. Since a closed polygon must have three sides at least, it follows at once that three edges at least must meet at each corner.

The edges of a figure can sometimes be combined together in different ways so as to make a variety of polygons. Only those

---

* The following references will be found useful. Maxwell, *On reciprocal figures and diagrams of forces*, Phil. Mag. 1864; Edin. Trans. vol. xxvi. 1870. The three examples mentioned in Arts. 347 and 349 are given by him. Maxwell was the first to give the theory with any completeness. Cremona, *Le figure reciproche nella statica grafica*, 1872; a French translation has been published and an English version has been given by Prof. Beare, 1890. Fleeming Jenkin, *On the practical application of reciprocal figures to the calculation of strains on frameworks and some forms of roofs*. He also notices that this method of calculating the stresses had been independently discovered by Mr Taylor, a practical draughtsman. He discusses the Warren girder, *Edin. Trans.* vol. xxv. 1869. Rankine's *Applied Mechanics*, eleventh edition, 1885. Maurice Lévy, *Statique Graphique*, second edition, 1886. He treats the subject at great length in several volumes. Culmann, *Die graphische statik*, Zurich, second edition, 1875. Major Clarke's *Principles of graphic statics*, second edition, 1888. Graham's *Graphic and analytic statics*, second edition, 1887. Eddy, *American Journal of Mathematics*, vol. i. 1878.

polygons which correspond to corners in the reciprocal figure are
to be regarded as faces. The figure is then said to be resolved
into its faces. The side of any face corresponds to an edge
terminated at the corresponding corner of the reciprocal figure.
Since an edge can have only two ends, it is clear that two faces
and only two must intersect in each edge.

341. Maxwell's Theorem. If the sides of a plane figure are the orthogonal
projections of the edges of a closed polyhedron, that plane figure has a reciprocal
which can be deduced by the following method.

Let one polyhedron be given and let its polar reciprocal be formed with regard
to the paraboloid $x^2+y^2=2hz$. Then we know that each face of either polyhedron
is the polar plane of the corresponding corner of the other. Smith's *Solid
Geometry*, Art. 152.

We shall now prove that the orthogonal projections of these two polyhedra on
the plane of xy are reciprocal figures with their corresponding sides at right angles.

The intersection of two faces is an edge of one polyhedron, and the straight line
joining the poles of these faces is an edge of the other. These edges correspond to
each other. Consider the edges which meet at a corner A of one polyhedron; the
corresponding edges of the second polyhedron lie in the polar plane of A and are
the sides of the face which corresponds to that corner. Thus *for every corner in
one polyhedron there corresponds a face with as many sides as the corner has edges.*

We shall next prove that *the projection of each edge of one polyhedron is at right
angles to the projection of the corresponding edge of the other.* To prove this we
write down the equations to the faces of one polyhedron which are the polar planes
of the two corners $(\xi\eta\zeta)$, $(\xi'\eta'\zeta')$ of the other. These are

$$h(z+\zeta)=x\xi+y\eta, \qquad h(z+\zeta')=x\xi'+y\eta'.$$

Eliminating z, we have the equation to the projection of an edge of the first
polyhedron, viz. $h(\zeta-\zeta')=x(\xi-\xi')+y(\eta-\eta')$. The equation to the projection of
the edge joining the two corners is $(y-\eta)(\xi-\xi')-(x-\xi)(\eta-\eta')=0$. These two
projections are evidently at right angles.

It is useful to notice that the pole of the plane $z=Ax+By+C$ is the point
whose coordinates are $\xi=hA$, $\eta=hB$, $\zeta=-C$.

Ex. Show that Maxwell's reciprocal is not altered (except in position) by
moving the paraboloid parallel to itself, and remains similar when the latus rectum
of the paraboloid is changed. What is the effect on the reciprocal figure of moving
the corners of the primitive polyhedron so that its projection is unchanged?

342. Cremona's Theorem. Another construction has been given by Cremona.
Let one polyhedron be given and let a second be derived from it by joining the
poles of the faces of the first. The Cremona-pole of a given plane is a certain
point which lies on the plane itself. If the edges of these two polyhedra are
orthogonally projected, these projections are reciprocal figures with their corresponding edges parallel.

Supposing the projection to be made on the plane of xy, the Cremona-pole may
be defined in any of the following ways. *Statically*, the Cremona-pole of a plane
is the nul point of that plane for a system of forces whose equivalent wrench is
situated in the axis of z and whose pitch is h. *Analytically*, the Cremona-pole of
the plane $z=Ax+By+C$ is the point $\xi=-hB$, $\eta=hA$, $\zeta=C$; see Art. 302.

Geometrically; let the plane intersect the axis of *z* in *C* and make an angle ϕ with that axis. The pole *O* lies on a straight line *CO* drawn in the given plane perpendicular to the axis of *z* so that $CO = h \cot \phi$.

We easily deduce Cremona's construction from that of Maxwell. If we turn Maxwell's reciprocal figure round the axis of *z* through a right angle, the coordinates of the pole used by him become $\xi = -hB$, $\eta = hA$, $\zeta = -C$. If we also change the sign of ζ, the coordinates become the same as those of the pole used in Cremona's construction. The effect of the rotation is that the corresponding lines in the projections of the two polyhedra become parallel, instead of perpendicular. The effect of the change of sign in ζ is that we replace the reciprocal polyhedron by its image formed by reflexion at the plane of *xy* as by a looking-glass. Since this last change does not affect the orthogonal projections on the plane of *xy*, it follows that the two constructions lead to the same reciprocal figures, except that the corresponding lines are in one case perpendicular to each other, in the other parallel.

343. *Example of a reciprocal figure.* The fig. 2 is composed of 8 corners, 18 edges and 12 triangular faces each having an angular point at *O* or *O'*. The hexagon enclosed by the six edges marked 1...6 not being included as a face, the figure may be regarded as the orthogonal projection of a polyhedron formed by placing two pyramids on a common base *ABCDEF* with their vertices on the same or on opposite sides. The figure therefore has a reciprocal.

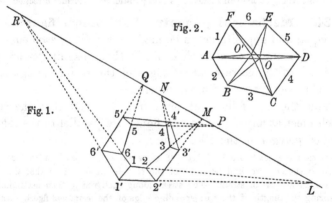

Fig. 1.   Fig. 2.

To construct this reciprocal we draw the two polar planes of *O*, *O'*; these intersect in some line *LMN*... whose orthogonal projection is by Maxwell's theorem at right angles to that of *OO'*. In fig. 1, the projection has been turned round through a right angle so that corresponding lines are parallel. Accordingly the projection of the intersection *LMN*... has been drawn parallel to that of *OO'*. Since 6 edges meet at *O* and *O'*, their polar planes give the two hexagons 1...6, 1'...6'. Since four edges meet at each of the other corners, the polar planes of these corners supply six quadrilateral faces to the reciprocal figure, the edges 11', 22', 33', &c. of fig. 1 being parallel to the edges 1, 2, 3, &c. of fig. 2.

The two edges 12, 1'2', lie in the planes of the two hexagonal faces and also in the planes of the quadrilaterals, they therefore intersect in the straight line LMN.

Fig. 1 will represent the general form, either of the reciprocal polyhedron, or its projection. The reciprocal figure thus constructed has 8 faces, 12 corners and 18 edges.

344. In the same way, when any plane figure is given, the polyhedron of which it is the projection can generally be found by erecting ordinates at the corners and joining the extremities. We must however take care that the faces thus constructed are planes. When the faces of the given figure are triangles, this condition is satisfied whatever be the lengths of the ordinates because a face bounded by three straight lines must be plane. It is also clear that when a figure is the projection of a polyhedron the area enclosed in that figure must be covered *twice* (or an even number of times) by the faces.

345. Reciprocal figures are usually constructed by drawing straight lines parallel to the edges of the given figure, assuming of course the properties already proved. To sketch fig. 1, we first draw from an assumed point L, the straight lines LMN, $L21$, $L2'1'$, parallel respectively to OO', OA, $O'A$. Assuming another point 2 on $L1$ we draw $22'$, $2M$ parallel to AB, OB, then in the figure of Art. 343 $2'M$ is parallel to $O'B$. The same is therefore true by similar figures (or by the properties of co-polar triangles) for all positions of the point 2 on $L1$. A point 3 being taken on $2M$ we draw $33'$, $3N$, $3'N$ parallel to BC, OC, $O'C$, and so on for the corners 4, 5, 6, the point 1 being known as the intersection of $R6$ and $L2$. If any one of these corners were chosen differently, say if 6 were moved nearer Q, we obtain a new triangle $R11'$ having its vertices on the straight lines LM, $L2$, $L2'$, and two sides $R1$, $R1'$, parallel to their former directions. Hence by the properties of co-polar triangles the third side $11'$ is also parallel to its former direction.

346. Mechanical property of reciprocal figures. Let two equal and opposite forces be made to act along each edge of a framework, one force at each end. If their magnitudes are proportional to the corresponding edges of the reciprocal figure, the forces at each corner are in equilibrium.

This theorem follows at once from the fact that the edges which meet at any corner in one figure are parallel to the sides of a closed polygon in the other figure.

For example, let figure 1 of Art. 343 represent a framework of 18 rods freely hinged at the corners, and let some of the rods be tightened so that the whole figure is in a state of strain. The stress along each rod is then determined by measuring the length of the corresponding edge of the reciprocal figure when that figure has been drawn. See also Art. 354.

347. Since each corner of a framework is in equilibrium under the action of the forces which meet at that corner, a corresponding polygon of forces can be drawn. There will thus be as many partial polygons as there are corners. When a reciprocal figure can be drawn, these polygons can be made to fit into each other so that every edge is represented once and once only in the complete force polygon. But if either of the conditions in Art. 340 were violated, so that a reciprocal diagram is impossible, the partial polygons may not fit completely into each other. The result would therefore be that one or more of

the forces would be represented by equal and parallel lines situated in different parts of the figure. Nevertheless some of the partial polygons may be made to fit, just as a portion of the framework may be regarded as the projection of a portion of some closed polyhedron. The force diagram thus imperfectly constructed may yet be of use to calculate the stresses.

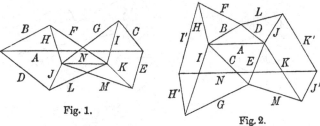

Fig. 1.

Fig. 2.

As an example of this, consider the framework represented in fig. 1, in which the rods F, G; L, M; &c. are supposed to cross without mutual action. If one rod is tightened, the resulting stresses along the others are determinate, yet a complete reciprocal figure cannot be constructed. The rod N forms an edge of four faces, viz. NFH, NGI, NJL, and NKM, so that if there could be a reciprocal figure, the line corresponding to N would have four extremities, which is impossible. In this case we can draw a diagram, represented in fig. 2, in which each of the forces H, I, J, K are represented by two parallel lines.

348. *External forces.* Let us remove the six bars which form the outer hexagon of fig. 1 in Art. 343 and also the connecting bars 11', 22', &c. We now apply at the corners 1...6 of the remaining hexagon forces $P_1...P_6$ to replace the stresses along the bars which have been removed. We thus have a framework consisting only of the bars 12, 23, &c. hinged at the corners and acted on by the now external forces $P_1...P_6$. This figure resembles the funicular polygon described in Art. 140, except that the forces which act at the corners are not necessarily vertical. When the external forces are given we modify the polygon in figure 2 to suit their magnitudes, see Art. 352. When therefore the stresses of a framework are caused by the action of external forces acting at the corners, these stresses can be graphically deduced when we can complete the figure in such a manner that a reciprocal can be drawn. It is however not usual actually to complete the figure, for the stresses which would exist in these additional bars if supplied are not required. It is sufficient to draw only so much of the figure as may be necessary to determine the stresses in the given framework.

349. A different mode of lettering the two figures is sometimes used, by which their reciprocity is more clearly brought into view. Since the lines which terminate in a corner of either figure correspond to lines which form a closed polygon in the other, it is obviously convenient to represent the corner in one figure and the polygon in the other by the same letter. In this way, the sides

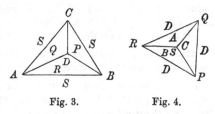

Fig. 3.   Fig. 4.

which meet in any corner A of fig. 3 are parallel to the sides which bound the space A in fig. 4, and the sides which bound the space P are parallel to those which meet at the corner marked P. Any side in one figure such as CD is bounded by the spaces P and Q and is therefore parallel to the straight line PQ in the other figure. This method of lettering the figures is called Bow's system. *On the economics of construction in relation to framed structures* (Spon, 1873).

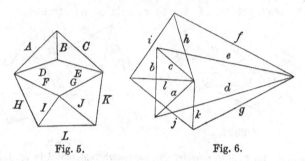

Fig. 5.   Fig. 6.

Another method of lettering the two figures has been used by Maxwell. Corresponding lines are represented by the same letter, but with some distinguishing mark; thus large letters may be used in one figure and small ones in the other. This method is illustrated in the diagram, which represents two reciprocal figures.

350. *A rectilinear figure being given, show how to find a reciprocal.* This may be best explained by considering an example. In the case of fig. 3 or 4, where all the faces are triangles, the reciprocal of either can be found by circumscribing circles about the faces. The straight lines which join the centres, two and two, are clearly perpendicular to the six sides of the given figure. One reciprocal figure having been thus constructed, any similar figure will also be reciprocal.

In more complicated cases such circles cannot be drawn. Let us consider how the reciprocal of fig. 5 in Art. 349 may be constructed. In drawing the reciprocal of a figure, it is generally convenient to begin with a corner at which three sides meet, for the reciprocal triangle corresponding to this corner will determine three lines of the reciprocal figure. By drawing the lines a, b, c parallel to A, B, C we construct the triangle reciprocal to the corner at which A, B, C meet. Through the intersection of b and c we draw a parallel e to E; because B and C form a triangle with E. In the same way d is drawn parallel to D through the intersection of a and b. We next notice that, since D, E, F, G form a polygon in one figure, the lines f and g may be constructed by drawing parallels to F and G through the intersection of e and d. Again the lines A, C, K, L, H form a closed polygon, hence the lines k, l, h must all pass through the intersection of a and c. The line i is drawn parallel to I through the intersection h, f. Lastly the line j is drawn parallel to J through the intersection g, k, and unless it passes through the intersection of l and i, a reciprocal figure cannot be formed. It follows however from the theorem in Art. 341 that this condition is satisfied.

Ex. 1. Two points are taken within a triangle, and the lines joining them to the corners are drawn. Construct the reciprocal figure.

Ex. 2. Three straight lines AA', BB', CC', if produced, meet in a point; AB, BC, CA, $A'B'$, $B'C'$, $C'A'$ are joined, thus forming three quadrilaterals and two triangles. Construct the reciprocal figure.

351. Let C be the number of corners in the given figure, E the number of sides or edges, F the number of faces or polygons. Let C', E', F' be the number of corners, edges and faces in the reciprocal polygon. It follows from the definition in Art. 340 that $E=E'$, $C=F'$, $F=C'$.

The sides of the reciprocal figure are formed by drawing straight lines parallel to those of the given figure. Taking any straight line AB parallel to one of the lines of the figure for a base, we construct two new sides by drawing through A and B parallels to the corresponding lines in the given figure. Continuing this process, every new corner is determined by the intersection of two new sides. As in Art. 151, the assumption of the first line AB determines two corners, and the remaining $C'-2$ corners are determined by drawing $2(C'-2)$ lines in addition to the assumed line AB. *Hence if* $E'=2C'-3$ *every corner is determined, and the figure is stiff. This is the condition that a diagram can be drawn in which the directions of the lines are arbitrarily given.* If E' is less than $2C'-3$, the form of the figure is indeterminate or deformable. If E' is greater than $2C'-3$, the construction is impossible unless $E'-2C'+3$ conditions among the directions of the lines are fulfilled.

In the first figure represented in Art. 349, there are four corners, four triangular faces and six edges; we have therefore in this figure $C+F=E+2$. Let another rectilinear figure be derived from this by drawing additional lines. The effect of drawing a line from a corner P to a point Q unconnected with the figure is to increase both C and E by unity. If we complete a new polygon by joining Q to another corner P', we increase both F and E by unity. If we divide any face into two parts by joining two points on its sides, we again increase equally $C+F$ and E. It follows, that if the relation $C+F=E+2$ hold for any one figure, the same relation * holds for all rectilinear figures derived from that one.

Considering both the given figure and the reciprocal, we have the relations

$$E=E', \qquad C=F', \qquad F=C', \qquad C+F=E+2, \qquad C'+F'=E'+2.$$

If the given figure is such that $C=F$, we have $E=2C-2$, $E'=2C'-2$. In this case the number of corners in either figure is equal to the number of faces, and each figure has one edge more than is necessary to stiffen it. That either figure may be possible, a geometrical condition for each must exist connecting the edges. When the given figure can be regarded as the projection of a polyhedron, it then follows from Maxwell's theorem that a reciprocal figure can be drawn. The conditions just mentioned must therefore be satisfied.

If $C<F$ as in Art. 343, we have $E>2C-2$, $E'<2C'-2$; on the same supposition the reciprocal figure is indeterminate. If $C>F$ we have $E<2C-2$, $E'>2C'-2$; in this case the construction of the reciprocal figure is impossible unless $C-F+1$ conditions are satisfied.

---

* This is the same as the relation (first given by Euler) which connects the number of corners, faces and edges of any simply connected polyhedron. We notice that in any polygon $C=E$ and $F=1$, so that $C+F=E+1$. Assuming any polygon as a base we construct the polyhedron by joining other polygons successively to the edges. It may easily be shown that, at each addition, we increase $C+F$ and E equally. *Hence the relation* $C+F=E+1$ *holds for unclosed polyhedrons.* When the final face is added, closing the figure, F is increased by unity, C and E remaining unchanged, *we therefore have* $C+F=E+2$ *for closed polyhedrons.* The limiting case of a polyhedron, all whose corners are in one plane, is a rectilineal figure having two faces only on each side. In such a figure Euler's relation must be true.

Statical view.

352. *The lines of action and the magnitudes of the forces P_1, P_2...P_5 being given, it is required to find their resultant.*

The magnitude and direction of the resultant can be found by constructing *a diagram or polygon of forces* in the manner explained in Art. 36. We draw straight lines parallel and proportional to the given forces and place them end to end in any order. The straight line closing the polygon, taken in the proper direction, represents the resultant. Let the forces P_1...P_5 be represented by the lines 1...5, the line 6 then represents the resultant in magnitude and reversed direction.

In constructing this polygon no reference has been made to the points of application of the forces, so that the forces are not fully represented. It will therefore be necessary to use a second diagram. This second figure is sometimes called *the framework* and sometimes *the funicular polygon.*

From any point O taken arbitrarily in the force diagram we draw radii vectores to the corners. These radii vectores divide the figure into a series of triangles, the sides of which are used to resolve the forces P_1 &c. in convenient directions by the use of the triangle of forces. The side joining O to any corner occurs in two triangles, and therefore represents two forces acting in opposite directions. No arrow has therefore been placed on that side. The arbitrary point O is usually called the *pole of the polygon.* The corners are represented by two figures; thus the intersection of the sides 1 and 2 is called the corner 12 and the straight line joining O to this corner is called the *polar radius* 12.

We are now in a position to construct the funicular polygon. Taking any arbitrary point L as the point of departure, we draw a straight line LA_1 parallel to the polar radius 61 to meet the line of action of P_1 in A_1. From A_1 we draw A_1A_2 parallel to the polar radius 12 to meet P_2 in A_2; then A_2A_3 is drawn parallel to the polar radius 23 to meet P_3 in A_3; then A_3A_4 and A_4A_5 are drawn parallel to the polar radii 34 and 45. Finally A_5A_6 is drawn parallel to 56 to meet A_1L (produced if necessary) in A_6. Then A_6 is the required point of application of the resultant force.

To understand this, we notice that the force P_1 at A_1 is resolved by one of the triangles of the force polygon into two forces acting along LA_1 and A_2A_1 respectively. The latter combined

with P_2 is equivalent to a force acting along A_3A_2. This combined
with P_3 is equivalent to one along A_4A_3, and so on. We thus see
that all the forces P_1, &c. P_5 are equivalent to two, one along LA_1
and the other along A_6A_5. These two must therefore intersect in
a point on the resultant force. In the figure P_6, drawn parallel
to the line 6, represents a force in equilibrium with $P_1 \ldots P_5$.

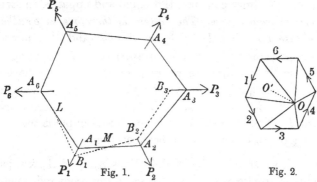

Fig. 1.   Fig. 2.

If we take some point, other than L, as a point of departure
we obtain a different funicular polygon having all its sides parallel
to those of $A_1A_2 \ldots A_6$. In this way by drawing two funicular
polygons we can obtain (if desired) two points on the line of action
of the resultant.

If we take some point other than O as the pole in the force
diagram, but keep the point of departure L unchanged, we obtain
another funicular polygon whose sides are not parallel to those
of $A_1A_2 \ldots A_6$. A few of these sides are represented by the dotted
lines. But the resulting point A_6 must still lie on the resultant.
We thus arrive at a geometrical theorem, that *for all poles with
the same force diagram the locus of A_6 is a straight line.*

353. Conditions of equilibrium. In this way we see that,
whenever the force polygon is *not closed*, the given system of forces
admits of a resultant whose position can be found by drawing any
one funicular polygon.

When the force polygon is closed the result is different. In
order to use the same two figures as before let us suppose that the
six forces $P_1 \ldots P_6$ form the given system. Taking any arbitrary
point L, we begin as before by drawing LA_1 parallel to the polar
radius 61. Continuing the construction for the funicular polygon,
we arrive at a point A_6 on the now given force P_6. To conclude

the construction we have to draw a straight line from A_6 parallel to the same polar 61 with which we began. This last straight line may be either coincident with, or parallel to, the straight line LA_1 with which we began the construction. The whole system of forces has thus been reduced to two equal and opposite forces, one along A_1L and the other along its parallel drawn from A_6.

If these two lines coincide, the equal and opposite forces along them cancel each other. *The system is therefore in equilibrium. In this case the funicular polygon drawn* (and therefore every funicular polygon which can be drawn) *is a closed polygon.*

If these two straight lines are parallel, the forces have been reduced to two equal, parallel, and opposite forces. *The system is therefore equivalent to a couple. In this case the funicular polygon is unclosed.* The moment of this resultant couple is the product of either force into the distance between them.

354. If we suppose the straight lines A_1A_2, A_2A_3, &c., joining the points of application of the forces to represent rods jointed at A_1, A_2, &c., the forces by which these press on the hinges act along their lengths, Art. 131. The figure has been so constructed that the reactions at each hinge balance the external force at that point. The combination of rods therefore forms a framework each part of which is in equilibrium under the action of the external forces, and the stresses in the several rods may be found by measuring the corresponding lines in the force diagram.

We notice that any set of forces acting at consecutive corners of the funicular polygon (such as P_4, P_5, P_6) are statically equivalent to the tensions or reactions along the straight lines at the extreme corners (viz. A_3A_4 and A_1A_6). These sides must therefore intersect in the resultant of the set of forces chosen. Hence, *whatever pole O is chosen and whatever point of departure L is taken, the locus of the intersection of any two corresponding sides of the funicular polygon* (such as A_3A_4 and A_1A_6) *is a straight line.* In a closed funicular polygon this straight line is the line of action of the resultant of either of the two sets of forces separated by the sides chosen. Thus the sides A_3A_4, A_1A_6 meet in the resultant either of P_4, P_5, P_6 or of P_3, P_2, P_1.

355. It may be noticed that fig. 1 does not admit of a reciprocal because the lines representing the forces $P_1...P_6$ do not form the edges of any face. Nevertheless a force diagram has been constructed. The reason is that fig. 1 is a part of a more complete figure which does admit of a reciprocal, Art. 343. It follows from

Art. 348 that if we complete the figure by drawing another funicular polygon corresponding to some other pole O, the whole figure becomes the projection of a polyhedron and therefore admits of a reciprocal. And so it will be found that the figures drawn to calculate the stresses of a framework are, in general, incomplete reciprocal figures. The parts essential to the problem in hand are sketched and the rest is omitted. The importance of the theory of reciprocal figures is that it enables us to investigate the relations of the several parts of the figure by pure geometry.

356. Parallel forces. When the forces are parallel, both the force diagram and the funicular polygon are simplified, see Art. 140. Thus let A_0A_1, A_1A_2, A_2A_3, A_3A_4 be light bars hinged together at A_1, A_2, A_3. Also let the weights P_1, P_2, P_3 act at A_1, A_2, A_3.

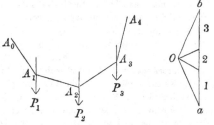

Here the force diagram is a straight line ab divided into segments representing the forces P_1, P_2, P_3. If Oa, Ob be parallel to the extreme bars A_0A_1, A_3A_4, then these lengths represent the tensions of these bars, and the lengths drawn from O to the corners 12, 23 represent the tensions of the intervening bars.

To find the resultant of three given forces P_1, P_2, P_3 we assume any arbitrary pole O in the force diagram and draw the corresponding funicular polygon $A_0A_1...A_4$. The extreme sides A_0A_1, A_4A_3 produced meet in a point on the line of action of the resultant. The magnitude is obviously the sum of the given forces, and its direction is parallel to those forces.

357. *The force polygon being given, and the point L of departure, let the pole move from any given position O along any straight line OO'. Prove (1) that each side of the funicular polygon turns round a fixed point, and (2) that all these fixed points lie in a straight line, which is parallel to the straight line OO'.* This theorem follows from the ordinary polar properties of Maxwell's reciprocal polyhedra, Art. 343. The following is a statical proof.

Referring to the figure of Art. 352, let L, M, N &c. be the points of intersection of corresponding sides of two polygons constructed with O, O' respectively as poles. Let (R_{61}, R_{21}) (R'_{61}, R'_{21}) be the reactions along the sides which meet on the force P_1 on the two polygons. Since these have a common resultant P_1, the four forces R_{61}, R'_{16}, R_{21} and R'_{12} are in equilibrium. Hence the resultant of R_{61}, R'_{16} acting at L must balance the resultant of R_{21}, R'_{12} acting at M. Each of these resultants must therefore act along LM. But looking at the force polygon, the forces R_{61}, R'_{61} are represented by the polar radii drawn from O, O' to the corner 61. Hence the resultant of R_{61}, R'_{16} is parallel to OO'. Similarly MN is parallel to OO'. Hence LMN is a straight line. [Lévy, *Statique Graphique.*]

Let a third funicular polygon be drawn corresponding to a third pole O'' situated on OO'. If this funicular polygon beginning at L intersect the first in M', N', &c., both LMN &c. and $LM'N'$ &c. are parallel to $OO'O''$, hence M coincides with M', N with N', and so on. The points M, N, &c. are therefore common to all the funicular polygons.

Find the locus of the pole O of a given force polygon that the corresponding funicular polygon starting from one given point M may pass through another given point N. The locus is known to be a straight line parallel to MN: the object is to construct the straight line.

Case 1. If the given points M, N lie between any two consecutive forces (say P_1, P_2), we may take MN as the initial side A_1A_2. The pole O must therefore lie on the straight line drawn through the corner 12 of the given force polygon parallel to the given line A_1A_2 (see Art. 352).

Case 2. Let the point M lie between any two forces (say P_1, P_2) and N between any other two (say P_3, P_4). We can remove the intervening force P_2, and replace it by two forces acting at M and N each parallel to P_2; let these be Q_2, Q_2', Art. 360. Similarly we can replace the other intervening force P_3 by two forces, each parallel to P_3, acting also at M and N; let these be Q_3, Q_3'. If we now adapt the given force polygon to these changes, the sides 2 and 3 only have to be altered. We have to draw forces parallel to Q_2, Q_3, Q_2', Q_3', beginning at the terminal extremity of the force 1 and ending (necessarily) at the initial extremity of the force 4. The points M, N now lie between the two consecutive forces $Q_3 Q_2'$, hence by Case 1 the locus of O is the straight line drawn parallel to MN through the intersection of these forces in the force diagram.   [Lévy, *Statique Graphique.*]

With given forces, show how to describe a funicular polygon to pass through any three given points L, M, N.

We first find the locus of the pole O when the funicular polygon has to pass through L and M, and then the locus when it has to pass through L and N. The intersection is the required point.

With given forces show how to describe a funicular polygon so that one side may be perpendicular to a given straight line.

Suppose the side A_1A_2 is to be perpendicular to a given straight line, then the polar radius 12 is also perpendicular to that line, Art. 352. Hence the pole O must lie on the straight line drawn through the corner 12 of the force polygon perpendicular to the given straight line.

Ex. Prove that, if the resultant of two of the forces is at right angles to the resultant of one of these and a third force of the system, a funicular polygon can be drawn with three right angles.   [Coll. Ex., 1887.]

358. *If we remove any set of consecutive forces from a funicular polygon, and replace them by other forces statically equivalent to them, show that the sides bounding this set of forces remain fixed in position and direction though not in length.* Suppose we replace P_4, P_5 by their resultant, then in the force diagram we replace the sides 4, 5 by the straight line joining 34 to 56. The polar radii 34 and 56 are therefore unaltered. But the bounding sides A_3A_4, A_5A_6 are drawn parallel to these bounding radii from fixed points A_3, A_6, hence they are unaltered in position and direction.

359. *If the forces are not in one plane, show that in general there is no funicular polygon.* Let the resultant of P_1, P_2,...P_n be required, and if possible let $A_1A_2...A_n$ be a funicular polygon. Then this polygon must satisfy two conditions; (1) since any one force P can be resolved into two components acting along the adjoining sides, each force and the two adjoining sides must lie in one plane, (2) the components of two consecutive forces along the side joining their points of application must be equal and opposite. When the forces lie in one plane, the first condition is satisfied already and the second condition alone has to be attended to, and this one condition suffices to find all the possible polygons.

If any one side $A_1 A_2$ of the polygon is chosen, the first condition in general determines all the other sides. To show this we notice that the plane through $A_1 A_2$ and P_2 must cut P_3 in A_3; thus $A_2 A_3$ is determined and so on round the polygon. Thus there are not sufficient constants left to satisfy the second condition, though of course in some special cases all the conditions might be satisfied together.

360. *Ex. 1.* Prove the following construction to resolve a given force P_2 acting at a given point A_2 into two forces, each parallel to P_2 and acting at two other given points A_1, A_3. Let a length ac represent P_2 in direction and magnitude on any *given scale*. Draw aO, cO parallel to $A_2 A_3$, $A_1 A_2$ respectively, and from their intersection O draw Ob parallel to $A_1 A_3$ to intersect ac in b. Then ab and bc represent the required components at A_3 and A_1.

Another construction. Produce P_2 to cut $A_1 A_3$ in N. Then $A_1 N$ and $N A_3$ represent the forces at A_3 and A_1 respectively on the same scale that $A_1 A_3$ represents the given force P_2. These would have to be reduced to the given scale by the method used in Euclid vi. 10.

Ex. 2. Show that a given force P can be resolved in only one way into three forces which act along three given straight lines, the force and the given straight lines being in one plane. Prove also the following construction. Let the given straight lines form the triangle ABC, and let the given force P intersect the sides in L, M, N. To find the force S which acts along any side AB, take Np to represent the force P in direction and magnitude, draw ps parallel to CN to intersect AB in s, then Ns represents the required force S. See Art. 120, Ex. 2.

Let Q, R, S be the forces which act along the sides. The sum of their moments about C must be equal to that of P. The moment of S about C is therefore equal to that of P. Since ps is parallel to CN, the areas CNp and CNs are equal, and therefore the moment of Ns about C is equal to that of P. Hence Ns represents S.

Ex. 3. Show how to resolve a couple by graphic methods into three forces which shall act along three given straight lines in a plane parallel to that of the couple. Prove also the following construction. Move the couple parallel to itself until one of its forces passes through the corner C of the given triangle, and let the other force intersect AB in N. Take Np to represent this second force, and draw ps parallel to CN to meet AB in A, then the required force along the side AB is represented by Ns.

361. *A light horizontal rod $A_0 A_5$ is supported at its two ends A_0, A_5 and has weights W_1, W_2, W_3, W_4, attached to any given points A_1, A_2, A_3, A_4. It is required to find by a graphical method the pressures on the points of support.*

Here all the forces are parallel, and the force diagram becomes a straight line. Let the line ab be divided into four portions representing the four weights $W_1 \dots W_4$, while bc and ca represent the pressures R' and R at A_5 and A_0. We have to determine the position of c.

Taking any pole O, we draw the polar radii joining O to the extremities of the lines which represent the forces. Drawing parallels beginning at A_0 we sketch a funicular polygon represented by $A_0B_1...B_5$. The polar radius Oc must be parallel to the line B_5A_0 closing the funicular. Thus c has been found and therefore the two pressures R, R'.

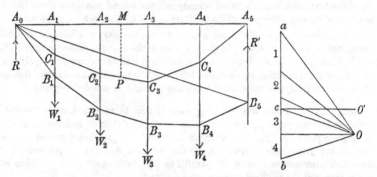

If the rod is heavy, the pressures R, R' are not affected by collecting the weight at the centre of gravity. Drawing any funicular, with this additional weight taken into account, the pressures on the points of support can be found as before.

362. *A light horizontal rod A_0A_5 being supported at its two ends and loaded with weights $W_1...W_4$ at the points $A_1...A_4$, it is required to find the stress couple at any point M.* Art. 145.

The pressures at the two ends having been determined, we describe a funicular polygon of these six forces, such that it passes through A_0 and A_5. We shall now prove that the stress couple at M is Hy, where y is the ordinate of the funicular at M and H is the horizontal tension.

Supposing the funicular polygon to be $A_0C_1...C_4A_5$, we notice that the system of rods represented by $A_0C_1, C_1C_2...C_4A_5$ are in equilibrium under the action of the weights $W_1...W_4$, the vertical pressures R, R', and the horizontal thrust H of A_1A_5, Art. 354. Taking moments about P, the extremity of the ordinate through M, for the portion $A_0...P$, we have Hy equal to the sum of the moments of the pressure R and the weights W_1, &c. on one side of P, i.e. Hy is the bending moment of the rod at M. Art. 143.

To draw the funicular polygon which passes through the points A_1 and A_5, we take a pole O' at any point on a horizontal line through the point c in the force diagram and then construct the polygon as before. Since cO is parallel to A_0B_5 it follows that, when O lies in cO', B_5 must coincide with A_5. It is evident that $O'c$ represents the horizontal tension.

If O' is moved along cO', the funicular polygon and therefore both the horizontal tension cO' and the ordinate MP change. The product however, being equal to the bending moment at M, is not altered; a result which may be independently verified.

If the rod is uniform and heavy, the moments about M of the weights of the portions A_0M, MA_5 are not altered by replacing those weights by half weights placed respectively at A_0, M and M, A_5, see Art. 134. If the stress couples at all the points $A_1...A_4$ are required, we can replace the weight of each segment by two half weights attached to its extremities. In this way the same funicular will determine all the stress couples.

363. Frameworks. *To show how the reactions along the bars of a framework may be found by graphical methods, the external forces being supposed to act at the corners.*

Let the given framework consist of a combination of three triangles, such as frequently occurs in iron roofs. Let any forces P_1, P_2, P_3, P_4, P_5 act at the corners A_1, A_2, A_3, A_4, A_5, and let the whole be in equilibrium. If these forces were parallel three

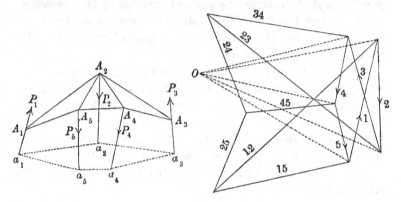

of them might represent weights placed at the joints, while the structure is supported on its two extremities A_1, A_3.

The five forces are in equilibrium, hence the five lines 1...5 which represent them in the force diagram form a closed pentagon. We shall now sketch the lines corresponding to the stresses of the framework.

The framework, as described above, does not admit of a reciprocal; let us assume for the present that it can be completed by drawing the pentagon $a_1...a_5$; Art. 355. The proper form of this addition to the figure is discussed in Art. 365 *.

The side A_1A_5 forms part of a quadrilateral $A_1A_5a_5a_1$. This quadrilateral corresponds to four lines in the reciprocal figure which meet in a point. Hence the reciprocal of the straight line

---

* If we do not refer to the theory of reciprocal figures the argument must be somewhat altered. As there are more than three forces at several corners of the framework, it will then require some attention to discover the force diagram, though when once known it can be drawn without difficulty to suit the numerical relations of the bars in any like structure.

To discover the line corresponding to A_1A_5 we notice that the forces at A_1 must be represented by a triangle two sides of which are parallel to P_1 and A_1A_5, those at A_5 by a quadrilateral two sides of which are parallel to P_5 and A_1A_5. As a trial construction we can satisfy these conditions by adopting the rule in the text. The success of the drawing will test the correctness of the hypothesis, Art. 347.

A_1A_5 is a straight line drawn through the intersection of the consecutive forces 1, 5 parallel to A_1A_5. The same argument applies to every bar of the frame $A_1A_2...A_5$; each is represented in the reciprocal by a straight line which passes through the junction of the consecutive forces at its extremities. This easy rule enables us to draw the reciprocal figure without difficulty. Thus the reciprocal of the side A_1A_2 is a straight line drawn parallel to A_1A_2 through the point of junction of the consecutive forces marked 1 and 2. These straight lines are marked in the force diagram with the suffixes of the straight lines to which they correspond in the framework.

The triangle representing the forces at A_1 having now been constructed, we turn our attention to those at the next corner A_5. These will be represented by a quadrilateral. Following the rule, we draw 45 parallel to A_4A_5 through the point of junction of the consecutive forces 4, 5. Thus three sides of the quadrilateral are known, viz. 5, 15, 45. Through the known intersection of 12 and 15 we draw a parallel to A_2A_5 completing the quadrilateral. The sides are 5, 15, 25, 45.

Turning our attention to the corner A_4, we draw 34 by the rule and again we know three sides of the corresponding quadrilateral, viz. 34, 4 and 45. The fourth side is completed by drawing 24 through the known intersection of 45 and 25. The four sides are 4, 45, 24, 34.

The triangle corresponding to the corner A_3 is completed by joining the known intersection of 34 and 24 to the point of junction of the consecutive forces 2, 3. By the rule this line should be parallel to the side A_2A_3. This serves as a partial verification of the correctness of the drawing.

Lastly the forces at the corner A_2 must be represented by a pentagon, but looking at the figure we find that all the sides of this pentagon, viz. 2, 23, 24, 25, 12, have been already drawn.

The magnitudes of the reactions along the bars of the given frame may now all be found by measuring the lengths of the different lines in the diagram.

364. The directions of the reactions along the bars of the framework are not usually marked by arrows in the force diagram because two equal and opposite forces act along each bar. It is more convenient to mark them as bars *in tension or in thrust.*

The former are called *ties* and the latter *thrusts*. Consider the corner A_1, the bars are parallel to the sides of the triangle 1, 12 and 15. The direction of the forces being known, those of 12 and 15 follow the usual rule for the triangle of forces. Hence at the point A_1 the forces act in the direction 15, 21. Therefore A_1A_2 is in a state of compression, i.e. it is a thrust, while A_1A_5 is in a state of tension and is a tie. We may represent these states by placing arrows in the framework at A_1, A_2 pointing *towards* A_1, A_2 respectively and arrows at A_1, A_5 pointing *from* A_1, A_5 respectively. Another method has been suggested by Prof. R. H. Smith in his work on Graphics. He proposes to indicate ties by the sign + and struts by −. These marks may be placed on either diagram.

365. We should notice that the figure thus constructed, though sufficient to find the stresses in the rods, is not a complete reciprocal figure. To enable us to complete the figure we must first draw such a polygon $a_1...a_5$, cutting the lines of action of the forces, that the whole figure may admit of a reciprocal. *Statically*, we see that this polygon must be a funicular of the given forces, for otherwise the forces at the corners $a_1...a_5$ would not be in equilibrium, Art. 354. *Geometrically*, the polygon should be such that the five quadrilaterals $a_1a_2A_1A_2$, &c. are the projections of plane faces of a polyhedron. This polyhedron is constructed by drawing ordinates at the corners. We know that, if we draw two funiculars $a_1...a_5$ and $b_1...b_5$ of the forces $P_1...P_5$, the five intersections of a_1a_2, b_1b_2; a_2a_3, b_2b_3; &c. lie in a straight line LMN, Art. 357. Referring to Art. 343 (where these funiculars are represented by 1...6 and 1'...6') we see that the five quadrilaterals $a_1a_2b_1b_2$, &c. may therefore be made the projections of plane faces. We construct the polyhedron by keeping $a_1...a_5$ fixed and erecting ordinates at $b_1...b_5$ proportional to their distances from LMN. Since the sides A_1A_2 &c. lie in the planes $a_1a_2b_1b_2$, &c. it follows that the five quadrilaterals $a_1a_2A_1A_2$, &c. are also the projections of plane faces. The ordinates at $A_1...A_5$ may then be drawn.

Taking $a_1...a_5$ to be a funicular polygon of the forces $P_1...P_5$ the corresponding lines on the force diagram are the dotted lines drawn from the corresponding pole O to the points of junction of the forces. It is evident that these lines are practically separate from the rest of the figure. Unless therefore we wish to assure ourselves that the forces $P_1...P_5$ are in equilibrium, it is unnecessary to draw either the funicular polygon $a_1...a_5$ or the corresponding lines in the force diagram. *It is usual to omit this part of the figure.*

366. Method of sections. We shall now show how the reactions are found by the method of sections. Let it be required to find the reactions along the rods A_2A_1, A_2A_5, A_5A_4. Let these reactions be called Q, R, S respectively. Draw a section cutting the frame along these rods, and let the points of intersection be B, C, D. If we imagine the whole structure on one side of this section to be removed, the remainder will stand if we apply the forces Q, R, S

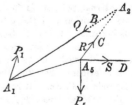

to the points B, C, D along the three rods respectively. Let us remove the structure on the right hand as being the more complicated, we have now to deduce the forces Q, R, S from the conditions of equilibrium of the remaining structure.

In our example *not more than three bars were cut by the section*. Since there are only three forces the problem is determinate. By Art. 360, Ex. 2, each force of any system can be replaced by three forces acting along three given straight lines, and *this resolution can be effected by a graphical construction*.

These reactions may also be easily found by the ordinary rules of analytical statics, as in Art. 120, where this problem is solved by taking moments about the intersections of these lines.

When the figure is so little complicated as the one we have just considered, either the method of the force diagram or the method of sections may be used indifferently. In general each has its own advantages. In the first we find all the reactions by constructing one figure with the help of the parallel ruler, but if there be a large number of bars the diagram may be very complicated. In the method of sections when only three reactions are required we find these without troubling ourselves about the others, provided these three and no others lie on one section.

367. In these frameworks, each rod, when its own weight can be neglected, is in equilibrium under the action of two forces, one at each extremity. These forces therefore act along the length of the rod, and thus the rods are only stretched or compressed. This is sometimes a matter of importance, for a rod can resist, without breaking, a tensional or compressing force when it would yield to an equal transverse force. The structure is therefore stronger than when rigidity at the joints is relied on to produce stiffness.

In actual structures some of the external forces may not act at a corner, for instance, the weight of any rod acts at its centroid. In such cases the resultant force on any bar must be found either by drawing a funicular polygon or by the rules of statics. This resultant is to be resolved into two parallel components acting one at each of the two joints to which the rod is attached.

This transformation of the forces which act on a rod cannot affect the distribution of stress over the rest of the structure, so that when these components are combined with the other forces which act at those joints the whole effect of the rest of the structure on each rod has been taken account of. So far as the rod itself is concerned, it is supposed to be able to support, without sensible bending, its own weight or any other forces which may act on it at points intermediate between its extremities.

368. Indeterminate Tensions. Let P_1, P_2,...P_n be a system of forces in equilibrium. Let $A_1...A_n$, $A'_1...A'_n$ be two funicular polygons of this system. Let the corresponding corners A_1, A'_1; A_2, A'_2 &c.

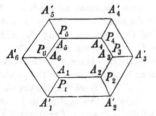

be joined by rods. Let us also suppose that the external polygon is formed of rods in a state of tension and the internal polygon of rods in thrust. It is clear from the properties of a funicular polygon that the framework thus constructed will be in equilibrium. It is also evident that the thrusts along the cross rods $A_1A'_1$ &c. will be equal respectively to the original forces P_1, P_2,...P_n. In this way a frame has been constructed with tensions along the rods apart from all external forces. See Art. 237. From the property of funicular polygons proved in Art. 357 *the corresponding sides of this frame intersect in points all of which lie in a straight line.*

If there are only three forces the polygons become triangles. Since the forces P_1, P_2, P_3 are in equilibrium the three straight lines $A_1A'_1$, $A_2A'_2$, $A_3A'_3$ which join the corresponding angular points must meet in a point. Such triangles are called co-polar. We see therefore that co-polar triangles admit of indeterminate tensions.

Levy's theorem, given in Art. 238, follows also from this proposition. Taking only six forces, because the figure has been drawn for a hexagon, let (P_1, P_4), (P_2, P_5), (P_3, P_6) be three sets of equal and opposite balancing forces. Let $A_1...A_6$ be any funicular polygon, but let the second funicular polygon be constructed so that A_1' coincides with A_4, and let the pole be so chosen that A_2' and A_3' coincide with A_5 and A_6, Art. 357. It then follows that the second funicular coincides throughout with the first. The cross bars A_1A_4, A_2A_5, A_3A_6 become the diagonals of the hexagon. Thus a frame of any even number of sides has been constructed in which the diagonals are in a state of thrust and the sides in tension.

369. The line of pressure. Let us suppose a series of connected bodies, such as the four represented in the figure, to be in equilibrium under the action of any forces, say the three P, Q, R. We suppose these bodies to be symmetrical about a plane which in the figure is taken to be the plane of the paper. The first body is hinged to some fixed support at A and also hinged at B to the body BCC'. This second body presses along its smooth plane surface CC' against a third body $CC'D$. This third body is hinged to a fourth body at D, and this last is hinged at E to a fixed point of support.

The pressure at A acts along some line Ap and intersects the force P at p. The resultant of these two must balance the action at the hinge B, and must therefore pass through B. This force acting at B intersects the force Q at q, and their resultant must balance the pressure at CC'. This resultant must therefore

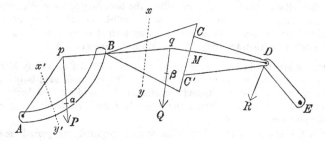

cut CC' at right angles in some point M. *Also the point M must lie within the area of contact, and the resultant must tend to press the surfaces at CC' together.* This pressure on the third body acts along qMD and intersects R at D. Finally the resultant of these two must pass through E.

It is evident that the line $ApqDE$ is a funicular polygon of the forces P, Q, R. When therefore such a series of bodies as we have here described rests in equilibrium with its extremities supported it is *sufficient and necessary for equilibrium that some one funicular polygon can be drawn which passes through all the hinges and cuts at right angles the surface of pressure.* This particular funicular polygon is called the line of pressure.

370. Let us take an ideal section, such as *xy*, which separates the whole system into two parts, and let it be required to find the resultant action across this section.

This action is really the resultant of the forces across each element of the sectional area. But since each portion of the system must act on the other portion in such a way as to keep that portion in equilibrium, we may also find the resultant from the general principle that it balances all the external forces which act on either of the two portions of the system : see also Art. 143. It immediately follows that the resultant action across *xy* is the force already described which acts along *pq*. Similar remarks apply to every section ; we therefore infer that *the resultant action across any section is the force which acts along the corresponding side of the line of pressure.*

If we move the section *xy* from one end *A* of the system to the other *B*, there may be some difficulty in determining which is the " corresponding side of the line of pressure " when the section passes the point of application of a force. Suppose for example *a* to be the point of application of *P*. If a section as *x'y'* is ever so little to the left of *a*, the corresponding side is *Ap*, but when the section is ever so little on the right of *a*, the corresponding side is *pq*. *If the section is parallel to the force P, the side corresponding to any section is the side of the line of pressure intersected by that section.* When therefore the forces are all vertical it will be found more convenient to consider the actions across vertical sections than across those inclined.

The resultant action across any section such as *x'y'* does not necessarily pass within the area of that section. The reason is that this action is the resultant of all the small forces across all the elements of area. As some of these elementary forces across the same sectional area may be tensions and some pressures, the line of action of the resultant may lie outside the area. If the forces all act in the same direction like those across the section *CC'* (where two bodies press against each other), the resultant must pass within the boundary of the section. Sometimes it is more useful to move the resultant parallel to itself and apply it at any convenient point within the boundary ; we must then of course *introduce a couple.* This is often done when the body *AB* is a thin rod. See Art. 142.

371. When the bodies are heavy we may find the action at any hinge or *boundary between two bodies* by the same rule. The weight of each body is to be collected at its centre of gravity and included in the list of external forces. The resultant action at any boundary is the force along the corresponding side of the funicular polygon.

But if the action across some section as *xy* is required, this partial funicular polygon will not suffice. We must now consider the body *BCC'* to be equivalent to two bodies separated by the plane *xy*. The weights of each of these portions may be collected at its own centre of gravity, and a funicular polygon may be drawn to suit this case. Thus, if *Q* is the weight of the body *BCC'* acting at its centre of gravity *β*, we remove *Q* and replace it by two weights acting at the respective centres of gravity of the portions *Bxy* and *xyCC'*. The funicular polygon will therefore have one more side than before. It also loses the corner on the force *Q* and gains two new corners which lie on the lines of action of these new weights. But since the action at *B* must still balance the external forces whose points of application are on the left of B, and the action at M must still balance the forces on the right of CC', it is clear that the sides pB and MD of the funicular polygon are not altered. Therefore the two corners of the new funicular polygon must lie respectively on Bq and qD. *Thus the new polygon is inscribed in the former partial funicular polygon.*

If we continue this process of separating the bodies into parts, we go on increasing the number of sides in the funicular polygon, but the side which passes through any real section is unchanged in position. Finally, when the bodies are subdivided into elements, the line of pressure becomes a curve. *This curve will touch all the partial polygons of pressure at each hinge and at each real surface of separation.*

EXAMPLES.

372. Ex. 1. A framework is constructed of eleven equal heavy bars. Nine of them form three equilateral triangles ABC, BDE, DFG with their bases AB, BD, DF hinged together in a horizontal straight line. The vertices C, E, G are joined by the remaining two bars. The Warren girder thus formed is supported at its two lower extremities A, F and loaded at the upper points C, E, G with weights w_1, w_2, w_1. Construct a force diagram showing the stresses in the bars.

Ex. 2. A horizontal girder has four bays AB, BC, CD, DE each 5 feet; it is stiffened by three vertical members BB', CC', DD' each 3 feet, by horizontal members $B'C'$, $C'D'$ and by oblique members AB', $B'C$, CD', $D'E$. Find by a graphical construction the tensions and thrusts produced in the members when a uniformly distributed load W is supported by the girder. [St John's Coll., 1893.]

Ex. 3. $ABCDEFG$ is a jointed frame in a vertical plane, constructed as follows. $ABCD$ and GFE are horizontal, A being vertically above G; $ABFG$, $BCEF$ are squares; CD is equal to CE; also BG, CF, DE are three diagonal stiffening bars. The frame is supported at the points A and G, while a weight is hung at D. Supposing the weights of each bar to act half at each of its ends, exhibit in a diagram the stresses in the various bars of the frame. Show that those in GF and BC are equal, likewise those in FE and CD, and determine which bars are struts and which are ties. The supporting force at A may be taken to be horizontal. [Coll. Ex., 1894.]

Ex. 4. A roof $ABCD$ is of the form of half a regular hexagon; it is stiffened by two cross-beams AC BD; and it rests on the walls at A and D. Find, by a stress diagram, the tensions and thrusts in its members produced by a uniform load of tiles. [St John's Coll., 1892.]

Ex. 5. A framework is composed of six light rods smoothly jointed so as to form a regular hexagon $ABCDEF$ whose centre is at O. The points BF, OA, OC, OE are also connected, without disturbing the regularity of the hexagon, by light rods of which the first two are to be regarded as having no contact with one another. If the framework be suspended from A and a weight W be attached to D, show by graphical methods that the thrust in BF will be $W\sqrt{3}$, and find the force along each of the other bars. [Trin. Coll., 1895.]

Ex. 6. A regular twelve-sided framework is formed by heavy loosely jointed rods and each angular point is connected by a light rod to a peg at the centre. The whole rests on the peg in a vertical plane with a diagonal vertical. Show that the stresses in the rods are indeterminate; and assuming that the horizontal rods are not under stress, draw a diagram in which lines are parallel to and proportional to the stress in each rod and calculate the stresses. [Coll. Ex., 1893.]

Ex. 7.   The lines of action of six forces in equilibrium are known.   One force is known, one other pair of the forces are in one known ratio, a second pair are in another known ratio.   Find a graphic construction determining the magnitudes of the five undetermined forces.   [Math. Tripos, 1895.]

Ex. 8.   *ABCD* is a rhombus of jointed rods, and *OB*, *OD* are two equal rods jointed to the rhombus at *B* and *D* and jointed at *O*.   Supposing all the joints smooth and parallel forces, not in the same line, applied to the framework at *O*, *A*, *C*; construct a force diagram.   Show that for equilibrium the directions of the forces must be parallel to *BD*.   [Math. Tripos, 1891.]

Ex. 9.   Four forces act in the sides *AB*, *BC*, *CD*, *DA* of a quadrilateral *ABCD*, and are proportional to those sides.   Construct the funicular, one of whose sides joins the middle points of *AB* and *BC*, when the *thrust* in that side is represented by *CA* on the same scale as the given forces are represented by the sides of the quadrilateral.   [St John's Coll., 1893.]

Ex. 10.   Prove that if the lines of action of $(n-1)$ forces be given, it is always possible to adjust their magnitudes so that the system of $(n-1)$ forces and their resultant reversed can hold in equilibrium a framework of jointed bars in the form of an equiangular polygon of n sides, a force acting at each corner.

[St John's Coll., 1890.]

Ex. 11.   Four points *A*, *B*, *C*, *D* are in equilibrium under forces acting between every two: prove the following construction for a force diagram of the system. With focus *D* a conic is described touching the sides of the triangle *ABC*, and *D′* is its second focus; *D′A′*, *D′B′*, *D′C′* are drawn perpendicular to the sides of the triangle *ABC*; then *D′A′B′C′* is a force diagram in which each side is perpendicular to the force it represents.   [Math. Tripos.]

Let *AD* cut *B′C′* in *P*; we notice (1) that *AD*, *AD′* make equal angles with the tangents drawn from *A*, hence the angles *PAC′*, *B′AD′* are equal; (2) that a circle can be described about *D′B′C′A′*, hence the angles *AC′P*, *AD′B′* are equal.   It follows that the triangles *PAC′*, *B′AD′* are equiangular.   Hence *AD* is perpendicular to *B′C′*.

Ex. 12.   Nine weightless rods are jointed together at their ends; six of them form the perimeter of a regular hexagon, and the other three each join one angular point to the opposite one; to each joint a weight *W* is attached, and the frame is hung in a vertical plane by strings attached to adjacent angles *A*, *B*, so that *AB* is horizontal, and the strings bisect the hexagon angles externally.   Find or show by a diagram the forces in all the rods.   [Coll. Ex., 1887.]

Ex. 13.   Two points *P*, *Q* are taken within a hexagon *ABCDEF*, the point *P* is joined to the corners *A*, *B*, *C*, *D*, and *Q* to the corners *D*, *E*, *F*, *A*.   Construct the reciprocal figure.

CHAPTER IX.

373. The centre of parallel forces. It has been proved in Art. 82 that the resultant of any number of parallel forces P_1, P_2, &c., acting at definite points A_1, A_2, &c., rigidly connected together, is a force ΣP.

Let the rigid system of points be moved about in any manner in space; let the forces P_1, P_2, &c. continue to act at these points, and let them retain unchanged their magnitudes and directions in space. It has also been proved that the line of action of the resultant always passes through a point fixed relatively to the points A_1, A_2, &c. This point is therefore regarded as the point of application of the resultant. It is called the centre of the parallel forces. The chief property of this point is its *fixity* relative to the system of points A_1, A_2, &c.

When the forces P_1, P_2, &c. are the weights of the particles of a body, the centre of parallel forces is called the centre of gravity. Thus the centre of gravity is a particular case of the centre of parallel forces.

374. *Definition of the centre of gravity.* We take as a system of parallel forces the weights of the several particles of a body. Each particle is supposed to be acted on by a force which is parallel to the vertical. This force is called gravity. The resultant of all these forces is the weight of the body. We infer from the theory of parallel forces that there is a certain point fixed in each body (or rigid system of bodies) such that in every position the line of action of the weight passes through that point. This point is called the centre of gravity *.

---

* The first idea of the centre of gravity is due to Archimedes, who flourished about 250 B.C. In his work on *Centres of gravity or aequiponderants* he determined the position of the centre of gravity of the parallelogram, the triangle, the ordinary rectilinear trapezium, the area of the parabola, the parabolic trapezium, &c. See the edition of his works in folio printed at the Clarendon Press, Oxford, 1792.

It is evident from this definition that if the centre of gravity of a body is supported the body will balance about it in *all* positions.

375. *A body has but one centre of gravity.* This is evident from the demonstration in the article already quoted. The following is an independent proof.

If possible let there be two such points, say A and B. As we turn the system into all positions, the resultant keeps its direction in space unaltered. Place the body so that the straight line AB is perpendicular to the direction of the resultant force. Then the line of action of that force cannot pass through both A and B.

376. Let (x_1, y_1, z_1), (x_2, y_2, z_2) &c. be the coordinates of the points of application of the parallel forces P_1, P_2, &c. respectively. Let these coordinates be referred to any axes, rectangular or oblique, but fixed in the system. By what has been already proved in Art. 80, the coordinates of the centre of parallel forces are

$$\bar{x} = \frac{\Sigma Px}{\Sigma P}, \qquad \bar{y} = \frac{\Sigma Py}{\Sigma P}, \qquad \bar{z} = \frac{\Sigma Pz}{\Sigma P}.$$

It is important to notice that, if all the forces were altered in the same ratio, the magnitude of the resultant would also be altered in the same ratio, but the coordinates of its point of application would not be changed.

377. When the weight of *any* two equal volumes of a substance are the same, the substance is said to be homogeneous or of uniform density. In such bodies the weights of different volumes are proportional to the volumes. The weight of any elementary volume dv may therefore be measured by the volume. Hence by Art. 376 we have

$$\bar{x} = \frac{\int dv \cdot x}{\int dv}, \qquad \bar{y} = \frac{\int dv \cdot y}{\int dv}, \qquad \bar{z} = \frac{\int dv \cdot z}{\int dv}.$$

We have here replaced the Σ by an integral, because the parallel forces we are considering are the weights of the elements of the body.

From these equations all trace of weight has disappeared. We might therefore call the point thus determined the *centre of volume*.

When the body is not homogeneous the weights of the elements are not proportional to their volumes. Let us represent the weight of a volume dv of the substance by ρdv. Here ρ will be different for each element of the body, and will be known as a function of the coordinates of the element when the structure of the body is given. For our present purpose the body is given when we know ρ as a function of x, y, z. We therefore have

$$\bar{x} = \frac{\int \rho dv \cdot x}{\int \rho dv}, \qquad \bar{y} = \frac{\int \rho dv \cdot y}{\int \rho dv}, \qquad \bar{z} = \frac{\int \rho dv \cdot z}{\int \rho dv}.$$

In these equations we may replace ρ by $\kappa \rho$, where κ is any quantity which is the same for all the elements of the body. All that is necessary is that ρdv should be proportional to the weight of dv.

We may therefore define ρ to be the limiting ratio of the weight of a small volume (enclosing the point (xyz)) to the weight of an equal volume of some standard homogeneous substance.

For the sake of brevity we shall speak of ρ as the density of the body. If the body is homogeneous the product of the density into the volume is called the mass. If heterogeneous, then ρdv is the mass of the elementary volume dv, and $\int \rho dv$ is the mass of the whole body. If we write $dm = \rho dv$, the equations become

$$\bar{x} = \frac{\int dm \cdot x}{\int dm}, \qquad \bar{y} = \frac{\int dm \cdot y}{\int dm}, \qquad \bar{z} = \frac{\int dm \cdot z}{\int dm}.$$

When we wish to regard the mass of an element as a quality of the body apart from its weight, we may speak of the point determined by these equations as the *centre of mass*.

378. Equations similar to these occur in other investigations besides those which relate to parallel forces. In such cases the quantity here denoted by P or m has some other meaning. Accordingly the point defined by these coordinates has had other names given to it, depending on the train of reasoning by which the equation has been reached. This may appear to complicate matters, but it has the advantage that the special name adopted in any case helps the reader to understand the particular property of the point to which attention is called.

We here arrive at the point as that particular case of the centre of parallel forces in which the forces are due to gravity. There may therefore be some propriety in using the term *centre of gravity*. There are also obvious advantages in using the short and colourless term of *centroid*. Another name, much used, is the *centre of inertia*. This expresses a dynamical property of the point which cannot be properly discussed in a treatise on statics.

379. The positions of the centres of gravity of many bodies are evident by inspection. Thus the centre of gravity of two equal particles is the middle point of the straight line which joins them. The centre of gravity of a uniform thin straight rod is at its middle point. The centre of gravity of a thin uniform circular disc is at its centre. Generally, if a body is symmetrical about a point, that point is the centre of gravity. If the body is symmetrical about an axis, the centre of gravity lies in that axis, and so on.

380. Working rule. To find the centre of gravity of any body or system of bodies, we proceed in the following manner. We divide the body or system into portions which may be either *finite in size or elementary.* But they *must be such that we know both the mass and position of the centre of gravity of each.* Let m_1, m_2, &c. be the masses of these portions, and let the coordinates of their respective centres of gravity be (x_1, y_1, z_1), (x_2, y_2, z_2), &c.

The weight of each portion is the resultant of the weights of the elementary particles, and may be supposed to act at the centre of gravity of that portion (Art. 82). We may therefore regard the whole body as acted on by a system of parallel forces whose magnitudes are proportional to m_1, m_2, &c., and whose points of application are the centres of gravity of m_1, m_2, &c. The position of the centre of gravity of the whole system is therefore found by substituting in the formulæ

$$\bar{x} = \frac{\Sigma mx}{\Sigma m}, \qquad \bar{y} = \frac{\Sigma my}{\Sigma m}, \qquad \bar{z} = \frac{\Sigma mz}{\Sigma m}.$$

381. In using this rule it is important to notice that *some of the masses may be negative.* Thus suppose one of the bodies is such that its mass and centre of gravity would be known if only *a certain vacant space were filled up.* We regard such a body as the difference of two bodies, one filling the whole volume of the body (including the vacant space) whose particles are acted on by gravity in the usual manner, the other filling the vacant space but such that its particles are acted on by forces equal and opposite to that of gravity. To represent this reversal of the direction of gravity it is sufficient to regard the mass of the latter body as negative. Since in the theory of parallel forces the forces may have any signs, it is clear that we may use the same formulæ to find the centre of gravity of this new system.

382. Ex. 1. A painter's palette is formed by cutting a small circle of radius b from a circular disc of radius a. It is required to find the distance of the centre of gravity of the remainder from the centre of the larger circle.

Let O and C be the centres of the larger and smaller circles respectively. Let $OC = c$. We take O as the origin and OC as the axis of x. The masses of the two circles are proportional to their areas; we therefore put $m_1 = \pi a^2$, $m_2 = -\pi b^2$. The latter is regarded as negative because its material has been removed from the larger circle. The centres of gravity of the two circles are at their centres, hence $x_1 = 0$, $x_2 = c$. We have therefore $\bar{x} = \dfrac{\Sigma mx}{\Sigma m} = \dfrac{\pi a^2 . 0 - \pi b^2 . c}{\pi a^2 - \pi b^2} = \dfrac{-b^2 c}{a^2 - b^2}$.

The negative sign in the result implies that the centre of gravity of the palette is on the side of O opposite to C.

Ex. 2. If any number of bodies have their centres of gravity on the same straight line, the centre of gravity of the whole of them lies on that straight line.

Take the straight line as the axis of x, then the y and z of each centre of gravity are both zero. Hence by Art. 380 $\bar{y}=0$, and $\bar{z}=0$.

Ex. 3. Two particles of masses m_1, m_2 are placed at A, B respectively. Prove that their centre of gravity G divides the distance AB inversely in the ratio of the masses. Art. 53, Ex. 1.

Ex. 4. Three particles are placed at the corners of a triangle; if their weights, w_1, w_2, w_3, vary so that they satisfy the linear equation $lw_1+mw_2+nw_3=0$, show that the locus of their centre of gravity is a straight line. What is the areal equation to the straight line? Art. 53, Ex. 2.

Ex. 5. Four weights are placed at four given points in space, the sum of two of the weights is given, and also the sum of the other two: prove that their centre of gravity lies on a fixed plane.　　　　　　　　　　　　　　　　[Math. Tripos, 1869.]

Ex. 6. Water is poured gently into a cylindrical cup of uniform thickness and density; prove that the locus of the centre of gravity of the water, the cup, and its handle, is a hyperbola.　　　　　　　　　　　　　　[Math. Tripos, 1859.]

Ex. 7. Water is gently poured into a vessel of any form; prove that, when so much water has been poured in that the centre of gravity of the vessel and water is in the lowest possible position, it will be in the surface of the water. [Math. T., 1859.]

Ex. 8. In the figure of Euclid, Book i. Prop. 47, if the perimeters of the squares be regarded as physical·lines uniform throughout, prove that the figure will balance about the middle point of the hypothenuse with that line horizontal, the lines of construction having no weight.　　　　　　　　[Math. Tripos, 1860.]

If we take the hypothenuse as the axis of x and its middle point as origin, it follows immediately that $\bar{x}=0$.

383. Area of a triangle. *To find the centre of gravity of a uniform triangular area ABC.*

Let us divide the area of the triangle into elementary portions or strips by drawing straight lines parallel to one side of the triangle. Bisect BC in D and join AD, and let AD intersect any straight line PNQ drawn parallel to BC in N. Then by similar triangles

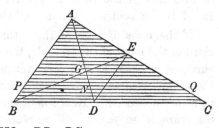

$$PN : NQ = BD : DC;$$

but $BD = DC$, hence PNQ is bisected in N. Thus every straight line drawn parallel to BC is bisected at its intersection with AD.

Since we can make each strip as narrow as we please, it follows that the centre of gravity of each (like that of a thin rod, Art. 379) is at its middle point. The centre of gravity of each strip therefore lies in AD. Hence the centre of gravity of the whole triangle lies in AD; see Art. 382, Ex. 2.

In the same way, if we draw BE from B to bisect AC in E, the centre of gravity lies in BE. The centre of gravity of the triangle is therefore at the intersection G of BE and AD.

Since D and E are the middle points of CB and CA, the triangle CED is similar to the triangle CAB. Hence ED is parallel to AB and is equal to one half of it. The triangles DEG, ABG are therefore also similar, and $DG : GA = ED : AB$. Thus DG is one half of AG, and therefore DG is one third of AD.

384. We have thus obtained two rules to find the centre of gravity of a uniform triangle.

(1) We may draw two median straight lines from any two angular points to bisect the opposite sides. The centre of gravity lies at their intersection.

(2) We may draw one median line from any one angular point, say A, to bisect the opposite side in D. The centre of gravity G lies in AD so that $AG = \frac{2}{3}AD$.

It will be found useful to observe that the centre of gravity of the area of the triangle is the same as that of three equal particles placed one at each angular point of the triangle.

Let the mass of each particle be m. The centre of gravity of the particles at B and C is the point D. The centre of gravity of all three is the same as that of $2m$ at D and m at A; it therefore divides AD in the ratio $1 : 2$ (Art. 382). But the point thus found is the centre of gravity of the triangle.

If the mass of each of these three particles is equal to one-third of the mass of the triangle, the resultant weight of the three particles is equal to the resultant weight of the triangle. And these two resultants have just been shown to have a common point of application. Hence *these three particles are equivalent to the triangle so far as all resolutions and moments of weights are concerned.*

Also, when we use the method of Art. 380 to find the centre of gravity of any figure composed of triangles, we may replace each of the triangles by three equivalent particles whose united mass is equal to that of the triangle. The centre of gravity of the whole figure may then be found by applying the rule to this collection of particles.

385. Ex. 1. The centre of gravity of the area of a triangle is the same as the centre of gravity of three equal particles placed one at each of the middle points of the sides.

Ex. 2. Lengths AP, BQ, CR are measured from the angular points of a triangle along the sides taken in order so that each length is proportional to the side along which it is measured. Show that the centre of gravity of three equal particles placed one at each of the points P, Q, R is the same as that of the triangle.

Prove also that the centres of gravity of the triangles APR, BQP, CRQ, lie on the sides of a fixed triangle, which is similar and equal to ABC.

Ex. 3. Lengths $AP, BQ,$ &c. are measured from the corners of a polygon along the sides taken in order so that each length is proportional to the side along which it is measured, the sides not being necessarily in one plane. Show that the centre of gravity of equal particles placed at $P, Q,$ &c. coincides with that of equal particles placed at the corners. Art. 79.

Ex. 4. Similar triangles $ABP, BCQ,$ &c. are described on the sides $AB, BC,$ &c. of a plane polygon taken in order. Show that the centre of gravity of equal weights placed at $P, Q,$ &c. coincides with that of equal weights placed at $A, B,$ &c.

Ex. 5. The perpendiculars from the angles A, B, C meet the sides of a triangle in P, Q, R: prove that the centre of gravity of six particles proportional respectively to $\sin^2 A, \sin^2 B, \sin^2 C, \cos^2 A, \cos^2 B, \cos^2 C$, placed at A, B, C, P, Q, R, coincides with that of the triangle PQR.   [Math. Tripos, 1872.]

Ex. 6. A point G is taken inside a tetrahedron $ABCD$. Find by a geometrical construction the plane section which having its corners on the edges DA, DB, DC, has its centre of gravity at G. Find also the limiting positions of G that the construction may be possible.

386. Perimeter of a triangle. Ex. 1. A triangle ABC is formed by three thin rods whose lengths are a, b, c. If H be the centre of gravity, prove that the areal coordinates of H are proportional to $b+c, c+a, a+b$.

Ex. 2. The centre of gravity of the perimeter of a triangle ABC is the centre of the circle inscribed in the triangle DEF, where D, E, F are the middle points of the sides of the triangle ABC.   [Lock's *Statics*.]

Ex. 3. If H be the centre of gravity of the perimeter of a triangle, G the centre of gravity of the area, I the centre of the inscribed circle, prove that H, G, I are in one straight line, and that GH is one half of IG. If O be the centre of the circumscribing circle, and P the orthocentre, show also that the triangles IGP, HGO are similar.

Ex. 4. The sides of a polygon are of equal weight. Prove that the centre of gravity of the perimeter coincides with that of equal particles placed at the corners. Art. 385, Ex. 3.

387. Quadrilateral areas. *To find the centre of gravity of any quadrilateral area $ABCD$.*

Using the rule in Art. 380, we replace the triangle ADC by three particles situated at A, D, C respectively, each equal to one-third of the mass of ADC. In the same way we replace the triangle ABC by three masses at A, B, C, each one-third of the mass of ABC. Each of the masses at A and C is therefore $\frac{1}{3}M$, if M be the mass of the whole quadrilateral.

Consider next the masses at B and D; call these m_1 and m_2. Their united mass is also $\frac{1}{3}M$, but this total mass is unequally divided between the particles in the ratio of the triangles $ABC : ADC$, i.e. in the ratio $BE : ED$. To obtain a more

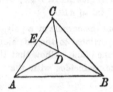 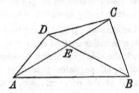

convenient distribution, let us replace these two masses by three others placed at B, D, and E. If the masses placed at B and D are each $\frac{1}{3}M$ and the mass placed at E is $-\frac{1}{3}M$, the sum of the masses is the same as before. It is also clear that their centre of gravity is the same as that of the masses m_1 and m_2. For by Art. 380 the distance of their centre of gravity from E is given by

$$\bar{x} = \frac{\Sigma mx}{\Sigma m} = \frac{\frac{1}{3}M \cdot BE - \frac{1}{3}M \cdot DE + \frac{1}{3}M \cdot 0}{\frac{1}{3}M}.$$

But the distance of the centre of gravity of the masses m_1, m_2 from E is given by

$$\bar{x} = \frac{m_1 \cdot BE - m_2 \cdot DE}{m_1 + m_2} = \frac{BE^2 - DE^2}{BE + DE},$$

which is the same as before.

The centre of gravity of the area of the quadrilateral is therefore the same as that of four equal particles, placed one at each angular point of the quadrilateral, together with a fifth particle of equal but negative mass, placed at the intersection of the diagonals.

We may put the result of this rule into an analytical form. Let (x_1, y_1), (x_2, y_2), &c. be the coordinates of the four angular points and of the intersection of the diagonals, then clearly

$$\bar{x} = \frac{1}{3}(x_1 + x_2 + x_3 + x_4 - x_5),$$

with a similar expression for $\bar{y}$. See the *Quarterly Journal of Mathematics*, vol. XI. 1871, p. 109.

The reader is advised to use the rule of equivalent points

partly because the analytical result follows at once, and partly because these equivalent points are used in rigid dynamics to enable us to write down the moments and products of inertia of a quadrilateral.

We may replace the four particles at the angular points by four others, equal to these, placed at the middle points of the sides, or in any of the equivalent positions described in Art. 385.

388. Ex. 1. Prove the following *geometrical* construction for the centre of gravity of a quadrilateral area. Let P, Q be points in BD, AC such that QA, PB are equal respectively to EC, ED; the centre of gravity of the quadrilateral coincides with that of the triangle EPQ. *Quarterly Journal of Mathematics*, vol. VI. 1864.

Ex. 2. A quadrilateral is divided into two triangles by one diagonal BD, and the centres of gravity of these triangles are M and N. Let MN cut BD in I, from the greater NI take NG equal to MI the lesser. Prove that G is the centre of gravity of the area of the quadrilateral.   [Guldin.]

Ex. 3. A trapezium has the two sides $AB=a$ and $CD=b$ parallel. Prove that the centre of gravity G of the quadrilateral area lies in the straight line joining the middle points M and N of AB and CD. Prove also that G divides MN so that $MG : GN = a + 2b : 2a + b$.   [Archimedes and Guldin.]

Notice that the ratio $MG : GN$ does not depend on the height of the trapezium but only on the lengths of the parallel sides.   [Poinsot.]

Ex. 4. Show that the centre of gravity of the quadrilateral area $ABCD$ coincides with that of four particles placed at the corners whose weights are respectively $\beta + \gamma + \delta$, $\gamma + \delta + a$, $\delta + a + \beta$, $a + \beta + \gamma$ where a, β, γ, δ are the reciprocals of EA, EB, EC, ED and E is the intersection of the diagonals.

[Caius Coll. 1877.]

Ex. 5. Any corner C of a pentagonal area $ABCDE$ is joined to the corners A, E, and the joining lines intersect EB, AD in F, G. Prove that the ordinate z of the centre of gravity of the pentagonal area is given by

$$3z = b + c + d - \frac{f + g - a - e}{1 - n}, \qquad n = \frac{(b-f)(d-g)}{(b-e)(d-a)}$$

where a, b, c, d, e, f, g are the ordinates of A, B, C, D, E, F, G, referred to any plane of xy.

389. Tetrahedron. *To find the centre of gravity of a tetra-hedron $ABCD$.*

Let us divide the tetrahedron into elementary slices by drawing planes parallel to one face. Let abc be one of these planes. Bisect BC in E and join DE, then, exactly as in the case of the triangle, DE will bisect all straight lines such as bc which are parallel to BC. Join AE and ae, then these are parallel to each other. Take $AF = \frac{2}{3}AE$, then F is the centre of gravity of the base ABC. Join DF and let it cut ae in f, then by similar triangles $af : AF = Da : DA = ae : AE$. Hence $af = \frac{2}{3}ae$, that is f

is the centre of gravity of the triangle abc. It therefore follows
that the centre of gravity of every elementary slice lies in DF.
Hence the centre of gravity of the whole tetrahedron lies in DF.
Thus *the centre of gravity of a tetrahedron lies in the straight line
which joins any angular point to the centre of gravity of the opposite
face.*

Let K be the centre of gravity of the face BCD; join AK.

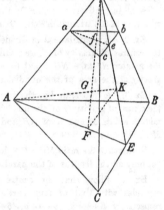

The centre of gravity also lies in
AK. Now both DF and AK lie
in the plane DAE, they therefore
intersect and the intersection G is
the required centre of gravity.

Exactly as in the corresponding
theorem for a triangle, we have FK
parallel to AD and $=\frac{1}{3}AD$. Hence
from the similar triangles AGD,
KGF, we see that $FG=\frac{1}{3}GD$. Thus
$DG=\frac{3}{4}DF$.

*To find the centre of gravity of
a tetrahedron we join any corner
(as D) to the centre of gravity (as F)
of the opposite face. The centre of gravity G lies in DF so that
$DG=\frac{3}{4}DF$.*

As in the case of a triangle, we may fix the position of the
centre of gravity of a tetrahedron by means of some equivalent
points. The *centre of gravity of a tetrahedron is the same as that
of four equal particles placed one at each angular point.* The
proof is exactly similar to that for a triangle.

390. Pyramid and Cone. *To find the centre of gravity of
the volume of a pyramid on a plane rectilinear base.*

Proceeding as in the case of the tetrahedron, we divide the
pyramid into elementary slices by drawing planes parallel to the
base. These sections are all similar to the base. The centre of
gravity of each slice, and therefore that of the whole pyramid, lies
in the straight line joining the vertex of the pyramid to the centre
of gravity of the base.

Next, we may divide the base into triangles. By joining the
angular points of these triangles to the vertex, we divide the whole
pyramid into tetrahedra having a common vertex. The centre

of each tetrahedron, and therefore that of the pyramid, lies in a plane parallel to the base such that its distance from the vertex is $\frac{3}{4}$ of the distance of the base.

Joining these two results together, we have the following rule to find the centre of gravity of a pyramid. *Join the vertex V to the centre of gravity F of the base and measure along VF from the vertex a length VG equal to three quarters of VF. Then G is the centre of gravity of the pyramid.*

When the base of the pyramid is curvilinear we regard the base as the limit of a polygon with an infinite number of elementary sides. We have therefore the following rule. *To find the centre of gravity of the volume of a cone on a circular or on an elliptic base; join the vertex V to the centre of gravity F of the base, and measure along VF from the vertex a length VG equal to three quarters of VF, then G is the centre of gravity of the cone.*

391. Ex. 1. A cone whose semivertical angle is $\tan^{-1} 1/\sqrt{2}$ is enclosed in the circumscribing sphere; show that it will rest in any position. [Math. T., 1851.]

Ex. 2. A pyramid, of which the base is a square, and the other faces equal isosceles triangles, is placed in the circumscribing spherical surface; prove that it will rest in any position if the cosine of the vertical angle of each of the triangular faces be $\frac{3}{4}$. [Math. Tripos, 1859.]

Ex. 3. A frustum of a tetrahedron is bounded by parallel faces ABC, $A'B'C'$. Prove that its centre of gravity G lies in the straight line joining the centres of gravity E, E' of the faces ABC, $A'B'C'$ and is such that $\dfrac{EG}{EE'} = \dfrac{1 + 2n + 3n^2}{4(1 + n + n^2)}$ where n is the ratio of any side of the triangle $A'B'C'$ to the corresponding side of the triangle ABC. [Poinsot.]

Ex. 4. A frustum of a tetrahedron $ABCD$ is bounded by faces ABC, $A'B'C'$ not necessarily parallel. Find its centre of gravity.

Let DA, DB, DC be regarded as a system of oblique axes, let the distances of A, B, C, A', B', C' from D be a, b, c, a', b', c'. Then

$$\bar{x} = \frac{3}{4}\frac{a^2bc - a'^2b'c'}{abc - a'b'c'}, \qquad \bar{y} = \frac{3}{4}\frac{ab^2c - a'b'^2c'}{abc - a'b'c'}, \qquad \bar{z} = \frac{3}{4}\frac{abc^2 - a'b'c'^2}{abc - a'b'c'}.$$

To prove these results, we regard the tetrahedra as the difference of two tetrahedra whose volumes are as $abc : a'b'c'$.

Ex. 5. The top of a right cone, semivertical angle a, cut off by a plane making an angle β with the axis, is placed on a perfectly rough inclined plane with the major axis of the base along a line of greatest slope of the plane; in this position the cone is on the point of toppling over: prove that the tangent of the inclination of the plane to the horizon has one of the values $\dfrac{4\sin 2a \pm \sin 2\beta}{\cos 2a - \cos 2\beta}$. [Math. T., 1876.]

392. Faces and edges of a tetrahedron. Ex. 1. Prove that the centre of gravity of the edges coincides with that of four weights placed at the corners equal respectively to the sum of the weights of the three edges which meet at that corner. Prove also that the same theorem is true if we read faces for edges, Arts. 79 and 86.

Ex. 2. The centre of gravity of the four faces of a tetrahedron is the centre of the sphere inscribed in a tetrahedron whose corners are the centres of gravity of the faces of the original tetrahedron.

Ex. 3. If H be the centre of gravity of the faces of a tetrahedron, G the centre of gravity of the volume, I the centre of the inscribed sphere, then H, G, I are in one straight line and HG is equal to one third of GI.

Ex. 4. The straight lines which join the middle points of opposite edges of a tetrahedron are called *the median lines*. Show that the medians pass through the centre of gravity G of the volume and are bisected by it.

Place particles of equal weight at the corners A, B, C, D. The centres of gravity of the particles at A, B and C, D are respectively at the middle points M, N of the edges AB, CD. Hence the centre of gravity of all four is at the middle point G of MN.

Ex. 5. A polyhedron circumscribes a sphere; show that the centres of gravity of the volume and of the surface, viz. G and H, and the centre O lie in the same straight line and that $OG = \frac{3}{4}OH$.   [Liouville's J., 1843.]

393. The isosceles tetrahedron. *An isosceles tetrahedron is one whose opposite edges are equal.* It follows from this definition that the sides of any two faces are equal each to each.

Ex. 1. Show that the following five points are coincident, viz. (1) the centre of gravity of the volume, (2) the centre of gravity of the six edges, (3) the centre of gravity of the four faces, (4) the centre of the circumscribing sphere, (5) the centre of the inscribed sphere. Let this point be called G.

Ex. 2. Show that the medians pass through G, are bisected by it and are perpendicular to their corresponding edges. Show also that the three medians are at right angles and form a system of three rectangular axes. See Casey's *Spherical Trigonometry*, 1889, Art. 127.

Let M, N, P, Q, R, S be the middle points of the edges AB, CD, BD, AC, AD, BC. Then PR, QS are parallel to AB and each is half AB; similarly PS, QR are parallel and equal to half CD. Since the opposite edges AB, CD are equal, it follows that $PQRS$ is a rhombus, and therefore that the diagonals or medians PQ, RS are at right angles. The median MN being perpendicular to the plane containing PQ, RS is perpendicular to PR, QS and therefore to the edge AB.

394. Double tetrahedra. *To find the centre of gravity of the solid bounded by six triangular faces, i.e. contained by two tetrahedra having a common face.*

Let the common base be ABC and D, D' the vertices. Join DD', and let it cut the base in E. We replace the tetrahedron $ABCD$ by four particles, each one-fourth its mass situated at the points A, B, C, D. Treating the other tetrahedron in the same way, we have at each of the points A, B, C a particle whose mass is equal to one-fourth of the solid, and at D, D' two particles whose united mass makes up the remaining fourth of the solid, and whose separate masses are in the ratio of the tetrahedra, i.e. in the ratio $DE : ED'$. Following exactly the steps of the reasoning in the case of a quadrilateral, it is easy to see that we can replace these two masses by two other

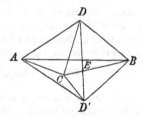

masses situated at D and D', and each one-fourth that of the whole solid, together with a third particle situated at E of the same mass but taken negatively. *The centre of gravity of the whole solid is the same as that of five equal particles placed at A, B, C, D, D' together with a sixth particle equal and opposite to any of the five placed at the intersection of DD' with the common face ABC.*

395. **Ex.** The centre of gravity of a pyramid on a plane quadrilateral base is the same as that of five equal particles placed at the five apices, and a sixth equal but negative particle placed at the intersection of the diagonals of the base. [To prove this draw a plane through the vertex and a diagonal of the base; the solid then becomes two tetrahedra joined together at a common face.]

396. Circular arc. *To find the centre of gravity of an arc of a circle.*

Let ACB be the arc, O its centre. Let the radius OC bisect the arc, let $OC = a$, and the angle $AOB = 2\alpha$. Let PQ be any element of the arc, and let the angle $POC = \theta$. Then in the fundamental formula of Art. 380 $m = ad\theta$, $x = a \cos \theta$. If $\bar{x}$ be the distance of the centre of gravity of the arc from O,

$$\bar{x} = \frac{\Sigma mx}{\Sigma m} = \frac{\int ad\theta . a \cos \theta}{\int ad\theta} = a \frac{\sin \alpha}{\alpha},$$

since the limits of θ are $\theta = -\alpha$ and $\theta = +\alpha$. As this result is frequently used, it will be convenient to put it into a form which will be convenient for reference.

$$\text{Distance of c. g. of arc from centre} \left.\right\} = \frac{\sin (\text{half angle})}{\text{half angle}} . \text{rad.} = \frac{\text{chord}}{\text{arc}} . \text{rad.}$$

This result was given by Wallis.

397. **Ex.** A series of $2n$ straight lines are inscribed in a circular arc, each straight line subtending an angle 2θ at the centre. Prove that the distance of their centre of gravity from the centre is $r \cos \theta \sin 2n\theta / 2n \sin \theta$. Thence deduce the centre of gravity of a circular arc of any angle.   [Guldin's Problem.]

398. Centre of gravity of any arc. The coordinates of the centre of gravity of the arc of any uniform plane curve are given by the formulæ

$$\bar{x} = \frac{\Sigma mx}{\Sigma m} = \frac{\int x ds}{\int ds}, \qquad \bar{y} = \frac{\int y ds}{\int ds},$$

where we write for the elementary arc ds its value given in the differential calculus. Thus we have

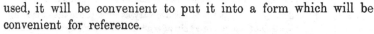

$$ds = \left\{1 + \left(\frac{dy}{dx}\right)^2\right\}^{\frac{1}{2}} dx \ \text{or} \ ds = \left\{r^2 + \left(\frac{dr}{d\theta}\right)^2\right\}^{\frac{1}{2}} d\theta,$$

according as the equation to the curve is given in the Cartesian form $y = f(x)$ or the polar form $r = F(\theta)$. If the curve be in three dimensions we have an expression for $\bar{z}$ similar to those written above. The corresponding expressions for ds are given in works on the differential calculus.

399. The process of finding the centre of gravity of an arc is merely that of substituting for ds from the given equation to the curve and then integrating. It seems unnecessary to give at length examples of what is merely integration, we shall therefore state only the results in a few cases likely to be useful.

Ex. 1. The coordinates of the centre of gravity of an arc of the catenary $y = \frac{c}{2} (e^{\frac{x}{c}} + e^{-\frac{x}{c}})$ from $x=0$ to $x=x$ are $\bar{x} = x - \frac{c(y-c)}{s}$, $\bar{y} = \frac{1}{2} \left(y + \frac{cx}{s} \right)$.

These admit of a geometrical interpretation. Let PQ be *any* arc of the catenary. Let the tangents at P and Q meet in T and the normals at P and Q meet in N. If $\bar{x}, \bar{y}$ be the coordinates of the centre of gravity of the arc PQ, then $\bar{x}=$abscissa of T, and $\bar{y}=$half the ordinate of N.

Ex. 2. Find the centre of gravity of the arc OP of a cycloid between the vertex O where $\phi=0$ and the point P, the equations to the curve being $x=2a\phi+a \sin 2\phi$, $y=a-a \cos 2\phi$, and the arc OP being $s=4a \sin \phi$.

Result $\bar{x}=2a\phi - \dfrac{2a}{3} \dfrac{(1-\cos\phi)^2 (2+\cos\phi)}{\sin\phi}$, and $\bar{y}=\frac{1}{3}y$.

Ex. 3. If G be the centre of gravity of any arc AP of the lemniscate $r^2=a^2 \cos 2\theta$, prove that OG bisects the angle AOP. One case of this is given in Walton's *Problems on Theoretical Mechanics*.

Ex. 4. The centre of gravity of any arc PQ of the curve $r^3 \sin 3\theta = a^3$ lies in the straight line joining the origin to the intersection of the tangents at P and Q.

Ex. 5. If the density at any point of the arc vary as r^{n-3}, prove that the centre of gravity of any arc PQ of the curve $r^n \sin n\theta = a^n$ lies in the straight line joining the origin to the intersection of the tangents at P and Q.

Ex. 6. The locus of the centre of gravity of an arc of given length of the lemniscate $r^2=a^2 \cos 2\theta$ is a curve which is the inverse of a concentric ellipse.

[R. A. Robert's theorem.]

400. Sectors of circles. *To find the centre of gravity of a sector of a circle.*

Let ACB be the arc of the sector, O its centre. As in Art. 396 let the radius OC bisect the arc, $OC = a$ and the angle $AOB = 2\alpha$.

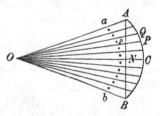

We divide the sector into elementary triangles of equal area. Let OPQ be any one of these triangles; following the rule of Art. 380 we collect its mass into its centre of gravity, i.e. into a point p where $Op = \frac{2}{3}OP$. Repeating this process for every triangle, we have a series of particles of equal mass arranged at equal distances along an arc ab of a circle. These are represented in the figure by the row of dots. In the limit when the triangles are infinitely small this becomes a homogeneous arc of a circle. The distance of the centre of gravity of the sector from O is therefore given by the result in Art. 396, viz.

$$\bar{x} = \frac{\sin \alpha}{\alpha} \tfrac{2}{3}a = \tfrac{2}{3}\frac{\text{chord } AB}{\text{arc } AB} . \text{radius } OC.$$

This result was given by Wallis.

401. Ex. *To find the coordinates of the centre of gravity of the area of a quadrant of a circle AOB.*

This is a particular case of the last article, viz. when $\alpha = \tfrac{1}{4}\pi$. If $\bar{x}, \bar{y}$ be the coordinates of G referred to OA, OB as axes, we have $\bar{x} = OG \cos \alpha = \dfrac{4a}{3\pi}$; $\bar{y} = \dfrac{4a}{3\pi}$.

402. Ex. The distance of the centre of gravity of the area of a segment of a circle measured from the centre is $\tfrac{2}{3} \dfrac{a \sin^3 \alpha}{a - \sin \alpha \cos \alpha}$, where a is the semiangle of the segment. [Guldin.]

403. Projection of areas. *If any plane area is orthogonally projected on any other plane, the centre of gravity of the projection is the projection of the centre of gravity of the primitive area.*

Let the plane on which the projection is made be the plane of xy, and let α be the inclination of the two planes. Let dS be any element of the area of the primitive, $d\Pi$ the area of its projection. Then by a known theorem in conics $d\Pi = dS \cos \alpha$. We also notice that the x and y coordinates of dS and $d\Pi$ are the same because the projection is orthogonal. The coordinates of the centre of gravity of either area are known from $\bar{x} = \dfrac{\Sigma mx}{\Sigma m}$, $\bar{y} = \dfrac{\Sigma my}{\Sigma m}$, where the m for one area is $d\Pi$ and for the other is dS. Since these are in a constant ratio, the values of $\bar{x}$ and $\bar{y}$ are the same for each area.

In order to use effectively the method of projections we join to it the two following well known theorems which are proved in books on conics: (1) the projections of parallel straight lines are parallel, (2) ratio of lengths of two parallel straight lines is unaltered by projection. *We then use the following rule.*

Suppose we had any geometrical relation between the lengths of lines in the primitive figure, and that we require the corresponding relation in the projected figure. We first express the given relation in the form of ratios of lengths of parallel straight lines. To do this it may be necessary to draw parallels to some of the lines in the primitive if there are no parallels to them mentioned in the given relation. Having put the geometrical relation into the form of ratios, the same relation is true for the projected figure.

404. Elliptic areas. Since an elliptic area is well known to be the orthogonal projection of a circle, we can deduce the centres of gravity of the various parts of an ellipse from those of the corresponding parts of a circle. The circle used for this purpose is sometimes called in conics the auxiliary circle.

405. *To find the centre of gravity of an elliptic area.*

The coordinates of the centre of gravity of a quadrant AOB of a circle, referred to OA, OB as axes, may be written in the form

$$\frac{\bar{x}}{OA} = \frac{\bar{y}}{OB} = \frac{4}{3\pi} \quad \dots\dots\dots\dots\dots\dots(1)$$

since OA, OB are both radii. But $\bar{x}$ and OA are parallel straight lines, and so also are $\bar{y}$ and OB. Hence these relations hold in the projected figure also.

If then OA, OB are the major and minor semiaxes of an ellipse, the coordinates of the centre of gravity of the area of the quadrant are given by (1).

If we make the plane on which we project intersect the quadrant of the circle in any straight line not one of the bounding radii the circular quadrant projects into an elliptic quadrant bounded by two conjugate diameters.

If then OA, OB are any two semiconjugates of an ellipse, the coordinates of the centre of gravity of the contained area are given by equations (1).

The position of the centre of gravity of a semi-ellipse was first found by Guldin.

406. Ex. 1. A chord PQ of an ellipse, centre C, passes always through a fixed point O. Prove that the locus of the centre of gravity of the triangle CPQ is a similar ellipse. [Coll. Exam.]

Ex. 2. The centre of gravity G of any elliptic sector bounded by the semi-diameters OP, OP' lies in the diameter OA' bisecting the chord PP', and is such that $\dfrac{OG}{OA'} = \dfrac{2}{3}\dfrac{\sin\theta}{\theta}$, where $\sin\theta$ is the ratio of half the chord PP' to the semiconjugate of OA'.

Ex. 3. The area A of any elliptic sector POP' is $A = \frac{1}{2} ab \, (\phi' - \phi)$, and the coordinates of the centre of gravity referred to the principal diameters, are

$$\frac{\bar{x}}{a} = \frac{2}{3} \frac{\sin \phi' - \sin \phi}{\phi' - \phi}, \qquad \frac{\bar{y}}{b} = \frac{2}{3} \frac{\cos \phi - \cos \phi'}{\phi' - \phi},$$

where ϕ, ϕ' are the eccentric angles of P and P'.

Ex. 4. Show that the centre of gravity G' of the elliptic segment bounded by any chord PP' is given by $OG' = \frac{2}{3} \dfrac{OA' \sin^3 \phi}{\phi - \sin \phi \cos \phi}$, where OA' is the conjugate of PP' and $\sin \phi$ is the ratio of PP' to the parallel diameter.

Ex. 5. The centre of gravity G of the area included between an ellipse and the two tangents drawn from any point T in the diameter OA' produced is given by

$$\frac{OG}{OA'} = \frac{1}{3} \frac{\tan^2 \phi \sin \phi}{\tan \phi - \phi}.$$

where $\sin \phi$ is the ratio of half the chord PP' of contact to the semiconjugate of OT.

Show also that the coordinates of G referred to the tangents TP, TP' as axes are

$$\frac{\bar{x}}{TP} = \frac{\bar{y}}{TP'} = \frac{1}{2} \frac{1}{\sin^2 \phi} \left(1 - \frac{1}{3} \frac{\tan \phi \sin^2 \phi}{\tan \phi - \phi} \right).$$

In the parabola, we have by rejecting the higher powers of ϕ, $\bar{x} = \frac{1}{5} TP$, $\bar{y} = \frac{1}{5} TP'$.

Ex. 6. The coordinates of the centre of gravity of the quadrilateral space bounded by arcs of four concentric and coaxial ellipses are

$$\bar{x} = \frac{2}{3} \frac{a^2{}_1 b_1 (\sin \phi_1' - \sin \phi_1) + a^2{}_2 b_2 (\sin \phi_2' - \sin \phi_2) + \&c.}{a_1 b_1 (\phi_1' - \phi_1) + a_2 b_2 (\phi_2' - \phi_2) + \&c.}$$

and a similar expression for $\bar{y}$.

407. Analytical Aspect of Projections. The geometrical method which has just been used in projecting the ellipse into the circle, or conversely, is really equivalent to a change of coordinates. We write $x = x'$, $y = gy'$, where g is a quantity at our disposal, which we so choose that the equation to the ellipse reduces to the simpler form of a circle. We can obviously extend this principle and apply it to any curve. Let us write $x = fx'$, $y = gy'$; we thus have *two* constants instead of one to choose as we please.

Geometrically this is equivalent to two successive projections. By writing $y = gy'$ we project the primitive on a plane passing through the axis of x, and then by writing $x = fx'$ we project the projection on another plane passing through the axis of y'. We may therefore in this generalized projection assume the two theorems of projection already mentioned, and transform all formulæ relating to ratios of parallel lengths from one figure to the other.

Analytically, let the equations to the several boundaries of any area A be changed into those of A' by writing $x = fx'$, $y = gy'$. Let $(\bar{x}, \bar{y})$, $(\bar{x}', \bar{y}')$ be the coordinates of the centres of gravity of A and A'. Then we have

$$A = \iint dx \, dy = fg \iint dx' \, dy' = fg A'.$$

In the same way $\bar{x} = f\bar{x}'$ and $\bar{y} = g\bar{y}'$. In these integrals the limits extend over corresponding areas.

Ex. Show that we may further generalize the method of projections by writing $x = a + bx' + cy'$, $y = e + fx' + gy'$. If A, A' be the areas of corresponding spaces, prove that $A = A' (bg - cf)$, $\bar{x} = a + b\bar{x}' + c\bar{y}'$, $\bar{y} = e + f\bar{x}' + g\bar{y}'$.

Notice that this is equivalent to a transformation to a new origin with oblique axes, followed by the projections.

408. The method of projection does not apply so conveniently to find the centres of gravity of hyperbolic areas because we have to use imaginary projections. By projecting the rectangular hyperbola instead of the circle we may find the centre of gravity of any hyperbolic area.

We may however infer from any *general* proposition proved for the ellipse the corresponding theorem for the hyperbola by using the law of continuity. For example, (see Ex. 2, Art. 406) the centre of gravity of a sector of an ellipse from $x=x$ to $x=a$ is given by $\bar{x}=\frac{2}{3} ak/\sin^{-1}k$, where k has been written for $(1-x^2/a^2)^{\frac{1}{2}}$ for the sake of brevity. This must be true also for the imaginary branches of the ellipse which originate in values of $x>a$. Put $k=k'\sqrt{-1}$ and use the formula in analytical trigonometry, $\theta\sqrt{(-1)}=\log(\cos\theta+\sqrt{-1}\sin\theta)$, where $\theta=\sin^{-1}k$; we find for the centre of gravity of a hyperbolic sector

$$\frac{x}{a}=\frac{2}{3}\frac{k'}{\log(k'+\sqrt{k'^2+1})}, \text{ where } k'=\left\{\left(\frac{x}{a}\right)^2-1\right\}^{\frac{1}{2}}.$$

409. Centre of gravity of any area.

After having obtained the fundamental formulæ of Art. 380 the discovery of the centres of gravity of any area is reduced to two processes. (1) We have to make a judicious choice of the element m, and (2) we have to effect the necessary integrations. The latter process is fully discussed in treatises on the integral calculus, in fact it is a part of that science rather than of statics. It will thus be unnecessary to do more here than make a few remarks on the choice of m with special reference to centres of gravity.

If the centre of gravity of the area bounded by two ordinates Aa, Bb be required, we put the equation of the curve into the form $y=f(x)$. We choose as our element the strip PQM. Here $PM=y$ and $m=ydx$. The coordinates of the centre of gravity of m are x and $\frac{1}{2}y$. Hence, Art. 380, the formulæ to be used are

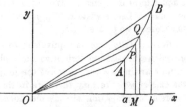

$$\bar{x}=\frac{\Sigma mx}{\Sigma m}=\frac{\int ydx.x}{\int ydx}, \qquad \bar{y}=\frac{\int ydx.\frac{1}{2}y}{\int ydx}.$$

If the centre of gravity of the sectorial area AOB is wanted, we put the equation into the form $r=f(\theta)$. We choose as our element the triangular strip POQ. Here $OP=r$, and $m=\frac{1}{2}r^2d\theta$. The Cartesian coordinates of the centre of gravity of m are $\frac{2}{3}r\cos\theta$ and $\frac{2}{3}r\sin\theta$. The formulæ to be used are

$$\bar{x}=\frac{\int \frac{1}{2}r^2d\theta.\frac{2}{3}r\cos\theta}{\int \frac{1}{2}r^2d\theta}, \qquad \bar{y}=\frac{\int \frac{1}{2}r^2d\theta.\frac{2}{3}r\sin\theta}{\int \frac{1}{2}r^2d\theta}.$$

Sometimes the equation to the curve is given with an auxiliary variable t, thus $x=\phi(t)$, $y=\psi(t)$. It is in this form for example that the equation to the cycloid is generally given. See Ex. 2, Art. 399. In this case when the polar area is required we quote from the differential calculus the formula $r\cdot d\theta=xdy-ydx$.

Substituting half of this for m in the standard expressions for $\bar{x}$ and $\bar{y}$, we have a convenient formula to find the centre of gravity.

410. If the figure whose centre of gravity is required is a triangle or quadrilateral whose sides are *curvilinear*, the proper choice for the element m will depend on the form of the curves.

If we join the angular points to the origin we have three or four sectors whose areas and centres of gravity may be separately found and thence, by Art. 380, the centre of gravity of the figure. Sometimes the bounding curves are of the same species so that when the process has been gone through for one sector the results for the other sectors may be inferred. In such cases the method is very advantageous. For example, we have already seen how the area and centre of gravity of a quadrilateral bounded by four elliptic arcs could be immediately deduced from the area and centre of gravity of an elliptic sector. See Ex. 6, Art. 406.

Putting this in an analytical form, we have for a curvilinear triangle whose sides are $r = f_1(\theta)$, $r' = f_2(\theta')$, $r'' = f_3(\theta'')$,

$$\Sigma mx = \tfrac{1}{3}\int_u^\beta r^3 \cos\theta \, d\theta + \tfrac{1}{3}\int_\beta^\gamma r'^3 \cos\theta' \, d\theta' + \tfrac{1}{3}\int_\gamma^a r''^3 \cos\theta'' \, d\theta''$$

$$\Sigma m = \tfrac{1}{2}\int_a^\beta r^2 d\theta + \tfrac{1}{2}\int_\beta^\gamma r'^2 d\theta' + \tfrac{1}{2}\int_\gamma^a r''^2 d\theta'',$$

where a, β, γ are the inclinations of the radii vectores of the angular points to the axis of x. *In forming these integrals we travel round the triangular figure taking the sides in order.*

It might appear at first sight that we are adding together all the three sectors instead of adding some together and subtracting the others. But it will be clear after a little consideration that in those sectors which should be subtracted from the others the $d\theta$ is made negative by taking the limits in the same order as we travel round the triangle.

Instead of joining the angular points to the origin we might draw perpendiculars on the axis of x. We then have

$$\Sigma mx = \int_a^b xy\,dx + \int_b^c x'y'\,dx' + \int_c^a x''y''\,dx'',$$

where a, b, c are the abscissæ of the angular points. As before, in taking the limits we travel round the sides in order.

411. Sometimes we may use *double integration*. Suppose we can express the equations to both the opposite sides of a curvilinear quadrilateral in one form by using an auxiliary quantity u. That is, let the one equation represent one boundary when $u = a$, and let the same equation represent the opposite boundary when $u = b$. Let this one equation be $\phi(x, y, u) = 0$. It is always possible to do this, for let $f_1(x, y) = 0$, $f_2(x, y) = 0$ be the boundaries, then

$$\phi = (u - a)f_1(x, y) + (u - b)f_2(x, y) = 0$$

represents one or the other according as $u = a$ or $u = b^*$. But this particular form is not always a convenient mode of expressing ϕ. In the same way let $\psi(x, y, v) = 0$ represent the other two boundaries when $v = e$ and $v = f$.

When this has been accomplished we have only to follow the rules of the integral calculus. By giving u and v all values between $u = a$ and $u = b$, $v = e$ and $v = f$, we obtain a double series of curves dividing the space into elements. Let m be the area of one of these elements and J the Jacobian determinant of x, y with regard to u, v, then $m = J\,du\,dv$. Hence

$$\bar{x} = \frac{\iint J\,du\,dv \cdot x}{\iint J\,du\,dv}, \qquad \bar{y} = \frac{\iint J\,du\,dv \cdot y}{\iint J\,du\,dv}.$$

---

* This is adapted from De Morgan's *Diff. Calc.* p. 392.

To find the Jacobian it *may* be necessary to solve the equations $\phi=0$, $\psi=0$, so as to express x, y in terms of u, v.　We then have $J=\dfrac{dx}{du}\dfrac{dy}{dv}-\dfrac{dx}{dv}\dfrac{dy}{du}$.　Unless we have been able in the first instance to express ϕ and ψ so conveniently that this Jacobian takes a simple form when expressed in terms of u, v, this method may lead to complicated analysis.　The advantage of the method is that the limits of integration $u=a$ to b, $v=e$ to f are constants, so that the integrations may be performed in any order or simultaneously.

412. Ex. 1.　An area is cut off from a parabola by a diameter ON and its ordinate PN: prove that $\bar{x}=\frac{3}{5}x$, $\bar{y}=\frac{3}{8}y$.

Ex. 2.　Two tangents TP, TP' are drawn to a parabola: show that the co-ordinates of the centre of gravity of the area between the curve and the tangents are $\bar{x}=\frac{1}{5}TP$, $\bar{y}=\frac{1}{5}TP'$ referred to TP, TP' as axes. Art. 406, Ex. 5.　[Walton.]

Regard the area as the difference between a triangle and a parabolic segment.

Ex. 3.　The equations of a cycloid are $x=a(1-\cos\theta)$, $y=a(\theta+\sin\theta)$.　Show that the centre of gravity of half the area is given by $\bar{x}=\frac{7}{6}a$, $\bar{y}=\dfrac{a}{2}\left(\pi-\dfrac{16}{9\pi}\right)$.

[Wallis.]

Ex. 4.　Find the centre of gravity of the half of either loop of the lemniscate $r^2=a^2\cos2\theta$ bounded by the axis.　The result is

$$\bar{x}=\frac{\pi a}{4\sqrt{2}}, \qquad \bar{y}=\frac{3\log(\sqrt{2}+1)-\sqrt{2}}{6\sqrt{2}}a.$$

Ex. 5.　Four parabolas whose equations are $y^2=a^3x$, $y^2=b^3x$, $x^2=e^3y$, $x^2=f^3y$ intersect and form a quadrilateral space.　Find the centre of gravity.

We take as the equations to the opposite sides $y^2=u^3x$ and $x^2=v^3y$.　Solving, we find $x=uv^2$, $y=u^2v$ and $J=3u^2v^2$.　This gives by substitution

$$\bar{x}=\frac{9}{20}\frac{(b^4-a^4)(f^5-e^5)}{(b^3-a^3)(f^3-e^3)}.$$

Ex. 6.　The centre of gravity of the space bounded by two ellipses and two hyperbolas all confocal lies in the straight line

$$-\frac{y}{x}=\frac{(a_2-a_1)(a_2'-a_1')(a_2^2+a_1a_2+a_1^2-a_2'^2-a_1'a_2'-a_1'^2)}{(b_2-b_1)(b_2'-b_1')(b_2^2+b_1b_2+b_1^2+b_2'^2+b_1'b_2'+b_1'^2)},$$

where the unaccented letters denote the semiaxes of the ellipse and the accented letters those of the hyperbola.

We take as the equation to the opposite sides $\dfrac{x^2}{u}+\dfrac{y^2}{u-h}=1$, $\dfrac{x^2}{v}+\dfrac{y^2}{v-h}=1$, where $u>h$ and $v<h$.　These give $hx^2=uv$, $-hy^2=(u-h)(v-h)$, as shown in Salmon's *Conics*.　The result then follows easily enough.

Ex. 7.　If the density at any point of a circular disc whose radius is a vary directly as the distance from the centre, and a circle described on a radius as diameter be cut out, prove that the centre of inertia of the remainder will be at a distance $\dfrac{6a}{15\pi-10}$ from the centre.　　　　　　　[Math. Tripos, 1875.]

Ex. 8.　A circular disc of radius r, whose density is proportional to the distance from the centre, has a hole cut in it bounded by a circle of diameter a which passes through the centre.　Show that the distance from the centre of the disc of the centre of gravity of the remaining portion is $\dfrac{6a^4}{15\pi r^3-10a^3}$.　　　　[Coll. Ex., 1888.]

Ex. 9. The curve for which the ordinate and abscissa of the centre of gravity of the area included between the ordinates $x=a$ and $x=x$ are in the same ratio as the bounding ordinate y and abscissa x is given by the equation $a^3y^3 - b^3x^3 = x^3y^3$.

[Math. Tripos, 1871.]

413. Pappus' Theorems. Before treating of the centres of gravity of surfaces or volumes it seems proper to discuss a method by which the centres of gravity of the arcs and areas already found may be used to find the surface or volume of a solid of revolution. The two following theorems were first given by Pappus at the end of the preface of his seventh book of *Mathematical Collections.*

Let any plane area revolve through any angle about an axis in its own plane, then

(1) *The area of the surface generated by its perimeter is equal to the product of the perimeter into the length of the path described by the centre of gravity of the perimeter.*

(2) *The volume of the solid generated by the area is equal to the product of the area into the length of the path described by the centre of gravity of the area.*

In both these theorems the axis is supposed not to intersect the perimeter or area.

414. Let AB be an arc of the curve, and let it lie in the plane xz. Let it revolve about the axis of z through any elementary angle $d\theta$. Any element $PQ = ds$ of the perimeter is thus brought into the position $P'Q'$, and the area traced out by PQ is $ds \cdot PP' = ds \cdot xd\theta$. The whole area or surface traced out by the finite arc AB is $d\theta \int x ds$. But this is $d\theta \cdot \bar{x}s$, if s be the arc AB and $\bar{x}$ the distance of its centre of gravity from the axis of z. If the arc now revolve again about Oz through a second elementary angle $d\theta$, an equal surface is again traced out. Hence, when the angle of rotation is θ, the area is $s \cdot \bar{x}\theta$. But $\bar{x}\theta$ is the length of the path traced out by the centre of gravity of the arc. The first proposition is therefore proved.

Next, let any closed curve in the plane of xz revolve as before about the axis of z through an angle $d\theta$. By this rotation any elementary area dA at R will describe a volume which may be regarded as an elementary cylinder. The base is dA, the altitude $xd\theta$, the volume is therefore $dA \cdot xd\theta$. The volume traced out by the whole area of the closed curve is $d\theta \int x dA$. But this is $d\theta . \bar{x} A$, if A be the area of the curve and $\bar{x}$ the distance of its centre of gravity from the axis of revolution. Integrating again for any finite value of θ, we find that the volume generated is $A . \bar{x} \theta$. This as before proves the theorem.

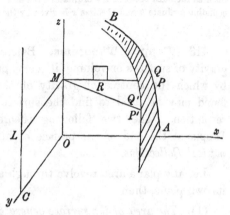

In both these proofs we have assumed that the whole of the curve lies on the same side of the axis of rotation. For suppose P_1 and P_2 were two points on the curve on opposite sides of the axis of z, then their abcissæ x_1 and x_2 would have opposite signs. Thus the elementary surfaces or volumes (having the factor $x d\theta$) would also have opposite signs. The integral gives the sum of these elementary surfaces or volumes taken with their proper signs. It follows that, when the axis cuts the curve, Pappus' two rules give the *difference of the surfaces or volumes traced out by the two parts of the curve on opposite sides of the axis of revolution.*

415. Ex. 1. Find the surface and volume of a tore or anchor-ring.

This solid may be regarded as generated by a complete revolution of a circle about an axis in its own plane. Let a be the distance of the centre from the axis, b the radius of the generating circle. Then $a > b$ if all the elements are to be regarded as positive. The arc of the generating circle is $2\pi b$, the length of the path described by its centre of gravity is $2\pi a$. *The surface is therefore $4\pi^2 ab$.* The area of the circle is πb^2, the length of the path described by its centre of gravity is $2\pi a$. *The volume is therefore $2\pi^2 ab^2$.*

Ex. 2. Find the volume of a solid sector of a sphere with a circular rim and also the area of its curved surface.

This solid may be regarded as generated by a complete revolution of a sector of a circle about one of the extreme radii. Let $2a$ be the angle of the sector, O its centre. The arc of the sector is $2a\alpha$. The length of the path described by its centre of gravity G is $2\pi . OG \sin \alpha$, where $OG = (a \sin \alpha)/\alpha$. *The spherical surface is therefore $4\pi a^2 \sin^2 \alpha$.* The area of the sector is $a^2 \alpha$. The length of the path of its centre of gravity G' is $2\pi . OG' \sin \alpha$, where $OG' = \frac{2}{3} OG$. *The volume is therefore $\frac{4}{3}\pi a^3 \sin^2 \alpha$.* It appears that both the surface and the volume vary as the versine of the sector.

Ex. 3. A solid is generated by the revolution of a triangle ABC about the side AB: prove that the surface is $\pi (a+b) p$ and the volume is $\frac{1}{3}\pi c p^2$, where p is the perpendicular from C on AB.

416. It should be noticed that for any elementary angle $d\theta$ the axis of rotation need only be an instantaneous axis. Suppose the plane area to move so as always to be normal to the curve described by the centre of gravity of the area. Then as the centre of gravity describes the arc ds, the area A may be regarded as turning round an axis through the centre of curvature of the path. Hence the elementary volume is Ads, and the volume described is the product of the area into the length of the path described by the centre of gravity of the area.

In the same way, if the area move so as always to be normal to the path described by the centre of gravity of the *perimeter*, the surface of the solid is the product of the arc into the length of the path of the centre of gravity of the perimeter.

417. *When the axis of rotation does not lie in the plane of the curve*, we can use a modification of Pappus' rule to find the *volume* generated by the motion of any area.

Let us suppose that the axis of rotation is parallel to the plane of the curve. Referring to the figure of Art. 414, let CL be the axis, and let RL be a perpendicular to it from any point R within the closed curve. The elementary area dA at R will now describe a portion of a thin ring whose centre is at L. The length of this portion is $\theta \cdot RL$. The area of the normal section of this ring is $dA \cos \phi$, where ϕ is the angle the normal RL to the ring makes with the area dA. The volume traced out is therefore $RL \cdot \cos \phi \cdot \theta dA$. But this is the same as $x\theta dA$. This is the same result as we obtained before when the axis of revolution was Oz.

If the element were to revolve round Oz it would trace out a ring of less radius than it actually does in its revolution round CL, and these rings would be differently situated in space. But the normal section of the larger ring is so much less than that of the smaller ring that the two volumes are equal.

We infer that Pappus' rule will apply to find the volume if we treat the *projection of the axis* on the plane of the curve as if it were the actual axis of rotation. The angle of rotation is to be the same for both axes.

If the area does not lie wholly on one side of the projection, it must be remembered that the volumes generated by the two parts on opposite sides of the projection will have opposite signs.

Ex. 1. If the axis of revolution is inclined to the plane of the area at an angle a, show that Pappus' rule will give the volume generated if we treat the projection of the axis on the plane as if it were the axis of revolution and regard the angle of rotation as $\theta \cos a$ instead of θ.

Ex. 2. A quadrant of a circle makes a complete revolution about an axis passing through its centre and making a right angle with one of its extreme radii and an angle a with the other. Show that the volume generated is $\frac{2}{3}\pi a^3 \cos a$.

Ex. 3. An arc A_1A_2 of a plane curve revolves about an axis perpendicular to its plane through an angle θ. Show that the area traced out is $\frac{1}{2}\theta (r_2{}^2 - r_1{}^2)$, where r_1, r_2 are the distances of A_1, A_2 from the axis.

It is supposed that the radius vector r is not a maximum or minimum at any point between A_1 and A_2. If it is either, the areas traced out by arcs on opposite sides of that point will have opposite signs.

Ex. 4. A solid is generated by the revolution of an area about the axis of z which lies in its own plane. The density D at any point P of the solid is a given function of z and ρ, where ρ is the distance of P from the axis. Prove that the mass may be found by Pappus' rule if we regard D as the surface density at any point P of the generating area where the coordinates of P are z and ρ.

418. Areas on the surface of a right cone. *To find the centre of gravity of the whole surface of a right cone excluding the base.* Guldin's Theorem.

Let O be the vertex, C the centre of the base, then OC is perpendicular to the plane of the base. The required centre of gravity lies in OC.

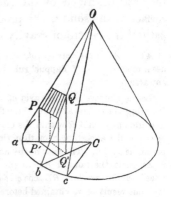

Divide the surface of the cone into elementary triangles by drawing straight lines from the vertex O to points a, b, c &c. in the base. The centre of gravity of each triangle lies in a plane parallel to the base and dividing the sides Oa, Ob &c. in the ratio $2 : 1$. The centre of gravity of the whole surface is therefore at the intersection of this plane with OC.

The centre of gravity of the surface of a right cone is two-thirds of the way from the vertex to the centre of the base.

Ex. Show that the same rule applies to find the centre of gravity of the whole curved surface of a right cone on an elliptic base or more generally on any base which is symmetrical about two diameters at right angles.

419. *To find the area and centre of gravity of a portion of the surface of a right cone on a circular base.*

Referring to the figure of Art. 418, let $PQ = dS$ be an element of the surface of the cone, $P'Q' = d\Pi$ its projection on the base. The angle between PQ and $P'Q'$ is the same as the angle between the triangle Oab and the plane of the base, and this angle is the complement of the semi-angle of the cone. We therefore have $d\Pi = dS \cdot \sin \alpha$, if α be the semi-angle of the cone. Since this is true for every element of area, it follows that to find the surface of any portion of a right cone we simply divide the area of its projection on a plane perpendicular to the axis by $\sin \alpha$.

If we take the axis of the cone for the axis of z, it is clear that dS and $d\Pi$ have the same coordinates of x and y. Hence, proceeding exactly as in Art. 403, we see that the projection of the centre of gravity of any portion of the surface of the cone on a plane perpendicular to the axis is the centre of gravity of the projection.

We have yet to find the z coordinate of the centre of gravity. Taking any plane perpendicular to the axis as the plane of xy, we have

$$\bar{z} = \frac{\Sigma mz}{\Sigma m} = \frac{\int dSz}{\int dS} = \frac{\int zd\Pi}{\int d\Pi};$$

thus the distance of the centre of gravity of any portion S of the surface from any plane perpendicular to the axis is equal to the volume of the cylindrical solid between S and its projection Π on that plane divided by the area Π.

These three results depend on the fact that the area of any element dS of the surface bears a constant ratio to its projection $d\Pi$ on the plane of xy. This again requires that every tangent plane to the surface should make a constant angle with the plane of xy. Other surfaces besides right cones and planes possess this property. Any developable surface which is the envelope of a system of planes making a given angle with the plane of xy will obviously satisfy the conditions.

Ex. 1. A cone of any form is intersected by a plane AB, and any straight line is drawn from the vertex to meet the section in H. Prove that the conical volume between the plane of the section and the vertex is equal to the product of $\frac{1}{3}OH$ into the projection of the area AB on a plane perpendicular to OH.

Ex. 2. A right cone, whose semi-angle is a, is intersected by a plane AB cutting the axis in H and making an angle β with the axis. Show that, (1) the surface S of the cone between the elliptic section AB and the vertex O is equal to the product of the area of the section AB into $\sin\beta\operatorname{cosec}a$;

(2) the centre of gravity of the surface S lies in a straight line drawn parallel to the axis of the cone from the centre C of the section AB;

(3) the distance of the centre of gravity of the surface S from $C = \frac{1}{3}OH$.

Since both the surface S and the section AB project into the same elliptic area $A'B'$, the two first results follow from what has been proved above.

To prove the third result we divide the surface into elementary triangles by drawing straight lines from the vertex O to the base AB. It follows, as in Art. 418, that the centre of gravity of the surface lies in a plane drawn parallel to the base through a trisection of OH.

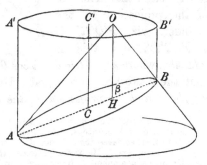

Ex. 3. A right cylinder stands on a plane base $A'B'$ of any form, and is intersected by any other plane AB. Show that (1) the surface of the cylinder between the plane AB and the base is equal to the product of the perimeter of the base into the ordinate (or altitude) of the plane at the centre of gravity of the perimeter, (2) the

volume of the cylinder between the plane AB and the base is equal to the product of the area of the base into the ordinate of the plane at the centre of gravity of the area.

By considering part of the perimeter of the base to be rectilinear and part curved, this gives the surface and volume of the portion of the cylinder cut off by two planes parallel to the axis and two transverse to the axis

Ex. 4. A right cylinder stands on the base $Ax^2 + By^2 = 1$, and is intersected by the plane $z = h + px + qy$. Prove that the coordinates of the centre of gravity of the volume are given by $4Ah\bar{x} = p$, $4Bh\bar{y} = q$, $2\bar{z} = h + p\bar{x} + q\bar{y}$.

420. Spherical Surfaces. There are two projections of the spherical surface which have been found useful. We can project any portion of the surface on the circumscribing cylinder and on a central plane. We shall consider these in order.

Let the origin be at the centre of the sphere, and let the rectangular axes x, y, z cut the surface in A, B, C. Let the polar coordinates of any point P be as usual $OP = a$, the angle $ZOP = \theta$ and the angle $NOA = \phi$. Let $PL = \rho$ be a perpendicular on the axis of z, then $OL = z$.

Let a cylinder circumscribe the sphere and touch it along the circle of which AB is a quadrant. Any point P on the sphere is projected on the cylinder by producing LP to meet the cylinder in P'. According to this definition any point P and its projection P' are so related that their z's and ϕ's are the same.

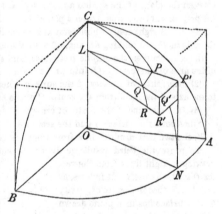

The area of any element PQR on the sphere is $PQ . QR$, and this is equal to $a \sin \theta d\phi . a d\theta$. The area of the projection on the cylinder, viz. $P'Q'R'$ is $P'Q' . Q'R'$, and this is $a d\phi . dz'$, where $z' = CL = a - a \cos \theta$. Substituting for z', we see that these two areas are equal. Hence *any elementary area on a sphere and its projection on the cylinder are equal**.

---

* The relation of the sphere to the cylinder in regard to their measurement was first discovered by Archimedes. He wrote two books on this subject. He investigated both their surfaces and volumes, whether entire or cut by planes perpendicular to their common axis. He was so pleased with these discoveries that he directed a cylinder enclosing a sphere to be engraved on his tombstone in commemoration of them.

It follows from this result that the area of any finite portion of the spherical surface is equal to the area of its projection on any circumscribing cylinder. *This rule enables us to find many areas on the sphere which are useful to us.* Thus the area cut off from the sphere by any two parallel planes whose distance apart is h is equal to the area of a band on the cylinder whose breadth is h. The area on the sphere is therefore $2\pi ah$. We notice that this result is independent of the position of the planes, except that they must be parallel. Thus the area of a segment of a sphere whose versed sine is h is $2\pi ah$.

421. This important theorem is used also in the construction of maps. The places on a terrestrial globe are projected in the manner just described on a circumscribing cylinder. The cylinder is then unrolled on a plane. In this way the whole earth may be represented on a map of a rectangular form. The *advantage of this construction is that any equal areas on the globe are represented by equal areas on the map.* This is true for large or small areas in whatever part of the globe they may be situated. The *disadvantage of the construction is that any small figure on the map is not similar to the corresponding figure on the globe.* If the figure is situated near the curve of contact of the cylinder, the similarity is sufficiently close for practical purposes, but if the figure is situated nearer the pole of this curve of contact, the dissimilarity is more striking. Thus a small circle very near the pole is represented by an elongated oval. In some other systems of making maps, as for example Mercator's, any small figure on the map is made similar to the corresponding figure on the globe, but in that case equal areas on the map do not correspond to equal areas on the globe.

Ex. A map is made on the following principle. Any point O on the surface of a globe of radius unity, and a corresponding point O' on a map being taken, the points P', Q' corresponding to the two points P, Q on the globe are found by taking the lengths $O'P' = a \tan \frac{1}{2}OP$, $O'Q' = a \tan \frac{1}{2}OQ$, the angle $P'O'Q'$ being made equal to POQ. Prove that *any infinitely small corresponding portions on the sphere and map are similar.* Show also that the scale of the map in the neighbourhood of any point P' varies as $a^2 + O'P'^2$.

If the tangents are replaced by sines in the relations given above, prove that *the areas of corresponding portions have a constant ratio.*

These are called the stereographic projection and the chordal construction.

422. *The altitude of the centre of gravity of any portion of the sphere above the plane of contact is equal to the altitude of the centre of gravity of its projection on the circumscribing cylinder.* To prove this it is sufficient to quote the formula $\bar{z} = \Sigma mz/\Sigma m$, and to remark that for the surface and its projection the m's and z's are equal, each to each.

From this we infer that the centre of gravity of the band on the sphere between any two parallel planes is the same as that for the corresponding band on the cylinder, and is therefore half way between the parallel planes, and lies on the perpendicular radius.

In the same way the *centre of gravity of a hollow thin hemisphere of uniform thickness bisects the middle radius.*

423. Ex. 1.   A segment of a sphere of height h rests on a plane base : show that the centre of gravity of the surface including the plane base is at a distance equal to $ah/(4a-h)$ from the base, where a is the radius of the sphere.

Ex. 2.   The distance of the centre of gravity of the surface of a lune from the axis is $\dfrac{\pi a}{4}\dfrac{\sin a}{a}$, where $2a$ is the angle of the lune.

Ex. 3.   A bowl of uniform thin material in the form of a segment of a sphere is closed by a circular lid of the same material and thickness, which is hinged across a diameter.   If it be placed on a smooth horizontal plane with one half of the lid turned back over the other half, show that the plane of the lid will make with the horizontal plane an angle ϕ given by $3\pi\tan\phi=4\tan\tfrac12 a$; a being the angle any radius of the lid subtends at the centre of the sphere.   [Math. Tripos, 1881.]

424.   *To find the centre of gravity of any spherical triangle.*

Let us begin by projecting *any* portion of the surface of the sphere on a central plane.   Let this be the plane of xy.   Let dS be any element of area, $d\Pi$ its projection, let θ be the angle the normal at dS makes with the axis of z.   Then

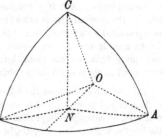

$$d\Pi = dS\cos\theta = dS\,.\,z/a.$$

Hence, integrating, we have $a\Pi = S\bar z$.

It follows that *the distance of the centre of gravity of any portion S of the surface of a sphere from a central plane* $=\dfrac{\Pi}{S}a$, where

Π *is the projection of S on that plane**.

This result follows from the equality $\cos\theta = z/a$.   Other surfaces besides spheres possess this property.   These surfaces are generated by the motion of a sphere of constant radius, whose centre moves in any manner in the plane of xy.   As an example an anchor ring or tore may be mentioned.

Let us now apply this Lemma to the spherical triangle.   Let A, B, C be the angles, a, b, c the sides, let O be the centre of the sphere, ρ its radius.   Let CN be a perpendicular from C on the plane AOB, let AN, BN be the two elliptic arcs which are the projections of the sides AC, BC of the spherical triangle.

By the lemma, $\bar z : \rho = $ area ANB : area ABC.   Also

$$\text{(area } ANB) = (\text{area } AOB) - (\text{area } AOC)\cos A - (\text{area } BOC)\cos B$$
$$= \tfrac12\rho^2(c - b\cos A - a\cos B).$$

If E be the spherical excess of the triangle, i.e. if $E = A + B + C - \pi$, we know by Spherical Trigonometry that the area $ABC = \rho^2 E$.   Hence

$$\frac{\bar z}{\rho} = \tfrac12\frac{c - b\cos A - a\cos B}{E}\,.$$

* We have here followed the method proposed by Prof. Giulio, chiefly because the lemma on which it depends is of general application and may be useful in other cases.   His memoir was published in the fourth volume of Liouville's *Journal de Mathématiques*.   An English version is also given in Walton's *Mechanical Problems*.

This formula gives the distance of the centre of gravity from the plane AOB containing any side AB of the triangle. The distances from the planes BOC, COA containing the other sides are expressed by similar formulae.

Ex. 1. If p, q, r be the perpendicular arcs from the angular points A, B, C on the opposite sides, and G the centre of gravity of the spherical triangle, prove that

$$\frac{\cos AOG}{a \sin p} = \frac{\cos BOG}{b \sin q} = \frac{\cos COG}{c \sin r} = \frac{1}{2E}.$$

This is equivalent to the result given in Moigno's *Statique*.

Ex. 2. A surface is generated by the revolution of the catenary about its axis. Let this be the axis of z and let the plane generated by the directrix be that of xy. Any portion S of its surface is projected orthogonally on the plane xy, and V is the volume of the cylindrical solid formed by the perpendiculars from the perimeter of S. Prove that the $\bar{x}$ and $\bar{y}$ of S and V are equal each to each, but the $\bar{z}$ of the first is double that of the second.   [Giulio, also Walton.]

425. Any surfaces and solids of revolution. A known plane curve revolves round an axis in its own plane which we shall take as the axis of z, and the angle of revolution is 2α. It is required to find the centres of gravity of the surface and volume thus generated.

It is clear that every point describes an arc of a circle whose centre is in the axis of z. Thus the whole solid is symmetrical about a plane passing through z and bisecting all these arcs. Let this be the plane of xz. The centres of gravity lie in this plane. Let PP' be half the arc described by P, the other half being behind the plane xz and not drawn in the figure.

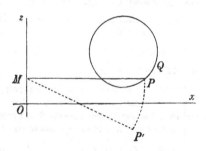

Let $PQ = ds$ be any arc of the generating curve, then the area of the elementary band described by ds is $m = 2x\alpha ds$ by Pappus' theorem. Its centre of gravity lies in MP at a distance from M equal to $(x \sin \alpha)/\alpha$. Hence the coordinates of the centre of gravity of the surface are

$$\bar{x} = \frac{\Sigma m x}{\Sigma m} = \frac{\int x^2 ds}{\int x ds} \cdot \frac{\sin \alpha}{\alpha}, \qquad \bar{z} = \frac{\int xz ds}{\int x ds}.$$

In the same way the coordinates of the centre of gravity of the volume are

$$\bar{x} = \frac{\Sigma m x}{\Sigma m} = \frac{\int x^2 d\sigma}{\int x d\sigma} \cdot \frac{\sin \alpha}{\alpha}, \qquad \bar{z} = \frac{\int xz d\sigma}{\int x d\sigma},$$

where $d\sigma$ is any element of the *area* of the given curve. We may write for $d\sigma$ either $dxdz$ or $rd\theta dr$ according as we choose to use Cartesian or polar coordinates, replacing the single integral sign by that for double integration.

It is evident that these integrals are those used in the higher Mathematics for the moments and products of inertia of the arcs and areas. When therefore we have once learnt the rules to find these moments of inertia, we seldom have to perform any integration; we simply quote the results as being well known. These rules are usually studied in connection with rigid dynamics, as a knowledge of them is essential for that science, but they are now given in some of the treatises on the integral calculus, for example in that by Prof. Williamson.

Ex. 1. A portion of an anchor ring is generated by the complete revolution of a quadrant of a circle (radius a) about an axis parallel to one of the extreme radii and distant b from it. Prove that the distances of the centres of gravity of the curved surface and volume from the plane described by the other extreme radius are

$$\frac{a\,(2b \pm a)}{\pi b \pm 2a} \text{ and } \frac{a\,(8b \pm 3a)}{2\,(3\pi b \pm 4a)}.$$

The axis of revolution is supposed not to cut the quadrant.

Ex. 2. A semi-ellipse revolves through one right angle about the bounding diameter. Show that the distance from the axis of the centre of gravity of the volume generated is $3ab/4\surd 2r$, where $2r$ is the length of the diameter.

Ex. 3. A triangular area makes a revolution through two right angles about an axis in its own plane. Prove that the distance of the centre of gravity of the volume from the axis is $\dfrac{2}{\pi}\,\dfrac{a^2+\beta^2+\gamma^2}{a+\beta+\gamma}$, where a, β, γ are the distances of the middle points of the sides from the axis.

Ex. 4. A circular area of radius a revolves about a line in its plane at a distance c from the centre, where c is greater than a. If $2a$ be the angle through which it revolves, find the volume generated and prove that the centre of gravity of the solid is at a distance from the line equal to $(4c^2+a^2)\sin a/4ca$.   [Coll. Ex., 1887.]

426. *To find the centre of gravity of a solid sector of a sphere with a circular rim.*

Referring to the figure of Art. 400, let OC be the middle radius of the solid sector, N the centre of the rim, G the centre of gravity of the sector, V its volume, V_0 the volume of the whole sphere, a the radius, then

$$OG = \tfrac{3}{4}\,\frac{ON+OC}{2}, \qquad V = V_0\,\frac{CN}{2a}. \qquad \text{[Wallis.]}$$

To prove this we follow the same method as that adopted to find the centre of gravity of a sector of a circle. Let PQ be an elementary area of the surface, then OPQ is a tetrahedron whose centre of gravity is at p where $Op = \frac{3}{4}OP$. Hence, if G' be the centre of gravity of the surface, $OG = \frac{3}{4}OG'$. But $OG' = \frac{1}{2}(ON + OC)$ by Art. 422. Hence the result follows. The volume V has been already found in Art. 415.

The centre of gravity of a solid hemisphere follows immediately from this result. Putting $ON = 0$, we see that *the centre of gravity of a solid hemisphere lies on the middle radius and is at a distance $\frac{3}{8}$ of that radius from the centre.*

The centre of gravity of a solid octant also follows at once. There are four octants on one side of any central plane and the centre of gravity of each of these is at the same distance from that plane. Hence the centre of gravity of all four must be also at the same distance, and this has just been proved to be $\frac{3}{8}a$. Hence, *for any octant, the distance of the centre of gravity from any one of the three plane faces is $\frac{3}{8}$ of the radius.*

427. Ex. 1.   The centre of gravity and volume of a solid segment of a sphere bounded by a plane distant z from the centre O are given by

$$OG = \frac{3}{4}\frac{(a+z)^2}{2a+z}, \qquad V = \frac{\pi}{3}(a-z)^2(2a+z).$$

Ex. 2.   Prove that in a sphere, whose density varies inversely as the distance from a point in the surface, the distance of the centre of gravity from that point bears to the diameter the ratio 2 : 5.   [Math. Tripos, 1867.]

Ex. 3.   Prove that the centre of gravity of a solid sphere, whose density varies inversely as the fifth power of the distance from an external point, is at the centre of the section of the sphere by the polar plane of the external point.   [Math. Tripos, 1872.]

428.   Centres of gravity of volumes connected with the ellipsoid.   In order to deduce the centre of gravity of any portion of an ellipsoid from that of the corresponding portion of a sphere, we shall use an extension of that method of projections by which we passed from the areas of circles to those of ellipses.

One point (xyz) is said to be projected into another $(x'y'z')$ when we write $x = ax'$, $y = by'$, $z = cz'$. The points are then said to correspond. Volumes V, V' correspond when their boundaries are traced out by corresponding points. If $(\overline{xyz})$, $(\overline{x'}\overline{y'}\overline{z'})$ be the centres of gravity of V, V' we have

$$V = \iiint dx\, dy\, dz = abc \iint dx'\, dy'\, dz' = abc\, V'.$$

In the same way $\bar{x} = a\bar{x}'$, $\bar{y} = b\bar{y}'$, $\bar{z} = c\bar{z}'$.

It appears from these equations that *any corresponding volumes have a constant ratio, and the centre of gravity of one corresponds to the centre of gravity of the other.*

We may also show* that (1) parallel straight lines correspond to parallels, and (2) the ratio of the lengths of parallel straight lines is unaltered by projection. Thus the rule already explained in Art. 403 for areas is true also for solids.

We may apply these principles to an ellipsoidal solid. The equation to an ellipsoid of semi-axes a, b, c is changed into that of a concentric sphere by writing $x = ax'$, $y = by'$, $z = cz'$. It follows that all projective theorems may be transferred from the sphere to the ellipsoid.

429. Ex. 1. Find the centre of gravity of a solid sector of an ellipsoid with an elliptic rim.

Let O and N be the centres of the ellipsoid and of the rim. Then ON is the conjugate diameter of the plane of the rim. Let it cut the ellipsoid in C. The corresponding theorem for a spherical sector is given in Art. 426. Since the values of OG and V there given depend on the ratios of parallel lengths, we may transfer them to the ellipsoid. The centre of gravity G of the ellipsoidal sector therefore lies in ON, and we have

$$OG = \tfrac{3}{4} \frac{ON + OC}{2}, \qquad V = \frac{CN}{2 \cdot OC} V_0.$$

Ex. 2. The coordinates of a solid octant of an ellipsoid bounded by three conjugate planes are $\bar{x} = \tfrac{3}{8} a$, $\bar{y} = \tfrac{3}{8} b$, $\bar{z} = \tfrac{3}{8} c$.

Ex. 3. The centre of gravity and volume of any solid segment of an ellipsoid are given by

$$OG = \tfrac{3}{4} \frac{(c+z)^2}{2c+z}, \qquad V = \frac{(c-z)^2 (2c+z)}{4c^3} V_0,$$

where $2c$ is the conjugate diameter of the plane of the segment, z its ordinate measured along c, and V_0 the volume of the whole ellipsoid.

430. Let us construct two concentric and coaxial ellipsoids forming between them a thin solid shell. Let (a, b, c), $(a + da, \&c.)$ be the semi-axes of these ellipsoids, p and $p + dp$ the perpendiculars on two parallel tangent planes. Then $t = dp$ is the thickness of the shell at any point. Let $d\sigma$ be an element of the surface of one ellipsoid, $d\Pi$ its projection on the plane of xy, then $d\Pi = d\sigma \frac{pz}{c^2}$.

Ex. 1. Show that the ordinate $\bar{z}$ of the centre of gravity of any portion of the shell is given by $\bar{z} V = c^2 \int \frac{t}{p} d\Pi$, where V is the volume of that portion of the shell.

Ex. 2. If the shell is bounded by similar ellipsoids, so that $\frac{da}{a} = \frac{db}{b} = \frac{dc}{c} = \frac{dp}{p}$, prove that $\bar{z} : c = \Pi dc : V$.

---

* Let the straight line AB project into $A'B'$ by writing $x = ax'$ leaving y, z unaltered. Geometrically we construct $A'B'$ by producing the abscissae (viz. LA, MB) of A and B in the given ratio $a : 1$. This gives $LA' = a \cdot LA$ and $MB' = a \cdot MB$. Repeating this process for a straight line CD parallel to AB, it is easy to see, by similar triangles, that $C'D'$ is also parallel to $A'B'$, and that the ratio $C'D' : A'B'$ = the ratio $CD : AB$. Having written $x = ax'$ we repeat the process by writing $y = by'$ and finally $z = cz'$. The theorems are obviously true after the third projection as well as after the first.

If two parallel planes cut off a portion from this *thin* shell, prove that its centre of gravity lies in the common conjugate diameter and is equidistant from the planes. Art. 428.

Ex. 3. If the shell is bounded by confocal ellipsoids, so that $ada = bdb = cdc = pdp$, prove that

$$\frac{\bar{z}}{c} = \frac{\Pi dc}{V}\left\{1 - \left(1 - \frac{c^2}{a^2}\right)\frac{k_2^2}{a^2} - \left(1 - \frac{c^2}{b^2}\right)\frac{k_1^2}{b^2}\right\},$$

where Πk_1^2 and Πk_2^2 are the moments of inertia of Π about the axes of x and y respectively, Art. 425.

Ex. 4. If the density of a shell bounded by concentric, similar, and similarly situated ellipsoids vary inversely as the cube of the distance from a point within the cavity, that point is the centre of gravity.

If the shell be thin, and the density vary inversely as the cube of the distance from an external point, the centre of gravity is in the polar plane of the point. At what point of the polar plane is the centre of gravity situated? [Math. T., 1880.]

Let the shell be thin, and let O be the point within the cavity. With O for vertex describe an elementary cone cutting off from the shell two elementary volumes. Let v and v' be these volumes, and r, r' their distances from O. By the properties of similar ellipsoids, we may show that $v/r^2 = v'/r'^2$. Let D, D' be the densities of these elements. Since $D = \mu/r^3$, $D' = \mu/r'^3$, we find $vDr = v'D'r'$, i.e. the centre of gravity of two elements is at O. It easily follows that the centre of gravity of the whole thin shell is at O. Joining many thin shells together, it also follows that the centre of gravity of a thick shell is at O.

Next, let O be an external point, and let the elementary cone whose vertex is at O intersect the polar plane of O in an element whose distance from O is ρ. Since ρ is the harmonic mean of r and r', we easily find $vDr + v'D'r' = (vD + v'D')\rho$, i.e. the centre of gravity of the two elementary volumes v and v' lies in the polar plane of O. It follows that the centre of gravity of the shell lies in the polar plane of O.

Lastly, let any number of particles m_1, m_2, &c., attract the origin according to the Newtonian law, and let the resultant attraction be a force X acting along the axis of x. If the coordinates of the particles be $(x_1 y_1 z_1)$ &c., we find by resolution

$$\Sigma \frac{mx}{r^3} = X, \qquad \Sigma \frac{my}{r^3} = 0, \qquad \Sigma \frac{mz}{r^3} = 0.$$

The two latter equations show that, if the masses m_1, m_2 &c. are divided by numbers proportional to the cubes of their distances from the origin, the centre of gravity of the masses so altered lies in the line of action of the force X. The first equation shows the distance of the centre of gravity from the origin.

In this way many propositions on attractions may be translated into propositions on centre of gravity, and vice versâ.

It will be shown in the chapter on attractions that the resultant attraction of a thin homogeneous shell bounded by similar ellipsoids at an external point O is normal to the confocal ellipsoid passing through O. The centre of gravity of the heterogeneous shell is the intersection of this normal with the polar plane of O.

431. Centres of gravity of the volume and surface of any solid. The fundamental formulae are in all cases those already found in Art. 380, viz.

$$\bar{x} = \frac{\Sigma mx}{\Sigma m}, \qquad \bar{y} = \frac{\Sigma my}{\Sigma m}, \qquad \bar{z} = \frac{\Sigma mz}{\Sigma m};$$

the differences we have to indicate arise only from the varying choice which we may make for the element m.

Let us first find the *centre of gravity of a volume*. For Cartesian coordinates we take $m = dx\,dy\,dz$, and replace the Σ by the sign of triple integration. We have then

$$\bar{x} = \frac{\iiint dx\,dy\,dz\,.\,x}{\iiint dx\,dy\,dz}, \qquad \bar{y} = \frac{\iiint dx\,dy\,dz\,.\,y}{\iiint dx\,dy\,dz}, \qquad \bar{z} = \frac{\iiint dx\,dy\,dz\,.\,z}{\iiint dx\,dy\,dz}.$$

These formulae evidently hold for oblique axes also.

For polar coordinates we take $m = r\,d\theta\,.\,dr\,.\,r \sin\theta\,d\phi$, and $x = r\sin\theta\cos\phi$, $y = r\sin\theta\sin\phi$, $z = r\cos\theta$, and replace Σ by the sign of triple integration. These relations are proved in treatises on the integral calculus. We find

$$\bar{x} = \frac{\iiint r^3\sin^2\theta\cos\phi\,dr\,d\theta\,d\phi}{\iiint r^2\sin\theta\,dr\,d\theta\,d\phi}, \quad \bar{y} = \frac{\iiint r^3\sin^2\theta\sin\phi\,dr\,d\theta\,d\phi}{\iiint r^2\sin\theta\,dr\,d\theta\,d\phi}, \quad \bar{z} = \frac{\iiint r^3\sin\theta\cos\theta\,dr\,d\theta\,d\phi}{\iiint r^2\sin\theta\,dr\,d\theta\,d\phi}.$$

For cylindrical coordinates we have $m = \rho\,d\phi\,.\,d\rho\,.\,dz$, and $x = \rho\cos\phi$, $y = \rho\sin\phi$. Hence

$$\bar{x} = \frac{\iiint \rho^2\cos\phi\,d\phi\,d\rho\,dz}{\iiint \rho\,d\phi\,d\rho\,dz}, \quad \bar{y} = \frac{\iiint \rho^2\sin\phi\,d\phi\,d\rho\,dz}{\iiint \rho\,d\phi\,d\rho\,dz}, \quad \bar{z} = \frac{\iiint \rho z\,d\phi\,d\rho\,dz}{\iiint \rho\,d\phi\,d\rho\,dz}.$$

Or again, if x, y, z be given functions of three auxiliary variables u, v, w, we can use the Jacobian form corresponding to that of Art. 411. We have then $m = J\,du\,dv\,dw$.

432. To find the *centre of gravity of the surface of a solid* we find the value of m suitable to the coordinates we wish to use.

If the equation to the surface is given in the Cartesian form $z = f(x, y)$, we project the element of surface on the plane of xy. The area of the projection is $dx\,dy$. If $(\alpha\beta\gamma)$ be the direction angles of the normal to the element, the area of the element must be $\sec\gamma\,dx\,dy$. This therefore is our value of m. We find

$$\bar{x} = \frac{\iint \sec\gamma\,dx\,dy\,.\,x}{\iint \sec\gamma\,dx\,dy}, \qquad \bar{y} = \frac{\iint \sec\gamma\,dx\,dy\,.\,y}{\iint \sec\gamma\,dx\,dy} \quad \&c.$$

Taking the equation to the normal, we find

$$\sec\gamma = \left\{ 1 + \left(\frac{dz}{dx}\right)^2 + \left(\frac{dz}{dy}\right)^2 \right\}^{\frac{1}{2}}.$$

In a similar way, if the equation to the surface is given in cylindrical coordinates $z = f(\rho, \phi)$, we find

$$m = \rho\,d\phi\,d\rho \left\{ 1 + \left(\frac{dz}{d\rho}\right)^2 + \left(\frac{dz}{\rho\,d\phi}\right)^2 \right\}^{\frac{1}{2}}.$$

If the surface is given in polar coordinates $r = f(\theta, \phi)$, we have

$$m = rd\theta d\phi \left\{ \left(\frac{dr}{d\phi}\right)^2 + \sin^2\theta \left(\frac{dr}{d\theta}\right)^2 + r^2 \sin^2\theta \right\}^{\frac{1}{2}}.$$

433. In some cases it is more advantageous to divide the solid into larger elements. We should especially try to choose as our element some thin lamina or shell whose volume and centre of gravity have been already found. Suppose, for example, we wish to find $\bar{x}$ for some solid. We take as the element a thin slice of the solid bounded by two planes perpendicular to x. If the boundary be a portion of an ellipse, triangle, or some other figure whose area A is known, we can use the formula

$$\bar{x} = \frac{\int A \, dx \, x}{\int A \, dx}.$$

In this method we have *only a single instead of a triple sign of integration*. If the centre of gravity of A is known as well as its area, we can find $\bar{y}$ and $\bar{z}$ by using the same element.

To take another example, suppose the solid heterogeneous. Then instead of the thin slice just mentioned we might take as the element a thin stratum of homogeneous substance. If the mass and centre of gravity of this stratum be known, a single integration will suffice to find the centre of gravity of the whole solid. *This method will be found useful whenever the boundary of the whole solid is a stratum of uniform density*, for in that case the limits of the integral will be usually constants.

434. Ex. 1. Find the centre of gravity of an octant of the solid

$$\left(\frac{x}{a}\right)^n + \left(\frac{y}{b}\right)^n + \left(\frac{z}{c}\right)^n = 1.$$

From the symmetry of the case it will be sufficient to find $\bar{z}$. It will also evidently simplify matters if we clear the equation of the quantities a, b, c; we therefore put $x = ax'$, $y = by'$, $z = cz'$, Art. 428.

If we take as our element a slice formed by planes parallel to xy, we shall require the area A of the section PMQ. This area is

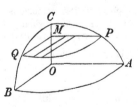

$$A = \int y' dx' = \int (1 - z'^n - x'^n)^{\frac{1}{n}} dx',$$

where the limits of integration are 0 to $(1 - z'^n)^{\frac{1}{n}}$. If we write $x'^n = (1 - z'^n) \, \xi$, this reduces to

$$A = (1 - z'^n)^{\frac{2}{n}} \frac{1}{n} \int (1 - \xi)^{\frac{1}{n}} \xi^{\frac{1}{n} - 1} \, d\xi = (1 - z'^n)^{\frac{2}{n}} B,$$

where the limits of the integral have been made 0 to 1, so that B can be expressed in gamma functions if required.

We have now, $\quad \dfrac{\bar{z}}{c} = \dfrac{\int A dz' \cdot z'}{\int A dz'} = \dfrac{\int (1 - z'^n)^{\frac{2}{n}} dz' \cdot z'}{\int (1 - z'^n)^{\frac{2}{n}} dz'}, \quad \begin{cases} z' = 0 \text{ to} \\ z' = 1 \end{cases}$.

If we put $z'^n = \xi$ and write m for $1/n$, this reduces to

$$\frac{\bar{z}}{c} = \frac{\int (1 - \xi)^{2m} \xi^{2m-1} d\xi}{\int (1 - \xi)^{2m} \xi^{m-1} d\xi} = \frac{\Gamma(2m+1)\,\Gamma(2m)}{\Gamma(4m+1)} \frac{\Gamma(3m+1)}{\Gamma(2m+1)\,\Gamma(m)};$$

using the equation $\Gamma(x+1) = x\Gamma(x)$, this becomes

$$\frac{\bar{z}}{c} = \tfrac{3}{4} \frac{\Gamma(2m)\,\Gamma(3m)}{\Gamma(m)\,\Gamma(4m)}, \quad \text{where } m = \frac{1}{n}.$$

Ex. 2. Find the centre of gravity of a hemisphere, the density at any point varying as the nth power of the distance from the centre.

Here we notice that any stratum of uniform density is a thin hemispherical shell, whose volume and centre of gravity are both known. We therefore take this stratum as the element. We have the further advantage that the limits are constants, because the external boundary of the solid is homogeneous.

Let the axis of z be along the middle radius, let $(r, r+dr)$ be the radii of any shell, and let the density $D = \mu r^n$. Then $m = 2\pi r^2 dr \cdot \mu r^n$, also the ordinate of its centre of gravity is $\frac{1}{2}r$, see Art. 422. Hence

$$\bar{z} = \frac{\int 2\pi r^2 dr \,\mu r^n \frac{1}{2} r}{\int 2\pi r^2 dr \,\mu r^n} = \frac{1}{2} \frac{n+3}{n+4} \frac{a^{n+4} - b^{n+4}}{a^{n+3} - b^{n+3}}.$$

The limits of the integral have been taken from $r = b$ to $r = a$, so that we have found the centre of gravity of a *shell* whose internal and external radii are b and a. For a hemisphere we put $b = 0$. If $n + 3$ is positive, we then have $\bar{z} = \dfrac{a}{2} \dfrac{n+3}{n+4}$. In other cases we find $\bar{z} = 0$. If either $n + 3$ or $n + 4$ is zero the integrals lead to logarithmic forms, but we still find $\bar{z} = 0$.

Ex. 3. Find the centre of gravity of the octant of an ellipsoid when the density at any point is $D = \mu x^l y^m z^n$.

To effect this we shall have to find the values of Σmz and Σm, which are both integrals of the form $\qquad \iiint x^l y^m z^n dx\,dy\,dz$

for all elements within the solid. To simplify matters, we write $(x/a)^2 = \xi$, &c. The limits of the integral are now fixed by the plane $\xi + \eta + \zeta = 1$. But these are the integrals known as Dirichlet's integrals, and are to be found in treatises on the Integral Calculus. The result is usually quoted in the form

$$\iiint \xi^{l-1} \eta^{m-1} \zeta^{n-1} d\xi\,d\eta\,d\zeta = \frac{\Gamma(l)\,\Gamma(m)\,\Gamma(n)}{\Gamma(l+m+n+1)},$$

though Liouville's extensions to ellipsoids and other surfaces are also given. Here $\Gamma(p+1) = 1 \cdot 2 \cdot 3 \ldots p$ when p is integral, and in all cases in which p is positive $\Gamma(p+1) = p\Gamma(p)$. Also $\Gamma(\frac{1}{2}) = \sqrt{\pi}$.

The result now follows from substitution; we find

$$\frac{\bar{z}}{c} = \frac{\Gamma\frac{1}{2}(n+2) \cdot \Gamma\frac{1}{2}(l+m+n+5)}{\Gamma\frac{1}{2}(n+1) \cdot \Gamma\frac{1}{2}(l+m+n+6)}.$$

When l, m, n are positive integers there is no difficulty in deducing the values of these gamma functions from the theorems just quoted.

In this way we can find Σmz and Σm and thence $\bar{z}$ whenever the density D is a function which can be expanded in a finite series of powers of x, y, z.

If the density at any point of an octant of an ellipsoid is $D = \mu xyz$, show that $\bar{z} = 16c/35$.

Ex. 4. If the density at any point of an octant of an ellipsoid vary as the square of the distance from the centre, show that $\bar{z} = \dfrac{5c}{16} \dfrac{a^2 + b^2 + 2c^2}{a^2 + b^2 + c^2}$.

Ex. 5. To find the centre of gravity of a triangular area whose density at any point is $D = \mu x^l y^m$.

To determine $\bar{x}$ and $\bar{y}$ we have to find Σm, Σmx and Σmy. All these are integrals of the form $\iint x^l y^m \, dx \, dy$. If y_1, y_2, y_3 are the ordinates of the corners of the triangle and Δ the area, it may be shown that

$$\iint y^n \, dx \, dy = \frac{2\Delta}{(n+1)(n+2)} \{ y_1{}^n + y_1{}^{n-1} y_2 + y_1{}^{n-1} y_3 + \ldots \} \ldots\ldots\ldots\ldots(1),$$

where the right hand side, after division by Δ, is the arithmetic mean of the homogeneous products of y_1, y_2, y_3. Thus when the density is $D = \mu y^n$ the ordinate $\bar{y}$ may be found by a simple substitution.

If we take $y + kx = 0$ as a new axis of x, (1) may be written in the form

$$\iint (y + kx)^n \, dx \, dy = \frac{2\Delta}{(n+1)(n+2)} \{ (y_1 + kx_1)^n + (y_1 + kx_1)^{n-1} (y_2 + kx_2) + \ldots \}.$$

Equating the coefficient of k on each side, we find

$$\iint n x y^{n-1} \, dx \, dy = \frac{2\Delta}{(n+1)(n+2)} \{ n x_1 y_1{}^{n-1} + (n-1) y_1{}^{n-2} y_2 x_1 + \&c. \}.$$

In general, if H_n be the arithmetic mean of the homogeneous products of y_1, y_2, y_3, we have

$$\iint x^p \frac{d^p}{dy^p} y^n \, dx \, dy = \Delta \left(x_1 \frac{d}{dy_1} + x_2 \frac{d}{dy_2} + x_3 \frac{d}{dy_3} \right)^p H_n.$$

One corner of a triangle is at the origin; if the density vary as the cube of the distance from the axis of x, show that $\bar{y} = \dfrac{2}{3} \dfrac{y_1{}^5 - y_2{}^5}{y_1{}^4 - y_2{}^4}$. Also write down the value of $\bar{x}$.

The same method may be used to find the centre of gravity of a quadrilateral, a tetrahedron or a double tetrahedron, when the density is $D = \mu x^l y^m z^n$. See a paper by the author in the *Quarterly Journal of Mathematics*, 1886.

435. Lagrange's two Theorems. *Def.* If the mass of a particle be multiplied by the square of its distance from a given point O, the product is called the moment of inertia of the particle about, or with regard to, the point O. The moment of inertia of a system of particles is the sum of the moments of inertia of the several particles.

436. *Lagrange's first Theorem.* The moment of inertia of a system of particles about any point O is equal to their moment of inertia about their centre of gravity together with what would be the moment of inertia about O of the whole mass if it were collected at its centre of gravity.

Let the particles m_1, m_2 &c. be situated at the points A_1, A_2 &c. Let $(x_1 y_1 z_1)$, $(x_2 y_2 z_2)$, &c. be the coordinates of A_1, A_2 &c.

referred to O as origin. Let $\bar{x}, \bar{y}, \bar{z}$ be the coordinates of the centre of gravity G. Also let $x = \bar{x} + x'$, $y = \bar{y} + y'$, &c. Now

$$\Sigma\,(m\,.\,OA^2) = \Sigma m\,\{(\bar{x} + x')^2 + (\bar{y} + y')^2 + (\bar{z} + z')^2\}$$
$$= \Sigma m\,.\,OG^2 + 2\bar{x}\Sigma mx' + 2\bar{y}\Sigma my' + 2\bar{z}\Sigma mz' + \Sigma\,(mGA^2).$$

Since the origin of the accented coordinates is the centre of gravity, we have $\Sigma mx' = 0$, $\Sigma my' = 0$, $\Sigma mz' = 0$. Hence putting $M = \Sigma m$, we have $\Sigma\,(m\,.\,OA^2) = M\,.\,OG^2 + \Sigma\,(m\,.\,GA^2)\ldots\ldots(A)$. This equation expresses Lagrange's theorem in an analytical form.

We notice that the moment of inertia of the body about any point O is least when that point is at the centre of gravity.

An important extension of this theorem is required in rigid dynamics. It is shown that, if $f(x, y, z)$ be any quadratic function of the coordinates of a particle, then

$$\Sigma mf(x, y, z) = Mf(\bar{x}, \bar{y}, \bar{z}) + \Sigma mf(x', y', z').$$

437. *Lagrange's second Theorem.* If m, m' be the masses of any two particles, AA' the distance between them, then the theorem may be analytically stated thus

$$\Sigma\,(mm'\,.\,AA'^2) = M\Sigma\,(m\,.\,GA^2)\ldots\ldots\ldots\ldots(B).$$

The sum of the continued products of the masses taken two together and the square of the distance between them is equal to the product of the whole mass by the moment of inertia about the centre of gravity.

This may be easily deduced from Lagrange's first theorem. We have by (A)

$$\Sigma m_a OA_a^2 = M\,.\,OG^2 + \Sigma m_a GA_a^2,$$

where Σ implies summation for all values of α. Putting the arbitrary point O successively at A_1, A_2 &c. we have

$$\Sigma m_a A_1 A_a^2 = M\,.\,A_1 G^2 + \Sigma m_a GA_a^2,$$
$$\Sigma m_a A_2 A_a^2 = M\,.\,A_2 G^2 + \Sigma m_a GA_a^2,$$
$$\&c. = \&c.$$

Multiplying these respectively by m_1, m_2 &c. and adding the products together, we have

$$\Sigma m_a m_\beta A_\beta A_a^2 = M\,\Sigma m_\beta A_\beta G^2 + \Sigma m_\beta\,.\,\Sigma m_a GA_a^2.$$

The Σ on the left hand side implies summation for all values of both α and β. Each term will therefore appear twice over, once in the form $m_\beta m_a\,.\,A_\beta A_a^2$, and a second time with α and β interchanged. If we wish to take each term once only, we must take half the right hand side. But the terms on the right hand side are the same. Hence

$$\Sigma m_\alpha m_\beta . A_\alpha A_\beta{}^2 = M\Sigma m_\alpha . GA_\alpha{}^2.$$

438. Ex. Let the symbol [ABC] represent the area of the triangle formed by joining the three points A, B, C. Let [ABCD] represent the volume of the tetrahedron formed by joining the four points in space A, B, C, D. We may extend the analytical expression for the area and volume to any number of points by the same notation. We then have the following extensions of Lagrange's two theorems

$$\Sigma m_\alpha OA_\alpha{}^2 = M . OG^2 + \Sigma m_\alpha GA_\alpha{}^2$$

$$\Sigma m_\alpha m_\beta [OA_\alpha A_\beta]^2 = M\Sigma m_\alpha [OGA_\alpha]^2 + \Sigma m_\alpha m_\beta [GA_\alpha A_\beta]^2$$

$$\Sigma m_\alpha m_\beta m_\gamma [OA_\alpha A_\beta A_\gamma]^2 = M\Sigma m_\alpha m_\beta [OGA_\alpha A_\beta]^2 + \Sigma m_\alpha m_\beta m_\gamma [GA_\alpha A_\beta A_\gamma]^2$$

&c. = &c.

$$\Sigma m_\alpha m_\beta A_\alpha A_\beta{}^2 = M\Sigma m_\alpha GA_\alpha{}^2$$

$$\Sigma m_\alpha m_\beta m_\gamma [A_\alpha A_\beta A_\gamma]^2 = M\Sigma m_\alpha m_\beta [GA_\alpha A_\beta]^2$$

$$\Sigma m_\alpha m_\beta m_\gamma m_\delta [A_\alpha A_\beta A_\gamma A_\delta]^2 = M\Sigma m_\alpha m_\beta m_\gamma [GA_\alpha A_\beta A_\gamma]^2$$

&c. = &c.

The first of each of these sets of equations is of course a repetition of Lagrange's equations. The remaining equations are due to Franklin.

[*American Journal of Mathematics*, Vol. x., 1888.]

439. Application to pure geometry. The property that every body has but one centre of gravity* may be used to assist us in discovering new geometrical theorems. The general method may be described in a few words. We place weights of the proper magnitudes at certain points in the figure. By combining these in several different orders we find different constructions for the centre of gravity. All these must give the same point. The following are a few examples.

Ex. 1. The two straight lines which join the middle points of the opposite sides of a quadrilateral and the straight line which joins the middle points of the two diagonals, intersect in one point and are bisected at that point. [Coll. Exam.]

Ex. 2. The centre of gravity of four particles of equal weight in the same plane is the centre of the conic which bisects the lines joining each pair of points.

[Only one chord of a conic is bisected at a given point, unless that point is the centre. Since, by the last example, three chords are bisected at the same point, that point is the centre.]   [Caius Coll.]

Ex. 3. Through each edge of a tetrahedron a plane is drawn bisecting the angle between the planes that meet in that edge and intersecting the opposite edge: prove that the three lines joining the points so determined on opposite edges meet in a point.   [St John's Coll., 1879.]

* In Milne's *Companion to the weekly problem papers* 1888, a number of examples will be found of the application of the "centroid" and of "force" to geometry.

Place weights at the corners proportional to the areas of the opposite faces. The centre of gravity of these four weights lies in each of the three straight lines.

440. The theorems on the centre of gravity are also useful in helping us to remember the relations of certain points, much used in our geometrical figures, to the other points and lines in the construction. For instance, when the results of Ex. 1 have been noticed, the distance of the centre of the inscribed conic from any straight line can be written down at once by taking moments about that line.

Ex. 1. The areal equation to the conic inscribed in the triangle of reference is $\sqrt{lx} + \sqrt{my} + \sqrt{nz} = 0$; show that the centre of the conic is the centre of gravity of three particles placed at the middle points of the sides, whose weights are proportional to l, m, n. It is also the centre of gravity of three particles whose weights are proportional to $m+n$, $n+l$, $l+m$, placed either at the points of contact or at the corners of the triangle.

Let the conic touch the sides in D, E, F, then D and E divide BC and AC in the ratios $m:n$ and $l:n$. Let ξ, η, ζ be the weights placed at A, B, C whose centre of gravity is the centre. Then ξ, η are respectively equivalent to $\xi(l+n)/n$ and $\eta(m+n)/n$ placed at E and D together with some weight at C, Art. 79. But since the straight line joining C to the centre O bisects DE, we see by taking moments about CO that the weights D and E are equal. Hence ξ and η are proportional to $m+n$ and $n+l$.

If the conic is a parabola $l+m+n=0$, because the weights must reduce to a couple. Hence the far extremity of the principal diameter, and therefore the far focus, is the centre of gravity of weights l, m, n placed at the corners A, B, C. Since the product of the perpendiculars from the foci on all tangents are equal, the near focus is the centre of gravity of three weights a^2/l, b^2/m, c^2/n placed at the corners.

Ex. 2. The areal equation to the conic circumscribed about a triangle is $lyz + mzx + nxy = 0$. Show that its centre is the centre of gravity of six particles, three placed at the corners whose weights are proportional to l^2, m^2, n^2, and three at the middle points of the sides whose weights are $-2mn$, $-2nl$, $-2lm$.

Ex. 3. Three particles of equal weight are placed at the corners of a triangle, and a fourth particle of negative weight is placed at the centre of the circumscribing circle. Show that the centre of gravity of all four is the centre of the nine-points circle or the orthocentre, according as the weight of the fourth particle is numerically equal to or double that of any one of the particles at the corners.

Ex. 4. The equation to a conic being $Ap^2 + Bq^2 + Cr^2 + 2Dqr + 2Erp + 2Fpq = 0$ in tangential coordinates, show that the centre of the conic is the centre of gravity of three weights proportional to $A+E+F$, $B+F+D$, $C+D+E$ placed at the corners. For other theorems see a paper by the author in the *Quarterly Journal*, Vol. VIII. 1866.

441. Theorems concerning the resolution and composition of forces may be used, as well as those relating to the centre of gravity, to prove geometrical properties.

Ex. 1. A straight line is drawn from the corner D of a tetrahedron making equal angles with the edges DA, DB, DC. Show that this straight line intersects the plane ABC in a point E such that AE/AD, BE/BD, CE/CD are proportional to the sines of the angles BEC, CEA, AED. Show also that $\dfrac{1}{AD} + \dfrac{1}{BD} + \dfrac{1}{CD} = \dfrac{3\cos\theta}{ED}$, where θ is the angle DE makes with any edge at D.

Ex. 2. *ABCD* is a quadrilateral, whose opposite sides meet in X and Y. Show that the bisectors of the angles X, Y, the bisectors of the angles B, D and the bisectors of the angles A, C intersect on a straight line, certain restrictions being made as to which pairs of bisectors are taken. See figure in Art. 132.

[Apply four equal forces to act along the sides of the quadrilateral, and find their resultant by combining them in different orders.]   [Math. Tripos, 1882.]

Ex. 3. Prove, by mechanical considerations, that the locus of the centres of all ellipses inscribed in the same quadrilateral is the straight line joining the middle points of any two diagonals.   [Coll. Exam.]

Let A, B, C, D be the corners taken in order. Apply forces along AB, AD, CB, CD proportional to these lengths. The tangents measured from each corner to the adjacent points of contact represent forces whose resultant passes through the centre. Since these eight forces make up the four forces AB, AD, CB, CD, the resultant passes through the centre. Again the resultant of AB, AD and also that of CB, CD bisect the diagonal BD. Similarly the resultant force bisects the other diagonal.

Ex. 4. If X, Y are the intersections of the opposite sides of a quadrilateral $ABCD$, prove that the ratio of the perpendiculars drawn from X and Y on the diagonal AC is equal to the ratio of the perpendiculars on the diagonal BD. Show also that each of these ratios is equal to the ratio of $AB \cdot CD \sin Y$ to $AD \cdot BC \sin X$. See figure of Art. 132.

CHAPTER X.

ON STRINGS.

442. The Catenary. The strings considered in this chapter are supposed to be perfectly flexible. By this we mean that the resultant action across any section of the string consists of a single force whose line of action is along a tangent to the length of the string. Any normal section is considered to be so small that the string may be regarded as a curved line, so that we may speak of its tangent, or its osculating plane.

The resultant action across any section of the string is called its tension, and in what follows will be represented by the letter T. This force may theoretically be positive or negative, but it is obvious that an actual string can only pull. The positive sign is given to the tension when it exerts a pull on any object instead of a push.

The weight of an element of length ds is represented by wds. In a uniform string w is the weight of a unit of length. If the string is not uniform, w is the weight of a unit of length of an imaginary string, such that any element of it (whose length is ds) is similar and equal to the particular element ds of the actual string.

443. *A heavy uniform string is suspended from two given points A, B, and is in equilibrium in a vertical plane. It is required to find the equation to the curve in which it hangs.* This curve is called the common Catenary*.

---

* The following short account of the history of the problem known under the name of the "Chaînette" is abridged from Montucla, Vol. ii., p. 468. The problem of finding the form of a heavy chain suspended from two fixed points was proposed by James Bernoulli as a question to the other geometers of that day. Four mathematicians, viz. James Bernoulli and his brother, Leibnitz and Huyghens, had the honour of solving it. They published their solutions in the *Actes de Leipsick*

Let C be the lowest point of the catenary, i.e. the point at which the tangent is horizontal. Take some horizontal straight line Ox as the axis of x, whose distance from C we may afterwards choose at pleasure. Draw CO perpendicular to it, and let O be the origin. Let ψ be the angle the tangent at any point P makes with Ox. Let T_0 and T be the tensions at C and P, and let $CP = s$. In the figure the axis of x, which is afterwards taken to represent the directrix, has been placed nearer the curve than it really is in order to save space.

The length CP of the string is in equilibrium under three forces, viz. the tensions T_0 and T acting at C and P in the directions of the arrows, and its weight ws acting at the centre of gravity G of the arc CP.

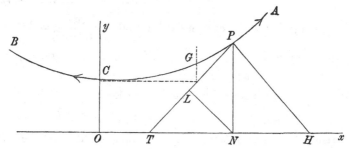

Resolving horizontally, we have $\qquad T \cos \psi = T_0 \ldots\ldots\ldots$ (1).

Resolving vertically, we have $\qquad T \sin \psi = ws \ldots\ldots\ldots$ (2).

Dividing one of these equations by the other,

$$\frac{dy}{dx} = \tan \psi = \frac{ws}{T_0} \ldots\ldots\ldots\ldots\ldots\ldots\ldots (3).$$

(Act. Erud. 1691) but without the analysis, apparently wishing to leave some laurels to be gathered by those who followed. David Gregory published a solution some years after in the *Phil. Trans.* 1697.

It is the custom of geometers to rise from one difficulty to another, and even to make new ones in order to have the pleasure of surmounting them. Bernoulli was no sooner in possession of the solution of his problem of the chaînette considered in its simplest case, than he proceeded to more difficult ones. He supposed next that the string was heterogeneous and enquired what should be the law of density that the curve should be of any given form, and what would be the curve if the string were extensible. He soon after published his solution, but reserved his analysis. Finally he proposed the problem, what would be the form of the string if it were acted on by a central force. The solutions of all these problems were afterwards given by John Bernoulli in his *Opera Omnia.* See also Ball's *Short History of Mathematics,* 1888.

Montucla remarks that the problem of the chaînette had excited the curiosity of Galileo, who had decided that the curve is a parabola. But this accusation is stated by Venturoli to be without foundation. Galileo had merely noticed the similarity between the two curves. See Venturoli, *Elements of Mechanics, translated by Cresswell,* p. 69, where the problem of the chaînette is discussed.

If the string is uniform w is constant, and it is then convenient to write $T_0 = wc$. To find the curve we must integrate the differential equation (3). We have

$$\left(\frac{ds}{dy}\right)^2 = 1 + \left(\frac{dx}{dy}\right)^2 = 1 + \frac{c^2}{s^2}.$$

$$\therefore \; dy = \pm \frac{sds}{\sqrt{(s^2 + c^2)}}; \qquad \therefore \; y + A = \pm \sqrt{(s^2 + c^2)}.$$

We must take the upper sign, for it is clear from (3) that, when x and s increase, y must also increase. When $s = 0$, $y + A = c$. Hence, if the axis of x is chosen to be at a distance c below the lowest point C of the string, we shall have $A = 0$. The equation now takes the form

$$y^2 = s^2 + c^2 \dots\dots\dots\dots\dots\dots\dots\dots(4).$$

Substituting this value of y in (3), we find $\dfrac{cds}{\sqrt{(s^2 + c^2)}} = dx$,

where the radical is to have the positive sign. Integrating,

$$c \log \{s + \sqrt{(s^2 + c^2)}\} = x + B.$$

But x and s vanish together, hence $B = c \log c$.

From this equation we find $\qquad \sqrt{(s^2 + c^2)} + s = ce^{\frac{x}{c}}.$
Inverting this and rationalizing the denominator in the usual

manner, we have $\qquad \sqrt{(s^2 + c^2)} - s = ce^{-\frac{x}{c}}.$
Adding and subtracting we deduce by (4)

$$y = \frac{c}{2}\left(e^{\frac{x}{c}} + e^{-\frac{x}{c}}\right), \qquad s = \frac{c}{2}\left(e^{\frac{x}{c}} - e^{-\frac{x}{c}}\right)\dots\dots\dots(5).$$

The first of these is the Cartesian equation to the common catenary. The straight lines which have here been taken as the axes of x and y are called respectively the *directrix* and the *axis of the catenary*. The point C is called the *vertex*.

Adding the squares of (1) and (2), we have by help of (4)

$$T^2 = w^2 (s^2 + c^2) = w^2 y^2;$$

$$\therefore \; T = wy \dots\dots\dots\dots\dots\dots\dots(6).$$

The equations (1) and (2) give us two important properties of the curve, viz. (1) *the horizontal tension at every point of the curve is the same and equal to wc*; (2) *the vertical tension at any point P is equal to ws, where s is the arc measured from the lowest point.* To these we join a third result embodied in (6), viz. (3) *the resultant tension at any point is equal to wy, where y is the ordinate measured from the directrix.*

444. Referring to the figure, let PN be the ordinate of P, then $T = w \cdot PN$. Draw NL perpendicular to the tangent at P, then the angle $PNL = \psi$. Hence

$$PL = PN \cdot \sin \psi = s \text{ by (2)},$$

$$NL = PN \cdot \cos \psi = c \text{ by (1)}.$$

These two geometrical properties of the curve may also be deduced from its Cartesian equation (5). By differentiating (3) we find

$$\frac{1}{\cos^2 \psi} \frac{d\psi}{ds} = \frac{1}{c}, \quad \therefore \ \rho = \frac{c}{\cos^2 \psi} \ \dots\dots\dots\dots (7).$$

We easily deduce from the right-angled triangle PNH, that the length of the normal, viz. PH, between the curve and the directrix is equal to the radius of curvature, viz. ρ, at P.

It will be noticed that these equations contain only one undetermined constant, viz. c; and when this is given the form of the curve is absolutely determined. Its position in space depends on the positions of the straight lines called its directrix and axis. This constant c is called *the parameter of the catenary.* Two arcs of catenaries which have their parameters equal are said to be arcs of equal catenaries.

Since $\rho \cos^2 \psi = c$, it is clear that c is large or small according as the curve is flat or much curved near its vertex. Thus if the string is suspended from two points A, B in the same horizontal line, then c is very large or very small compared with the distance between A and B according as the string is tight or loose.

The relations between the quantities y, s, c, ρ, ψ and T in the common catenary may be easily remembered by referring to the rectilineal figure $PLNH$. We have $PN=y$, $PL=s$, $NL=c$, $PH=\rho$, $T=w \cdot PN$ and the angles LNP, NPH are each equal to ψ. Thus the important relations (1), (2), (3), (4), and (7) follow from the ordinary properties of a right-angled triangle.

445. Since the three forces, viz., the tensions at A and B and the weight are in equilibrium, it follows that their lines of action must meet in a point. Hence the *centre of gravity G of the arc must lie vertically over the intersection of the tangents at the extremities of the arc.* This is a statical proof of one part of the more general theorem given in Art. 399, Ex. 1, where it is also proved that *the vertical ordinate of the centre of gravity is half that of the intersection of the normals at the extremities of the arc.*

446. Ex. 1. Show that it is impossible to pull a heavy string by forces at its extremities so as to make it quite straight unless the string is vertical.

If it be straight let ψ be the inclination to the horizon, W its weight. Then, resolving perpendicularly to its length, $W \cos \psi = 0$, which gives ψ equal to a right angle. This proof does not require the string to be uniform.

Ex. 2. If a string be suspended from any two points A and B not in the same vertical, and be nearly straight, show that c is very large.

Let ψ, ψ' be the inclinations at A and B, and l the length of the string. Then $l = s - s' = c (\tan \psi - \tan \psi')$. Since ψ and ψ' are nearly equal, c is large compared with l.

Ex. 3. A heavy uniform string AB of length l is suspended from a fixed point A, while the other extremity B is pulled horizontally by a given force $F = wa$. Show that the horizontal and vertical distances between A and B are $a \log \dfrac{l + \sqrt{(l^2 + a^2)}}{a}$ and $\sqrt{(l^2 + a^2)} - a$ respectively.

Ex. 4. The extremities A and B of a heavy string of length $2l$ are attached to two small rings which can slide on a fixed horizontal wire. Each of these rings is acted on by a horizontal force $F = wl$. Show that the distance apart of the rings is $2l \log (1 + \sqrt{2})$.

Ex. 5. If the inclination ψ of the tangent at any point P of the catenary is taken as the independent variable, prove that

$$x = c \log \tan \left(\frac{\pi}{4} + \frac{\psi}{2} \right), \quad y = \frac{c}{\cos \psi}, \quad s = c \tan \psi, \quad \rho = \frac{c}{\cos^2 \psi}.$$

If $\bar{x}$, $\bar{y}$ be the coordinates of the centre of gravity of the arc measured from the vertex up to the point P, prove also that $\bar{x} = x - c \tan \dfrac{\psi}{2}$, $\quad \bar{y} = \dfrac{1}{2} \left(\dfrac{c}{\cos \psi} + x \cot \psi \right)$.

447. If the position in space of the points A and B of suspension and the length of the string or chain are given, we may obtain sufficient equations to find the parameter c of the catenary, and the positions in space of its directrix and axis.

Let the given point A be taken as an origin of coordinates, and let the axes be horizontal and vertical. Let (h, k) be the coordinates of B referred to A, and let l be the length of the string AB. These three quantities are therefore given. Let (x, y), $(x + h, y + k)$ be the coordinates of A, B referred to the directrix and axis of the catenary. Then x, y, c are the three quantities to be found. By Art. 443

$$y = \frac{c}{2} (e^{\frac{x}{c}} + e^{-\frac{x}{c}}), \qquad y + k = \frac{c}{2} (e^{\frac{x+h}{c}} + e^{-\frac{x+h}{c}}) \quad \ldots\ldots\ldots\ldots(A).$$

Also by Art. 443, since l is the algebraic difference of the arcs CA, CB,

$$l = \frac{c}{2} (e^{\frac{x+h}{c}} - e^{-\frac{x+h}{c}}) - \frac{c}{2} (e^{\frac{x}{c}} - e^{-\frac{x}{c}}) \quad \ldots\ldots\ldots\ldots(B).$$

If C lie between A and B, x will be negative.

These three equations are sufficient to determine x, y and c. They cannot however be solved in finite terms. We may eliminate x, y in the following manner.

Writing $u = e^{\frac{x}{c}}$, $v = e^{\frac{h}{c}}$, we find from (A) and (B)

$$\left. \begin{aligned} k &= \frac{c}{2} \left(u - \frac{1}{uv} \right) (v - 1) \\ l &= \frac{c}{2} \left(u + \frac{1}{uv} \right) (v - 1) \end{aligned} \right\} \quad \ldots\ldots\ldots(C).$$

We notice that v contains only c and the known quantity h. Hence, subtracting the squares of these equations in order to eliminate u, we find

$$\pm \sqrt{(l^2 - k^2)} = c \left(e^{\frac{h}{2c}} - e^{-\frac{h}{2c}}\right) \dots\dots\dots\dots\dots\dots\dots\dots\text{(D)}.$$

This agrees with the equation given by Poisson in his *Traité de Mécanique*.

The value of c has to be found from this equation. It gives two real finite values of c, one positive and the other negative but numerically equal. A negative value for c would make y negative and would therefore correspond to a catenary with its concavity downwards. It is therefore clear that the positive value of c is to be taken.

To analyse the equation (D), we let $c = 1/\gamma$, and arrange the terms of the equation in the form
$$z = e^{m\gamma} - e^{-m\gamma} - a\gamma = 0\dots\dots\dots\dots\dots\dots\dots\dots\text{(E)},$$
so that a and m are both positive. We have $a^2 = l^2 - k^2$, and $2m = h$. Since the length l of the string must be longer than the straight line joining the points of suspension, it is clear that a must be greater than $2m$. By differentiation,

$$\frac{dz}{d\gamma} = m \left(e^{m\gamma} + e^{-m\gamma}\right) - a.$$

Thus $dz/d\gamma$ is negative when $\gamma = 0$, so that, as γ increases from zero, z is at first zero, then becomes negative and finally becomes positive for large values of γ. There is therefore some one value of γ, say $\gamma = i$, at which $z = 0$. If there could be another, say $\gamma = i'$, then $dz/d\gamma$ must vanish twice, once between $\gamma = 0$ and $\gamma = i$, and again between $\gamma = i$ and $\gamma = i'$. We shall now show that this is impossible. By differentiating twice we have

$$\frac{d^2z}{d\gamma^2} = m^2 \left(e^{m\gamma} - e^{-m\gamma}\right);$$

thus $d^2z/d\gamma^2$ is positive when γ is greater than zero. Hence $dz/d\gamma$ continually increases with γ from its initial value $2m - a$ when $\gamma = 0$. It therefore cannot vanish twice when γ is positive. It appears from this reasoning that the equation gives only one positive value of c.

The solitary positive value of c having been found from (D), we can form a simple equation to find u by adding one of the equations (C) to the other. In this way we find one real value of x. The value of y is then found from the first of the equations (A). Thus it appears that, *when a uniform string is suspended from two fixed points of support, there is only one position of equilibrium.*

The equation (D) can be solved by approximation when h/c is so small that we can expand the exponentials and retain only the first powers of h/c which do not disappear of themselves. This occurs when c is large, i.e. when the string is nearly tight. In such cases, however, it will be found more convenient to resume the problem from the beginning rather than to quote the equations (D) or (E).

448. Ex. 1. A uniform string of length l is suspended from two points A and B in the same horizontal line, whose distance apart is h. If h and l are nearly equal, find the parameter of the catenary.

Referring to the figure of Art. 443, we see that $s = \frac{1}{2}l$, $x = \frac{1}{2}h$. Hence using one of the equations (5) of that article, we have $l = c \left(e^{\frac{h}{2c}} - e^{-\frac{h}{2c}}\right)$.

Whatever the given values of h and l may be, the value of c must be found from this equation. When h and l are nearly equal, we know by Art. 446, Ex. 2, that h/c

is small. Hence, expanding the exponentials and retaining only the lowest powers of h/c which do not disappear, we have $\qquad c^2 = \dfrac{h^3}{24\,(l-h)}$.

Since the string considered in this problem is nearly horizontal, the tension of every element is nearly the same. If the string be slightly extensible, so that the extension of any element is some function of the tension, the stretched string will still be homogeneous. The form will therefore be a catenary, and its parameter will be given by the same formula, provided l represents its stretched length.

In order to use this formula, the length l of the string and the distance h between A and B must be measured. But measurements cannot be made without error. To use any formula correctly it is necessary to estimate the effects of such errors. Taking the logarithmic differential we have $\qquad \dfrac{2\delta c}{c} = \dfrac{3\delta h}{h} - \dfrac{\pm\delta l \pm \delta h}{l-h}$.

Here δh and δl are the errors of h and l due to measurement. We see that the error in c might be a large proportion of c if either h or $l-h$ were small. In our case $l-h$ is small. Hence to find c we must so make our measurements that the error of $l-h$ is small compared with the small quantity $l-h$, while the length h need be measured only so truly that its error is within the same fraction of the larger quantity h. Thus greater care must be taken in measuring $l-h$ than h.

Suppose, for example, that $h=30$ feet and $l=31$ feet, with possible errors of measurement either way of only one thousandth part of the thing measured. The value of c given by the formula is 33·5 feet, but its possible error is as much as one thirtieth part of itself.

Ex. 2. A uniform measuring chain of length l is tightly stretched over a river, the middle point just touching the surface of the water, while each of the extremities has an elevation k above the surface. Show that the difference between the length of the measuring chain and the breadth of the river is nearly $\dfrac{8}{3}\dfrac{k^2}{l}$.

Ex. 3. A heavy string of length $2l$ is suspended from two fixed points A, B in the same horizontal line at a distance apart equal to $2a$. A ring of weight W can slide freely on the string, and is in equilibrium at the lowest point. Find the parameter of the catenary and the position of the weight.

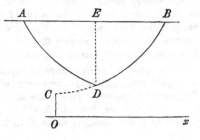

Let D be the position of the heavy ring, then BD and AD are equal portions of a catenary. Produce BD to its vertex C, and let Ox, OC be the directrix and axis of the catenary DB. Let x be the abscissa of D. Then since l is the difference of the arcs

CB, CD, we have $\qquad l = \dfrac{c}{2}\left(e^{\frac{x+a}{c}} - e^{-\frac{x+a}{c}}\right) - \dfrac{c}{2}\left(e^{\frac{x}{c}} - e^{-\frac{x}{c}}\right)$ ....(1).

Also, since the weight of the ring is supported by the two vertical tensions of the string, $\qquad\qquad W = 2w\,\dfrac{c}{2}\left(e^{\frac{x}{c}} - e^{-\frac{x}{c}}\right)$ ....(2).

The equations (1) and (2) determine x and c. Thence the ordinates of D and B may be found, and therefore the depth of D below AB.

If the weight of the ring is much greater than the weight of the string, each string is nearly tight. Thus a/c is small, but x/c is not necessarily small, for the vertex C may be at a considerable distance from D. If we expand the terms containing the exponent a/c and eliminate those containing x/c, we find

$$c = Wa/2w \sqrt{(l^2 - a^2)} \text{ nearly.}$$

The contrary holds if the weight of the ring is much smaller than the weight of the string. If W were zero the two catenaries BD and DA would be continuous, and the vertex would be at D. Hence when W is very small, the vertex will be near D and therefore x/a will be small. But a/c is not necessarily small. Expanding the terms with small exponentials, we find from (2) that $x = W/2w$. Then

(1) gives
$$l = \frac{c}{2}(e^{\frac{a}{c}} - e^{-\frac{a}{c}}) + \frac{W}{2w}\{\tfrac{1}{2}(e^{\frac{a}{c}} + e^{-\frac{a}{c}}) - 1\}.$$

If the weight W were absent this equation would reduce to the one already discussed above. If γ be the change produced in the value of c there found by adding the weight W, we find, by writing $c + \gamma$ for c in the first term on the right hand side, that $\left(l - \dfrac{ak}{c}\right)\gamma + \dfrac{W}{2w}(k - c) = 0$, where k is the ordinate of B before the addition of W.

If the *weight W had been attached to any point D of the string* not its middle point, AD, BD would still form catenaries, whose positions could be found in a similar manner. We may notice that, however different the two strings may appear to be, *the catenaries have equal parameters*. For consider the equilibrium of the weight W; we see by resolving horizontally that the wc of each catenary must be the same.

If the *string be passed through a fine smooth ring fixed in space* through which it could slide freely, the two strings on each side must have their tensions equal. Hence the *two catenaries have the same directrix*. The parameters are not necessarily equal, for the difference between the horizontal tensions of the two catenaries is equal to the horizontal pressure on the ring, which need not be zero.

Ex. 4. A heavy string of length l is suspended from two points A, A' in the same horizontal line, and passes through a smooth ring D fixed in space. If DN be a perpendicular from D on AA' and $NA = h$, $NA' = h'$, $DN = k$, prove that the parameters c, c' may be obtained from

$$4c^2 = l^2 \left\{\cosh\frac{h'}{2c'}\cosech\left(\frac{h}{2c} + \frac{h'}{2c'}\right)\right\}^2 - k^2\left(\cosech\frac{h}{2c}\right)^2,$$

and a similar equation with the accented and unaccented letters interchanged.

Ex. 5. A portion AC of a uniform heavy chain rests extended in the form of a straight line on a rough horizontal plane, while the other portion CB hangs in the form of a catenary from a given point B above the plane. The whole chain is on the point of motion towards the vertical through B. If l be the length of the whole chain and h be the altitude of B above the plane, show that the parameter c of the catenary is equal to $\mu(l + \mu h) - \mu\sqrt{\{(\mu^2 + 1)h^2 + 2\mu hl\}}$.

Ex. 6. A heavy string hangs over two small smooth fixed pegs. The two ends of the string are free, and the central portion hangs in a catenary. Show that the free ends are on the directrix of the catenary. If the two pegs are on the same level and distant $2a$ apart, show that equilibrium is impossible unless the length of the string is equal to or greater than $2ae$.   [Coll. Exam.]

Ex. 7. A heavy uniform chain is suspended from two fixed points A and B in the same horizontal line, and the tangent at A makes an angle $45°$ with the horizon.

Prove that the depth of the lowest point of the chain below AB is to the length of the chain as $\sqrt{(2)} - 1 : 2$.

Ex. 8. A uniform heavy chain is fastened at its extremities to two rings of equal weight, which slide on smooth rods intersecting in a vertical plane, and inclined at the same angle a to the vertical: find the condition that the tension at the lowest point may be equal to half the weight of the chain ; and, in that case, show that the vertical distance of the rings from the point of intersection of the rods is $l \cot a \log (\sqrt{2} + 1)$, where $2l$ is the length of the chain.   [Math. Tripos, 1856.]

Ex. 9. A heavy string of uniform density and thickness is suspended from two given points in the same horizontal plane. A weight, an nth that of the string, is attached to its lowest point ; show that, if θ, ϕ be the inclinations to the vertical of the tangents at the highest and lowest points of the string, $\tan \phi = (1 + n) \tan \theta$.
          [Math. Tripos, 1858.]

Ex. 10. If a, β be the angles which a string of length l makes with the vertical at the points of support, show that the height of one point above the other is

$$l \cos \tfrac{1}{2} (a + \beta) / \cos \tfrac{1}{2} (a - \beta).$$ [Pet. Coll., 1855.]

Ex. 11. A heavy endless string passes over two small smooth fixed pegs in the same horizontal line, and a small smooth ring without weight binds together the upper and lower portions of the string : prove that the ratio of the cosines of the angles which the portions of the string at either peg make with the horizon, is equal to that of the tangents of the angles which the portions of the string at the ring make with the vertical.   [Math. Tripos, 1872.]

Ex. 12. A and B are two smooth pegs in the same horizontal line, and C is a third smooth peg vertically below the middle point of AB ; an endless string hangs upon them forming three catenaries AB, BC, and CA : if the lowest point of the catenary AB coincides with C, prove that the pegs AB divide the whole string into two parts in the ratio of $2w + w'$ to $2w - w'$, where w and w' are the vertical components of the pressures on A and C respectively.   [Math. Tripos, 1870.]

Ex. 13. An endless uniform chain is hung over two small smooth pegs in the same horizontal line. Show that, when it is in a position of equilibrium, the ratio of the distance between the vertices of the two catenaries to half the length of the chain is the tangent of half the angle of inclination of the portions near the pegs.
          [Math. Tripos, 1855.]

Ex. 14. A heavy uniform string of length $4l$ passes through two small smooth rings resting on a fixed horizontal bar. Prove that, if one of the rings be kept stationary, the other being held at any other point of the bar, the locus of the position of equilibrium of that end of the string which is the further from the stationary ring may be represented by the equation $x = 2\sqrt{(ly)} \log \dfrac{l}{y}$.   [Coll. Ex.]

Ex. 15. A heavy uniform string is suspended from two points A and B in the same horizontal line, and to any point P of the string a heavy particle is attached. Prove that the two portions of the string are parts of equal Catenaries.

Prove also that the portion of the tangent at A intercepted between the verticals through P and the centre of gravity of the string is divided by the tangent at B in a ratio independent of the position of P.

If θ, ϕ be the angles the tangents at P make with the horizon, a and β those made by the tangents at A and B, show that $\dfrac{\tan \theta + \tan \phi}{\tan a + \tan \beta}$ is constant for all positions of P.   [St John's Coll.]

Ex. 16. A heavy uniform string hangs over two smooth pegs in the same horizontal line. If the length of each portion which hangs freely be equal to the length between the pegs, prove that the whole length of the string is to the distance between the pegs as $\sqrt{(3)}$ to $\log \sqrt{(3)}$. Compare also the pressures on each peg with the weight of the string.

Ex. 17. A uniform endless string of length l is placed symmetrically over a smooth cube which is fixed with one diagonal vertical. Prove that the string will slip over the cube unless the side of the cube is greater than $\frac{1}{6} l \sqrt{2} \log (1 + \sqrt{2})$.
[Emm. Coll., 1891.]

Ex. 18. An endless inextensible string hangs in two festoons over two small pegs in the same horizontal line. Prove that, if θ be the inclination to the vertical of one branch of the string at its highest point, the inclination of the other branch at the same point must be either θ or ϕ, where ϕ has only *one* value and is a function of θ only. If $\cot \frac{1}{2}\theta = e^{\sec \theta}$, then $\phi = \theta$. [Coll. Ex.]

Ex. 19. Four smooth pegs are placed in a vertical plane so as to form a square, the diagonals being one vertical and one horizontal. Round the pegs an endless chain is passed so as to pass over the three upper and under the lower one. If the directions of the strings make with the vertical angles equal to a at the upper peg, β and γ at each of the middle and δ at the lower peg, prove the following relations:
$$\sin \beta \log \cot \tfrac{1}{2}a \tan \tfrac{1}{2}\beta = \sin \gamma \log \cot \tfrac{1}{2}\gamma \tan \tfrac{1}{2}\delta,$$
$$\sin \beta \sin \delta + \sin a \sin \gamma = 2 \sin a \sin \delta. \qquad \text{[Caius Coll.]}$$

Ex. 20. A bar of length $2a$ has its ends fastened to those of a heavy string of length $2l$, by which it is hung symmetrically over a peg. The weight of the bar is n times, and the horizontal tension $\frac{1}{2}m$ times the weight of the string. Show that
$$m^2 + n^2 = \left\{ (n+1) \operatorname{cosech} \frac{a}{ml} - n \coth \frac{a}{ml} \right\}^2. \qquad \text{[Coll. Ex., 1889.]}$$

Ex. 21. One end of a heavy chain is attached to the extremity of a fixed rod, the other end is fastened to a small smooth ring which slides on the rod: prove that in the position of equilibrium $\log \{ \cot \frac{1}{2}\theta \cot (\frac{1}{4}\pi - \frac{1}{2}\psi) \} = \cot \theta (\sec \psi - \operatorname{cosec} \theta)$, θ being the inclination of the rod to the horizon, and ψ that of the chain at its highest point. [Coll. Ex.]

Ex. 22. A string of length πa is fastened to two points at a distance apart equal to $2a$, and is repelled by a force perpendicular to the line joining the points and varying inversely as the square of the distance from it. Show that the form of the string is a semi-circle. [Coll. Ex., 1882.]

Ex. 23. A chain, of length $2l$ and weight $2W$, hangs with one end A attached to a fixed point in a smooth horizontal wire, and the other end B attached to a smooth ring which slides along the wire. Initially A and B are together. Show that the work done in drawing the ring along the wire till the chain at A is inclined at an angle of 45° to the vertical is $Wl(1 - \sqrt{2} + \log \overline{1 + \sqrt{2}})$. [Coll. Ex., 1883.]

Ex. 24. Determine if the catenary is the only curve such that, if AB be any arc whose centre of gravity is G, and AT, BT tangents at A and B, then GT is always parallel to a fixed line in space.

Ex. 25. A uniform heavy chain of length $2a$ is suspended from two points in the same horizontal line; if one of these points be moveable, find the equation of the locus of the vertex of the catenary formed by the string; and show that the area cut off from this locus by a horizontal line through the fixed point is $\frac{1}{4}a^2 (\pi^2 - 4)$. [Math. Tripos, 1867.]

449. Stability of equilibrium. Some problems on the equilibrium of heavy strings may be conveniently solved by using the principle that the depth of the centre of gravity below some fixed straight line is a maximum or minimum, Art. 218. If the curve of the string be varied from its form as a catenary, the use of this principle will require the calculus of variations. But if we restrict the arbitrary displacements to be such that the string retains its form as a catenary, though the parameter c may be varied, the problem may be solved by the ordinary processes of the differential calculus.

This method presents some advantages when we desire to know whether the equilibrium is stable or not. We know, by Art. 218, that *the equilibrium will be stable or unstable according as the depth of the centre of gravity below some fixed horizontal plane is a true maximum or minimum.*

Ex. 1. A string of length $2l$ hangs over two smooth pegs which are in the same horizontal plane and at a distance $2a$ apart. The two ends of the string are free, and its central portion hangs in a catenary. Show that equilibrium is impossible unless l be at least equal to ae; and that, if $l > ae$, the catenary in the position of stable equilibrium for symmetrical displacements will be defined by that root of $ce^{\frac{a}{c}} = l$ which is greater than a. [Math. Tripos, 1878.]

Let $2s$ be the length of the string between the pegs. Taking the horizontal line joining the pegs for the axis of x, we easily find (Art. 399) that the depth $\bar{y}$ of the centre of gravity of the catenary and the two parts hanging over the pegs is given by
$$2l\bar{y} = sy - ca + (l-s)^2.$$

Substituting for y and s their values in terms of c, we find
$$2l\frac{d\bar{y}}{dc} = \left(c - \frac{l}{\rho}\right)\frac{\rho^2(c-a)-(c+a)}{c},$$

where ρ stands for $e^{\frac{a}{c}}$. It is easy to see that the second factor on the right hand side is negative for all positive values of c. Equating $d\bar{y}/dc$ to zero, we find that the possible positions of equilibrium are given by $l = c\rho$. To find the least value of l given by this equation we put $dl/dc = 0$; this gives $c = a$, so that l must be equal to or greater than ae.

For any value of l greater than ae there are two possible values of c, one greater and the other less than a. To determine which of these two catenaries is stable, we examine the sign of the second differential coefficient, Art. 220. We easily find,
when $l = c\rho$, $$2l\frac{d^2\bar{y}}{dc^2} = (c-a)\frac{\rho^2(c-a)-(c+a)}{c^2}.$$

In order that the equilibrium may be stable, this expression must be negative. This requires that c should be greater than a.

Ex. 2. A heavy string of given length has one extremity attached to a fixed point A, and hangs over a small smooth peg B on the same level with A, the other extremity of the string being free. Show that, if the length of the string exceed a certain value, there are two positions of equilibrium, and that the one in which the catenary has the greater parameter is stable.

450. Heterogeneous chain. *A heavy heterogeneous chain is suspended from two given points A and B. Find the equation to the catenary.*

This problem may be solved in a manner similar to that used in Art. 443 for a homogeneous chain. Since the equations (1) and (2) of that article are obtained by simple resolutions, they will be true with some slight modifications when the string is not uniform. In our case the weight of the string measured from the lowest point is $\int wds$ between the limits $s = 0$, $s = s$, Art. 442. We have therefore by the same resolutions

$$T \cos \psi = T_0 \quad \dots \dots (1), \qquad T \sin \psi = \int wds \quad \dots \dots (2).$$

Dividing one of these by the other as before, we find

$$\int wds = T_0 \tan \psi, \qquad \therefore \ w = \frac{T_0}{\rho \cos^2 \psi} \quad \dots \dots \dots (3),$$

substituting for ρ and $\tan \psi$, their Cartesian values

$$w = T_0 \frac{d^2y}{dx^2} \frac{dx}{ds} = T_0 \left\{ 1 + \left(\frac{dy}{dx} \right)^2 \right\}^{-\frac{1}{2}} \frac{d^2y}{dx^2} \quad \dots \dots \dots (4).$$

Conversely, when the law of density is known, say $w = f(s)$, the equation (3) gives a relation between s and dy/dx which we may write in the form $dy/dx = f_1(s)$. We easily deduce from this

$$x = \int \{1 + (f_1(s))^2\}^{-\frac{1}{2}} ds, \qquad y = \int \{1 + (f_1(s))^2\}^{-\frac{1}{2}} f_1(s) \, ds,$$

whence x and y can be expressed in terms of an auxiliary variable which has a geometrical meaning.

Ex. 1. Prove that the tension at any point P of the heterogeneous catenary is equal to the weight of a uniform chain whose length is the projection of the radius of curvature on the vertical and whose density is the same as that of the catenary at P.

Ex. 2. A straight line BR is drawn through any fixed point B in the axis of y parallel to the normal at P to the curve, cutting the axis of x in R. Prove that (1) the tension at P is (T_0/c) times the length BR and (2) the weight of the arc OP, measured from the lowest point O, is (T_0/c) times the length OR, where $OB = c$ and T_0 is the horizontal tension; Art. 35.

451. Cycloidal chain. A heterogeneous chain hangs in the form of a cycloid under the action of gravity: find the law of density.

In a cycloid we have $\rho = 4a \cos \psi$ and $s = 4a \sin \psi$, where a is the radius of the rolling circle. Substituting, we find $w = \dfrac{T_0}{4a} \sec^3 \psi = \dfrac{16a^2 T_0}{(16a^2 - s^2)^{\frac{3}{2}}}$.

It appears from this result that all the lower part of the chain is of nearly uniform density; thus the density at a point whose distance from the vertex measured along the arc is equal to the radius of the rolling circle is about ten ninths of the density at the vertex. The density increases rapidly higher up the chain and is infinite at the cusp. If then the chain when suspended from two points in the same horizontal line is not very curved, the chain may be regarded as nearly uniform.

The chief interest connected with this chain is that, when slightly disturbed from its position of equilibrium, it makes small oscillations whose periods and amplitudes can be investigated.

Ex. Drawing the usual figure for a cycloid, let O be the lowest point of the curve, B the middle point of the line joining the cusps. Let the normal at any point P of the curve intersect the line joining the cusps in M, and let BR be drawn through B parallel to MP to intersect the horizontal through O in R. Prove that the centre of gravity of the arc OP is the intersection of BR with the vertical through M. We find $\bar{x} = 2a\psi$, $\bar{y} = 2a\psi \cot \psi$, if B is the origin.

452. Parabolic chain. A heavy chain AOB is suspended from another chain DCE by vertical strings, which are so numerous that every element of AOB is attached to the corresponding element of DCE. If the weights of DCE and of the vertical strings are inconsiderable compared with that of AOB, find the form of the chain DCE that the chain AOB may be horizontal in the position of equilibrium.

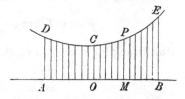

The tensions at O, M of the chain AOB being equal and horizontal, the weight of the length OM is supported by the tensions at C and P of the chain DCE. Thus DCE may be regarded as a heterogeneous heavy chain, such that the weight of any length PC is mx. Resolving horizontally and vertically for this portion of the chain, we have

$$T \cos \psi = T_0, \qquad T \sin \psi = mx.$$

Dividing one of these by the other,

$$mx = T_0 \tan \psi = T_0 \, dy/dx, \qquad \therefore \tfrac{1}{2}mx^2 = T_0 (y - c).$$

The form of the chain DCE is therefore a parabola.

One point of interest connected with this result is that the chain AOB might be replaced by a uniform heavy bar to represent the roadway of a bridge. The tensions of the chains due to the weight of the bridge would not then tend to break or bend the roadway. It is only necessary that the roadway should be strong enough to bear without bending the additional weights due to carriages. But this would not be true if the light chain DCE were not in the form of a parabola.

The results are more complicated if the weight of the chain DCE is taken into account, and if the chains of support, instead of being vertical, are arranged in some other way.

This problem was first discussed by Nicolas Fuss, *Nova Acta Petropolitanæ*, Tom. 12, 1794. It was proposed to erect a bridge across the Neva suspended by vertical chains from four chains stretched across the river. He decided that the chains of his day could not support the necessary tension without breaking.

Ex. 1. Prove that in the parabolic catenary the tension at any point P is $(T_0/2a)$ times the length of the normal between P and the axis of the parabola, where $2a$ is the semi-latus rectum. Prove also that the line density w at P is T_0 divided by the length of the normal.

Ex. 2. Prove that the weight of the chain OP measured from the lowest point O of the curve is $(T_0/2a)$ times the distance of P from the axis of the parabola; and deduce $T_0 = 2am$.

Ex. 3. The centre of gravity G of an arc bounded by any chord lies in the diameter bisecting the chord, and $PG = \frac{1}{3}PN$ where the diameter cuts the parabola in P and the chord in N.

Ex. 4. Referring to the figure, we notice that, since the tensions at C and P support the weight of the roadway OM, the tangents at C and P must intersect in a point vertically over the centre of gravity of OM. Thence deduce that the curve CP is a parabola.

Ex. 5. If the weight of any element ds of the string $DCPE$ is represented by $w\,(ds + n\,dx)$, show that the catenary is given by $x = \displaystyle\int \frac{c\,dz}{n + \sqrt{(1 + z^2)}}$, where z is the tangent of the inclination of the tangent to the horizon, and c is a constant. [Fuss.]

Ex..6. Prove that the form of the curve of the chain of a suspension bridge *when the weight of the rods is taken into account*, but the weight of the rest of the bridge neglected, is the orthogonal projection of a catenary, the rods being supposed vertical and equidistant.   [Math. Tripos, 1880.]

453. The Catenary of equal strength. A heavy chain, suspended from two fixed points, is such that the area of its section is proportional to the tension. Find the form of the chain.

If wds be the weight of an element ds, the conditions of the question require that $T = cw$, where c is some constant. The equations (1) and (2) of Art. 450 now become
$$T \cos \psi = T_0, \qquad T \sin \psi = \frac{1}{c} \int T ds.$$

Substituting in the second equation the value of T given by the first, we have $c \tan \psi = \int \sec \psi ds$. Differentiating, we find $c \sec^2 \psi = \sec \psi \, ds/d\psi$ and $\therefore \rho \cos \psi = c$.

This result also easily follows from the intrinsic equation of equilibrium (2) given in Art. 454. We have $Tds/\rho = wds \cos \psi$. But when the string is equally strong throughout $T = cw$, hence $\rho \cos \psi = c$. *The projection of the radius of curvature on the vertical is therefore constant and equal to c.*

To deduce the Cartesian equation we substitute for ρ and $\cos \psi$,
$$\left\{ 1 + \left(\frac{dy}{dx} \right)^2 \right\}^{-1} \frac{d^2 y}{dx^2} = \frac{1}{c}, \qquad \therefore \tan^{-1} \frac{dy}{dx} = \frac{x}{c} + A.$$

If the origin be taken at the lowest point, the constant A is zero. We then find
$$y = c \log \sec \frac{x}{c}.$$

Tracing this curve, we see that the ordinate y increases from zero as x increases from zero positively or negatively, and that there are two vertical asymptotes given by $x = \pm \frac{1}{2}\pi c$. When x lies between $\frac{1}{2}\pi c$ and $\frac{3}{2}\pi c$, the ordinate is imaginary; when x lies between $\frac{3}{2}\pi c$ and $\frac{5}{2}\pi c$, the curve is the same as that between $x = \pm \frac{1}{2}\pi c$. For greater values of x, the ordinate is again imaginary and so on. The curve therefore consists of an infinite number of branches all equal and similar to that between $x = \pm \frac{1}{2}\pi c$. This is therefore the only part of the curve which it is necessary to consider. Since the ordinates of the bridge must be finite, the values of x are restricted to lie between $\pm \frac{1}{2}\pi c$. *The span therefore cannot be so great as πc.*

Let O be the lowest point of the curve, C the centre of curvature at any point P, and PH a perpendicular on the vertical through C. Then $CH = c$. The sides of the triangle PCH are perpendicular and proportional to the forces which act on the arc OP, viz. the tension at P, the weight of OP and the horizontal tension T_0 at O. It follows that (1) *the tension at P is (T_0/c) times the length of the radius of* *curvature and* (2) *the weight of the arc OP is* (T_0/c) *times the projection of the radius of curvature on the horizontal.*

This curve was called the catenary of equal strength by Davies Gilbert, who invented it on the occasion of the erection of the suspension bridge across the Menai Straits. See *Phil. Trans.* 1826, part iii., page 202. In the first volume of *Liouville's Journal*, 1836, there is a note by G. Coriolis on the "chaînette" of equal resistance. Coriolis does not appear to have been aware that this form of chain had already been discussed several years before.

Ex. 1.   Prove (1) $x = c\psi$, (2) $s = c \log \tan \frac{1}{4}(\pi + 2\psi)$.

Ex. 2.   Prove that the depth of the centre of gravity of any arc below the intersection of the normals at its extremities is constant and equal to c. Prove also that its abscissa is equal to that of the intersection of the tangents at the same points.

Ex. 3.   The distance between the points of support of a catenary of uniform strength is a, and the length of the chain is l. Show that the parameter c must be found from $\tanh \dfrac{l}{4c} = \tan \dfrac{a}{4c}$. Show also that this equation gives a positive value of c greater than a/π.

Ex. 4.   Show that the horizontal projection of the span is in every case less than π times the greatest length of uniform chain of the same material that can be hung by one end. Assume the strength of any part of the chain to be proportional to the mass per unit of length.   [Kelvin, Math. Tripos, 1874.]

If L be the length of uniform chain spoken of, the tension at the point of support is its weight, i.e. wL. Again, the tension at any point of the heterogeneous chain is cw, hence c must be less than L. Hence the span must be less than πL.

454. String under any Forces. *To form the general intrinsic equations of equilibrium of a string under the action of any forces.* Let A be any fixed point of reference on the string, $AP = s$, $AQ = s + ds$. Let T be the tension at P; then, since T is a function of s, $T + dT$ is the tension at $Q*$.

Let the impressed forces on the element PQ be resolved along the tangent, radius of curvature, and binormal at P. Thus Fds is the force on ds resolved along the tangent in the direction in which s is measured; Gds is the force on ds resolved along the radius of curvature ρ in the direction in which ρ is measured, i.e. inwards; Hds is the force on ds resolved perpendicular to the plane of the curve at P, and estimated positive in either direction of the binormal. These three directions are called the principal directions or principal axes of the curve at P.

Let $d\psi$ be the angle between the tangents at P and Q. Hence also the angle $PCQ = d\psi$. The element ds is in equilibrium under

---

* It should be noticed that, if s were measured from B towards A, so that $BQ = s$, then T would be the tension at Q, $T + dT$ that at P.

the forces T, $T + dT$ acting along the tangents at P, Q and the forces Fds, Gds, Hds. Resolving along the tangent at P,

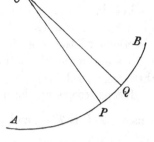

$$(T + dT)\cos d\psi - T + Fds = 0,$$

which reduces to

$$dT + Fds = 0\ldots\ldots\ldots(1).$$

Resolving along the radius of curvature at P, we have

$$(T + dT)\sin d\psi + Gds = 0,$$

$$\therefore T\frac{ds}{\rho} + Gds = 0 \ldots\ldots\ldots\ldots\ldots\ldots (2).$$

We have now to resolve perpendicular to the osculating plane at P of the curve. Since two consecutive tangents to a curve lie in the osculating plane, the tensions have no component perpendicular to this plane. We have therefore

$$Hds = 0\ldots\ldots\ldots\ldots\ldots\ldots\ldots(3).$$

The three equations (1), (2), (3) are the general intrinsic equations of equilibrium.

The density of the string is supposed to be included in the expressions Fds, Gds, Hds for the forces on the element. The equations of equilibrium therefore apply, whether the string is uniform, or whether its density varies from point to point.

From these equations we infer that the tensions T and $T + dT$, acting at the extremities of any element, are equivalent to two other forces, viz. dT and $T\dfrac{ds}{\rho}$, acting respectively along the tangent to, and the radius of curvature of, the curve at either extremity of the element. In problems on strings it is often convenient to replace the tensions by these two forces. The advantage of this change is that the direction cosines of the tangent and of the radius of curvature are known by the differential calculus. When therefore we form the equations of statics, we can easily resolve these two forces and the given impressed forces in any directions we may find convenient.

Ex. Show that the form of the string is such that at every point the resultant of the applied forces lies in the osculating plane, and makes with the principal normal to the string an angle $\tan^{-1}\dfrac{d\log T}{d\psi}$.

455. *To form the general Cartesian equations of equilibrium of a string*.*

Let ds be the length of any element PQ of the string. Let the forces on this element when resolved parallel to the positive directions of the axes be $X ds$, $Y ds$, $Z ds$. The element is in equilibrium under the action of the tensions at P and Q and these three impressed forces.

Let us resolve all these parallel to the axis of x. The resolved tension at P is $T \dfrac{dx}{ds}$, and pulls the element PQ towards the left hand. At Q, s has become $s + ds$, the horizontal tension at Q is therefore

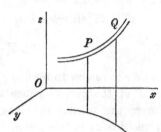

$$\left(T\frac{dx}{ds}\right) + \frac{d}{ds}\left(T\frac{dx}{ds}\right)ds,$$

and this pulls the element PQ towards the right hand side. Taking both these and the force $X ds$, we have

$$\frac{d}{ds}\left(T\frac{dx}{ds}\right)ds + X ds = 0.$$

Treating the other components in the same way, we find

$$\left.\begin{aligned}\frac{d}{ds}\left(T\frac{dx}{ds}\right) + X = 0 \\[1mm] \frac{d}{ds}\left(T\frac{dy}{ds}\right) + Y = 0 \\[1mm] \frac{d}{ds}\left(T\frac{dz}{ds}\right) + Z = 0\end{aligned}\right\}.$$

456. Ex. 1. Show that the polar equations of equilibrium of a string in one plane under forces $P ds$, $Q ds$, acting along and perpendicular to the radius vector, are

$$\frac{d}{ds}(T\cos\phi) - \frac{T}{r}\sin^2\phi + P = 0, \qquad \frac{d}{ds}(T\sin\phi) + \frac{T}{r}\sin\phi\cos\phi + Q = 0,$$

where $\cos\phi = dr/ds$ and $\sin\phi = r d\theta/ds$. Thence deduce the equations of equilibrium of a string in space of three dimensions, referred to cylindrical coordinates.

* The equations of equilibrium of a string under the action of any forces in two dimensions were given in a Cartesian form by Nicolas Fuss, *Nova Acta Petropolitanæ*, 1796. He gives two solutions, one by moments, and another by considering the tension. In this second solution, after resolving parallel to the axes, he deduces algebraically equations equivalent to those obtained by resolving along the tangent and normal. He goes on to apply his equations to the chaînette and other similar problems.

Ex. 2. A string is in equilibrium in the form of a helix, and the tension is constant throughout the string. Show that the force on any element tends directly from the axis of the helix.

Ex. 3. The extremities of a string of given length are attached to two given points, and each element ds of the string is acted on by a repulsive force tending directly from the axis of z and equal to $2\mu r ds$. If $(r\theta z)$ be the cylindrical coordinates of any point, prove that

$$T = A - \mu r^2,$$

$$\frac{d\theta}{dz} = \frac{B}{r^2}, \qquad \left(\frac{dr}{dz}\right)^2 = C\left(1 - \frac{\mu}{A}r^2\right)^2 - \frac{B^2}{r^2} - 1.$$

Show how the five arbitrary constants are determined. Explain how a helix is, in certain cases, the solution.

Ex. 4. A heavy chain is suspended from two points, and hangs partly immersed in a fluid. Show that the curvatures of the portions just inside and just outside the surface of the fluid are as $D - D'$ to D, where D and D' are the densities of the chain and fluid.   [St John's Coll.]

The weights of the elements just above and just below the surface of the fluid are proportional to Dds and $(D - D')\,ds$. If T be the tension, the resolved parts of these weights along the normal must be Tds/ρ and Tds/ρ'. Hence $D/(D - D') = \rho'/\rho$.

Ex. 5. A heavy string is suspended from two fixed points A and B, and the density is such that the form of the string is an equiangular spiral. Show that the density at any point P is inversely proportional to $r\cos^2\psi$, where r is the distance of P from the pole, and ψ is the angle which the tangent at P makes with the horizon.

[Trin. Coll., 1881.]

Ex. 6. A heavy string, which is not uniform, is suspended from two fixed points. Prove that the catenary formed of a given uniform string which touches at any point the curve in which the string hangs and has the same tension at that point will be of invariable dimensions.

457. Constrained Strings. *A string rests on a curve of any form in one plane, and is acted on by forces at its extremities. It is required to find the conditions of equilibrium and the tension at any point.*

There are four cases of this proposition which are of considerable importance ; we shall consider these in order.

Let us first suppose that the weight of the string is so slight that it may be neglected compared with the forces applied at the two extremities of the string. Let us also suppose that the curve is perfectly smooth. The forces on an element ds are merely the tensions at its ends and the reaction or pressure of the curve. Let Rds be this pressure, then R is the *pressure per unit of length of the string.* For the sake of brevity this is usually expressed by saying that R is the pressure *at* the element. It is usual to estimate the pressure of the curve on the string as positive when it acts in the direction opposite to that in which the radius of curvature is measured.

Resolving along the tangent and normal to the string, we have by Art. 454,

$$dT = 0, \qquad T\frac{ds}{\rho} - Rds = 0.$$

We infer from these equations that, *when a light string rests on a smooth curve, the tension is constant, and the pressure at any point varies as the curvature.*

458. This theorem has a wider range than would perhaps appear at first sight. Since the curve may be of any form, the result includes the case of a string in equilibrium under any forces which are at every point normal to the curve. Supposing the normal forces given, the form of the curve can be found from the result just proved, viz. that at every point the curvature is proportional to the normal force.

As an example we may consider Bernoulli's problem; *to find the form of a rectangular sail,* two opposite sides of which are fixed so as to be parallel to each other and perpendicular to the direction of the wind. The weight of the sail is neglected compared with the pressure produced by the wind. Let us enquire what is the curve formed by a plane section of the sail drawn perpendicular to the fixed sides.

Two answers may be given to this question according as the wind after acting on the sail immediately finds an issue, or remains to press on the sail like a gas in equilibrium. On the former hypothesis we assume as the law of resistance, that the pressure of the wind on any element of the sail acts along the normal to the element and is proportional to the square of the resolved velocity of the wind. We have therefore $R = w \cos^2 \psi$, where ψ is the angle the normal to the section of the sail makes with the direction of the wind, and w is a constant. This gives $c/\rho = \cos^2 \psi$. By Art. 444 we infer that the *curve is a catenary*, whose axis is in the direction of the wind, and whose directrix is vertical.

If the air presses on the sail like a gas in equilibrium, the pressure on each side of the sail is equal in all directions by the laws of hydrostatics, but the pressure is greater on one side than on the other. We have therefore R equal to this constant difference, hence also ρ is constant, and the *required curve is a circle.*

Ex. 1. A "square sail" of a ship is fastened to the mast by two yard-arms, and is such that when filled with wind every section by a horizontal plane is a straight line parallel to the yards. Show that, assuming the ordinary law of resistance, it will have the greatest effect in propelling the ship when $3 \sin (\alpha - 2\phi) - \sin \alpha = 0$, where α is the angle between the direction from which the wind comes and the ship's keel, and ϕ is the angle between the yard and the ship's keel.   [Caius Coll.]

Ex. 2. A light string has one end fixed at the vertex of a smooth cycloid; prove that as the string, while taut, is wound on the curve, the line of action of the resultant pressure on the cycloid envelopes another cycloid of double parameter.   [Coll. Ex., 1890.]

[The resultant pressure of the curve on an arc of the string balances the tensions at the extremities of the arc. It therefore passes through the intersection of the tangents at those extremities and bisects the angle between them.]

459. Heavy smooth string. *Let us next suppose that the weight of the string cannot be neglected.* Let wds be the weight of the element ds. Let ψ be the angle the tangent PK at P makes with the horizontal.

The element PQ is in equilibrium under the action of wds along the ordinate PN, Rds along the normal PG, and the tensions at P and Q. Resolving along the tangent and normal at P, we have

$$dT - wds \sin \psi = 0 \quad \Big\} \dots\dots(1),$$

$$T\frac{ds}{\rho} - wds \cos \psi - Rds = 0 \Big\} \dots\dots(2).$$

Since $\sin \psi = dy/ds$, the first equation gives by integration

$$T = wy + C \dots\dots\dots\dots\dots\dots\dots\dots (3).$$

Hence, if T_1, T_2 be the tensions at two points whose ordinates are y_1, y_2, $T_2 - T_1 = w(y_2 - y_1)$.

This important result may be stated thus, *If a heavy string rest on a smooth curve, the difference of the tensions at any two points is equal to the weight of a string whose length is the vertical distance between the points.*

460. It may be remarked that this result has been obtained solely by resolving along the tangent to the string, and is altogether independent of the truth of the second equation. If then the whole length of the string does not lie on the curve, but if

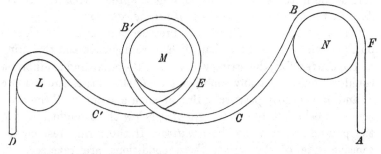

part of it be free and stretch across to and over some other curve, the theorem is still true. Thus if the string $ABCD$ stretch round the smooth curves L, M, N, as indicated in the figure, the tension at any point B or C exceeds that at A by the weight of a string whose length is the vertical distance of B or C above A.

Since the tensions at A and D are zero, it follows that the free extremities of a heavy chain are in the same horizontal line.

In the same way the tension is a maximum at the highest point. Also no point of the string, such as C or C', can be beneath the horizontal line joining the free extremities.

To determine the pressure at any point P (see fig. of Art. 459) we write the equation (2) in the form

$$R\rho = T - w\rho \cos \psi,$$

where the pressure R of the curve on the string, when positive, acts outwards, i.e. in the direction opposite to that in which the radius of curvature ρ is measured, Art. 457. If T_1 be the tension at any fixed point A, and z the altitude of any point P above A, we have by (3) $T = T_1 + wz$. It therefore follows that

$$R\rho = T_1 + w(z - \rho \cos \psi).$$

If we measure a length $PS = \rho$ along the normal at P *outwards*, the point S may be called the *anti-centre*. It is clear that $z - \rho \cos \psi$ is the altitude of S above A. Hence, if a heavy string rest on a smooth curve, *the value of $R\rho$ at any point P exceeds the tension at A by the weight of a string whose length is the altitude of the anti-centre of P above A.*

If the extremity A be free, as in the figure of this article, then $R\rho$ at any point B is equal to w multiplied by the altitude of the anti-centre of B above A. If part of the string is free, as at C and C', the pressure R is zero. Hence the anti-centres of curvature all lie in the straight line joining the free extremities A and D. *This is the common directrix of all the catenaries.*

In these equations Rds is the pressure outwards of the curve on the string. It is clear that, if R were negative and the string on the convex side, the string would leave the curve and equilibrium could not exist. At any such point as B, the anti-centre is above B and R is clearly positive. But at such a point as E the anti-centre is below E, and if it were *also* below the straight line AD, the pressure at E would be *negative*. If the string rest on the concave side of the curve, these conditions are reversed. In general, it is necessary for equilibrium that $R\rho$ should be positive or negative according as the string is on the convex or concave side of the curve.

Summing up the results arrived at in this article, we see that a horizontal straight line can be drawn such that the *tension at each point P of the string is wy*, where y is the altitude of P above the straight line. *This straight line may be called the statical* *directrix of the string.* No part of the string can be below the statical directrix, and the free ends, if there are any, must lie on it. *If R be the outward pressure of the curve on the string, Rρ is equal to wy',* where y' is the altitude of the anti-centre of P above the directrix. It is therefore necessary that at every point of the string the anti-centre should be above or below the directrix according as the string is on the convex or concave side of the curve.

Ex. 1. Show that the locus of the anti-centre of a circle is another circle.

Ex. 2. Show that the coordinates of the anti-centre at any point P of an ellipse referred to its axes are given by $ax = 2a^2 \cos \phi - c^2 \cos^3 \phi$ $by = 2b^2 \sin \phi + c^2 \sin^3 \phi$, where $c^2 = a^2 - b^2$, and ϕ is the eccentric angle of P.

Ex. 3. If S be the anti-centre at any point P of a curve, show that the normal to the locus of S makes with PS an angle θ given by $\tan \theta = \frac{1}{2} d\rho / ds$.

461. It should be noticed that at the points where the string leaves the constraining curve, both the curvature of the string and the pressure R may change abruptly. Thus in the figure of Art. 460 at a point a little below F the radius of curvature of the string is infinite and R is zero. At a point a little above F the curvature of the string is the same as that of the body N, and the pressure R is equal to T/ρ. At such a point as E the abrupt change if any in the value of the product $R\rho$ (in accordance with the rule of Art. 460) is equal to the weight of a string whose length is the vertical distance between the anti-centres on each side of the point.

When the external forces which act on the string are such that their magnitudes per unit of length are finite, an abrupt change of tension cannot occur. If the tensions on each side of any point could differ by a finite quantity, an infinitesimal length of string containing the point would be in equilibrium under the influence of two unequal forces acting in opposite directions. In the same way there can be no abrupt change in the direction of the tangent *except at a point where the tension is zero,* for if the tangents on each side of any point made a finite angle with each other, the element of string at that point would be in equilibrium under the action of two finite tensions not opposed to each other.

462. Ex. 1. A heavy string (length $2l$) passes completely round a smooth horizontal cylinder (radius a) with the two ends hanging freely down on each side. The parts of the string on the upper semi-circumference are close together, so that the whole string may be regarded as lying in a vertical plane perpendicular to the

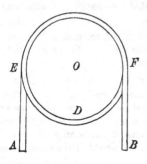
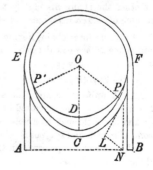

axis of the cylinder. Find the position of rest and the least length of string consistent with equilibrium.

First, let us suppose that the string is in contact with the circle along the lower semi-circumference as well as the upper. Then a length $l - \frac{3}{2}\pi a$ hangs vertically on each side. Let D be the lowest point of the circle, the anti-centre of D is at a depth $2a$ below the centre O of the circle. Hence, unless $l - \frac{3}{2}\pi a > 2a$, the string cannot rest in contact with the circle.

Secondly, let us suppose that a portion of the string hangs freely in the form of a catenary. Let P' be one of the points of contact of the catenary with the circle. Let P be any point on the catenary, drawn in the figure merely to show the triangle PLN, Art. 444. Let the angle $P'OD = \psi$, so that ψ is the inclination of the tangent at P' to the horizon. Let x, y be the coordinates of P', $s = CP'$. By examining the triangle PLN, we see that $y = c \sec \psi$, $s = c \tan \psi$. Since $x = a \sin \psi$, we have by (5) of Art. 443

$$\sec \psi + \tan \psi = e^{\frac{a \sin \psi}{c}} \quad \text{...............................(1).}$$

As already explained, the free extremities A, B of the string are on a level with the directrix, Art. 460. Hence $BF = y + a \cos \psi$; also the arc $FE = \pi a$, $EP' = (\frac{1}{2}\pi - \psi) a$, and $P'C = s$. The sum of these four quantities is l,

$$\therefore \ c (\sec \psi + \tan \psi) + a \cos \psi - a\psi + \tfrac{3}{2}\pi a = l \quad \text{....................(2).}$$

Putting $v = \frac{1}{2} \log \dfrac{1 + \sin \psi}{1 - \sin \psi}$, we find from (1) and (2)

$$c = \frac{a \sin \psi}{v} \qquad \frac{l}{a} = \sqrt{\frac{1 + \sin \psi}{1 - \sin \psi}} \left(\frac{\sin \psi}{v} + 1 - \sin \psi \right) - \psi + \tfrac{3}{2}\pi.$$

The second of these equations gives the length of the string corresponding to any given position of equilibrium.

To find the least value of l consistent with equilibrium, we equate to zero the differential coefficient of l. As this leads to some rather long reductions, the results only are here stated. Noticing that $dv/d\psi = \sec \psi$, we find

$$\frac{1}{a} \frac{dl}{d\psi} = \frac{(1 - v)\,(v \cos^2 \psi - \sin \psi)}{v^2 (1 - \sin \psi)} = 0.$$

By expanding v in powers of $\sin \psi$, we may show that $(v \cos^2 \psi - \sin \psi)$ is negative and does not vanish for any value of $\sin \psi$ between zero and unity. Equating to zero the factor $(1 - v)$, we find that $\sin \psi = (e^2 - 1)/(e^2 + 1)$. As $dl/d\psi$ changes sign from $-$ to $+$ as $\sin \psi$ increases, we see that l is a minimum. Effecting the numerical calculations, we have $\psi = \cdot 86$, and $l - \frac{3}{2}\pi a = (e - \psi) a$, which reduces to $(1 \cdot 85)\, a$.

For any given value of l, greater than this minimum, there are two positions of equilibrium. In one a portion of the string hangs freely in the form of a catenary; in the other the string fits closely to the cylinder or hangs free according as the given value of $l - \frac{3}{2}\pi a$ is greater or less than $2a$.

Ex. 2. A uniform chain, having its ends fastened together, is hung round the circumference of a vertical circle. If a be the radius of the circle, $2a\gamma$ the arc which the string touches, and l the whole length, prove

$$(l - 2a\gamma)\,\{\log(-\cos \gamma) - \log(1 + \sin \gamma)\} = 2a \sin^2\gamma \sec \gamma. \qquad \text{[May Exam.]}$$

Ex. 3. A uniform inextensible string of given length hangs freely from two fixed points. It is then enclosed in a fine fixed tube which touches no part of the string, and is cut through at a point where the tangent makes an angle γ with the horizon. Prove that at a point where the tangent makes an angle ψ with the horizon the ratio of the pressure on the tube to the weight of the string per unit of length becomes $\cos^2 \psi \sec \gamma$. \qquad [Math. Tripos, 1886.]

463. Rough curve, light string. *To consider the case in which the weight of the string is inconsiderable, but the curve is rough.* Referring to the figure of Art. 459, we shall suppose the extremities A and B to be acted on by unequal forces F, F'. Our object is to find the conditions of limiting equilibrium; let us then suppose the string is on the point of motion in the direction AB. The friction on every element PQ is equal to $\mu R ds$, where μ is the coefficient of friction. This force acts in the direction opposite to motion, viz. from B to A.

Introducing this force into the equations obtained in Art. 459 by resolving the forces along the tangent and normal, and omitting the terms containing the weight of the element, we have

$$dT - \mu R ds = 0 \ldots\ldots(1), \qquad T\frac{ds}{\rho} - R ds = 0 \ldots\ldots(2).$$

Eliminating R, we find, $\qquad \dfrac{dT}{T} = \mu\,\dfrac{ds}{\rho} = \mu d\psi\,;$

$$\therefore \ \log T = \mu\psi + A, \qquad \therefore \ T = Be^{\mu\psi},$$

where A and B are undetermined constants. If T_1, T_2 be the tensions at two points at which the tangents make angles ψ_1, ψ_2 with the axis of x, this equation gives

$$T_2 = T_1 e^{\mu(\psi_2 - \psi_1)} \ldots\ldots\ldots\ldots\ldots\ldots\ldots (3).$$

It will be found useful to put the result in the form of a rule. *If a light string rest on a rough curve in a state bordering on motion, the ratio of the tensions at any two points is equal to e to the power of μ times the angle between the tangents or between the normals at those points.*

The sign to be given to μ in this equation depends on the direction in which the friction acts. In using the rule, however, no difficulty arises from this ambiguity; for (1) it is evident that that tension is the greater of the two which is opposed to the friction, and (2) it must be the ratio of the greater tension to the lesser (not the lesser to the greater) which is equal to the exponential with the positive index.

To determine the angle between the tangents; let a straight line, starting from a position coincident with one tangent, roll on the string until it coincides with the other tangent; the angle turned round by this moving· tangent is the angle required.

The pressure at any point is given by (2), and we see that $R\rho$ at any point is equal to the tension at that point.

464. If the forces F, F' which act at the extremities A, B are given, and if the length l of the string is also given, we may find the limiting positions of equilibrium in the following manner. Put the equation to the curve in the form $\psi = f(s)$. Let s be the required arc-coordinate of A, then $s + l$ is that of B. The ψ's of A and B are therefore $f(s)$ and $f(s + l)$. Hence, by taking the logarithms of equation (3),

$$\log F_2 - \log F_1 = \mu \{ f(s + l) - f(s) \}.$$

From this equation s must be found. The other limiting position may be found by writing $-\mu$ for μ.

465. It should be noticed that the equation (3) of Art. 463 is independent of the size of the curve. Suppose a *heavy string to pass through a small rough ring or over a small peg*, and to be in a state bordering on motion; the weight of the portion of string on the pulley may sometimes be neglected compared with the tensions of the string on either side. If the strings on either side make a finite angle with each other, the pressures and therefore the frictions will not be small, and cannot be neglected. We infer that, *when a heavy tight string passes through a small rough ring, the ratio of the tensions on each side is given by the same rule as that for a light string.*

466. Ex. 1. A rope is wound twice round a rough post, and the extremities are acted on by forces F, F'. Find the ratio of $F : F'$ when the rope is on the point of slipping. [Here the angle between the tangents is 4π, hence the ratio of the greater force to the other is $e^{4\pi\mu}$.]

Ex. 2. A circle has its plane vertical, and is pressed against a vertical wall by a string fixed to a point in the wall above the circle. The string sustains a weight P, the coefficient of friction between the string and circle is μ, and the wall is perfectly rough. When the circle is on the point of sliding, prove that, if W be the weight of the circle and θ the angle between the string and the wall, $P(1 + \cos \theta) e^{\mu\theta} = W + 2P$. [Coll. Exam.]

Ex. 3. A light string is placed over a rough vertical circle, and a uniform heavy rod, whose length is equal to the diameter of the circle, has one end attached to each end of the string, and rests in a horizontal position. Find within what points on the rod a given mass may be placed, without disturbing the equilibrium of the system : and show that the given mass may be placed anywhere on the rod, provided the ratio of its weight to that of the rod does not exceed $\frac{1}{2}(e^{\mu\pi} - 1)$, where μ is the coefficient of friction between the string and the circle. [Coll. Exam., 1880.]

Ex. 4. A string, whose weight is neglected, passes over a rough fixed horizontal cylinder and is attached to a weight W; P is the weight which will just raise W, and P' the weight which will just sustain W; show that, if R, R' are the corresponding resultant pressures of the string on the cylinder, $P : P' :: R^2 : R'^2$. [Math. T., 1880.]

Ex. 5. A band without weight passes tightly round the circumference of two unequal rough wheels. One wheel is fixed while the other is made to turn slowly round its centre. Show that the band will slip first on the smaller wheel.

Ex. 6. On the top of a rough fixed sphere (radius c) is placed a heavy particle, to which are tied two equally heavy particles by light strings each of length $c\theta$; show that, when the latter particles are as near together as possible, the planes of the strings make with one another an angle ϕ, where $2 \sin(\theta - \lambda) \cos \dfrac{\phi}{2} = \sin \lambda \cdot e^{\theta \tan \lambda}$, and λ is the angle of friction between the particles and the sphere, and between the strings and the sphere.   [Coll. Exam., 1887.]

Ex. 7. A uniform heavy string of length $2l$ passes through two given small fixed rings A, B in the same horizontal line. Supposing the string to be on the point of slipping inwards at both A and B, find the position of equilibrium.

If $2s$ be the portion of the string between the pegs, y the ordinate of the catenary at either peg, the tensions at the two sides of either ring are proportional to y and $l-s$. Referring to the triangle PLN in the figure of Art. 443, we see that the angle through which the string has been turned is the supplement of the least angle whose sine is c/y. Hence we have by (3) $\log \dfrac{y}{l-s} = \left(\pi - \sin^{-1} \dfrac{c}{y} \right) \mu$. Also if $2a$ be the known distance between the rings, we have $x = a$. Substituting for y and s their values in terms of x or a given in Art. 443, we have an equation to find c. Hence y and s may be found.

Ex. 8. A, B, C are three rough points in a vertical plane; P, Q, R are the greatest weights which can be severally supported by a weight W when connected with it by strings passing over A, B, C, over A, B, and over B, C respectively. Show that the coefficient of friction at B is $\dfrac{1}{\pi} \log \dfrac{QR}{PW}$.   [Math. Tripos, 1851.]

Let α, β, γ be the angles through which the string is bent at ABC, their sum is π. By Art. 463 $\log P/W$, $\log Q/W$, $\log R/W$ are respectively equal to $\mu\alpha + \mu'\beta + \mu''\gamma$, $\mu\alpha + \mu'(\beta + \gamma)$, $\mu'(\alpha + \beta) + \mu''\gamma$. The result follows by substitution. It is supposed that B lies between the verticals through A and C.

Ex. 9. A string, whose length is l, is hung over two rough pegs at a distance a apart in a horizontal line. If one free end of the string is as much as possible lower than the other, the inclination to the vertical of the tangent to the string at either peg is given by the equation $\dfrac{l}{a} \sin \theta \cdot \log \cot \dfrac{\theta}{2} = \cos \theta + \cosh \mu (\pi - \theta)$.   [St John's Coll., 1881.]

Ex. 10. An endless uniform heavy chain is passed round two rough pegs in the same horizontal line, being partly supported by a smooth peg situated midway in the line between the other pegs, so that the chain hangs in three festoons. If α, β are the angles which the tangents at one of the rough pegs make with the vertical, and μ is the coefficient of friction, prove that the limiting values of α and β are given by the equation $e^{\pm\mu(\pi - \alpha + \beta)} = 2 \dfrac{\sin \alpha \log \cot \frac{1}{2}\alpha}{\sin \beta \log \cot \frac{1}{2}\beta}$.   [Math. Tripos, 1879.]

467. Rough curve, heavy string. *We shall now consider the general case in which both the weight of the string and the roughness of the curve are taken account of.*

Referring to the figure of Art. 459, and introducing both the weight and the roughness into the equations (1) and (2), we have

$$dT - wds \sin \psi - \mu Rds = 0 \ldots\ldots\ldots\ldots\ldots (1),$$

$$\frac{Tds}{\rho} - wds \cos \psi - Rds = 0 \ldots\ldots\ldots\ldots\ldots (2).$$

In applying these equations to other forms of the string we must remember that the friction is μ times the pressure taken positively. Thus as the string is heavy it might lie on the concave side of the curve. We must then change the sign of R in the second equation, but not in the first.

We shall presently have occasion to write $\rho = ds/d\psi$. If the figure is not so drawn that s and ψ increase together, we shall have $\rho = -ds/d\psi$. To solve these equations, we eliminate R,

$$\therefore \quad \frac{dT}{d\psi} - \mu T = w\rho\,(\sin\psi - \mu\cos\psi) \quad\ldots\ldots\ldots\ldots(3).$$

This is one of the standard forms in the theory of differential equations. According to rule we multiply by $e^{-\mu\psi}$ and integrate;

$$\therefore \quad Te^{-\mu\psi} = \int w\rho\,(\sin\psi - \mu\cos\psi)\,e^{-\mu\psi}\,d\psi + C\ldots\ldots\ldots(4).$$

We cannot effect this integration until the form of the curve is given. By using the rules of the differential calculus we first express ρ as a function of ψ. Then substituting and integrating, we find

$$Te^{-\mu\psi} = f(\psi) + C\ldots\ldots\ldots\ldots\ldots\ldots\ldots(5).$$

The value of T having been found by this equation, R follows from either (1) or (2). *It should be noticed that we have not assumed that the string is necessarily uniform.*

The pressure at any point is given by the equation

$$R\rho = T - w\rho\cos\psi.$$

It may be noticed that this is the same as the corresponding equation for a heavy string on a smooth curve, Art. 460.

If the string is not on the point of motion, we replace the term $-\mu R ds$ in (1) by $-F ds$, where F is the friction per unit of length.

Ex. If the string is uniform and of finite length, and if the extremities are acted on by forces P_1, P_2, prove that the whole friction called into play is $\int F ds = P_2 - P_1 - wz$, where $z = y_2 - y_1$, so that z is the vertical distance between the extremities of the string.

468. It appears from the last article that the determination of the circumstances of the equilibrium of a heavy string on a rough curve depends on the integral

$$I = \int w\rho e^{-\mu\psi}\,(\sin\psi - \mu\cos\psi)\,d\psi.$$

This integral can be found in several cases.

If the curve is a circle and the string homogeneous, we have $\rho = a$. We easily find

$$I = \frac{wa}{\mu^2 + 1}\left\{(\mu^2 - 1)\cos\psi - 2\mu\sin\psi\right\}e^{-\mu\psi}.$$

If the curve is an equiangular spiral and the string homogeneous, we have $r = ae^{\theta\cot a}$. Since $\rho\sin a = r$ and $\psi = \theta + a$, the integral may be obtained from the last by writing $\mu - \cot a$ for μ, and $ae^{-a\cot a}\operatorname{cosec} a$ for a.

If the curve is a cycloid with its base inclined to the horizon at any angle, we have $\rho = 4a \cos(\psi - a)$, where a is the radius of the generating circle. More generally, if the curve is such that $w\rho$ can be expanded in a series of *positive integral powers* of $\sin \psi$ and $\cos \psi$, we can express $w\rho (\sin \psi - \mu \cos \psi)$ in a series of sines and cosines of multiple angles. In this case the integral can be found by a method similar to that used for the circle.

If the curve is a catenary we have $\rho \cos^2 \psi = c$ and $I = wc \sec \psi e^{-\mu\psi}$. More generally, if the curve is such that $\rho = a \cos^n \psi$, where n is a *positive* or *negative integer*, we may find I by a formula of reduction. We easily see that

$$\{\mu^2 + (n+1)^2\} I_n - (n-1)(n+2) I_{n-2}$$
$$= wa (\cos \psi)^{n-1} e^{-\mu\psi} \{n+2 - \mu(n+2)\sin\psi\cos\psi - (n+1-\mu^2)\cos^2\psi\}.$$

469. Ex. 1. A heavy string occupies a quadrant of the upper half of a rough vertical circle in a state bordering on motion. Prove that the radius through the lower extremity makes an angle a with the vertical given by $\tan(a - 2\epsilon) = e^{-\frac{1}{2}\mu\pi}$, where $\mu = \tan \epsilon$.

Ex. 2. A heavy string, resting on a rough vertical circle with one extremity at the highest point, is on the point of motion. If the length of the string is equal to a quadrant, prove that $\frac{1}{2}\pi \tan \epsilon = \log \tan 2\epsilon$. [Coll. Ex., 1881.]

Ex. 3. A single moveable pulley, of weight W, is just supported by a power P, which is applied at one end of a cord which goes under the pulley and is then fastened to a fixed point; show that, if ϕ be the angle subtended at the centre by the part of the string in contact with the pulley, ϕ is given by the equation

$$P(1 - 2e^{\mu\phi} \cos \phi + e^{2\mu\phi})^{\frac{1}{2}} = W. \qquad \text{[Coll. Ex., 1882.]}$$

Ex. 4. If a heavy string be laid on a rough catenary, with its vertex upwards and its axis vertical, so that one extremity is at the vertex, the string will just rest if its length be equal to the parameter of the catenary, provided the coefficient of friction be $(2 \log 2)/\pi$. [Coll. Ex., 1885.]

Ex. 5. A heavy string AB is placed on the concave side of a rough cycloidal curve whose base is inclined at an angle a to the horizon, with one extremity A at the lowest point and the other B at the vertex. Prove that the string will be in a state bordering on motion if $\dfrac{\tan \epsilon - 2\tan a}{\tan \epsilon + (1 - 3\cos^2 \epsilon)\tan a} = e^{a\tan \epsilon}$, where $\tan \epsilon$ is the coefficient of friction.

Ex. 6. A heavy string rests on a rough cycloid with its base horizontal and plane vertical. The normals at the extremities of the string make with the vertical angles each equal to a, which is also the angle of friction between string and cycloid. If, when the cycloid is tilted about one end till the base makes an angle a with the horizontal, the string is on the point of motion, show that

$$3 - 2\sec^2 a = e^{-2a\tan a}.$$

[It is assumed that no part of the string hangs freely.] [Coll. Ex., 1883.]

Ex. 7. A heavy uniform flexible string rests on a smooth complete cycloid, the axis of which is vertical and vertex upwards, the whole length of the string exactly coinciding with the whole arc of the cycloid; prove that the pressure at any point of the cycloid varies inversely as the curvature. [Math. Tripos, 1865.]

Ex. 8. A heavy string AB is laid on a rough convex curve in a vertical plane, and the friction at every point acts in the same direction along the curve. Show that it will rest if the inclination of the chord AB to the horizon be less than $\tan^{-1}\mu$, where μ is the coefficient of friction.   [June Ex., 1878.]

470. The following proposition will be found to include a number of problems which lead to known integrals.

Let the form be known in which a heterogeneous unconstrained string, supported at each end, rests in equilibrium in one plane under the action of any forces. Let this known curve be $y = f(x)$. Let us now suppose this string to be placed in the same position on a *rough curve fixed in space* whose equation is also $y = f(x)$. If the extremities of the string be acted on by forces such that the string is on the point of slipping, then

$$(T + G\rho)\,e^{-\mu\psi} = C, \qquad\qquad R\rho\,e^{-\mu\psi} = C \dots\dots\dots\dots\dots\dots\dots\dots(1),$$

where C is constant throughout the length of the string. Here, as in Art. 454, Gds is the resolved normal force inwards on the element ds. The standard case is the same as that taken in Art. 467. The string is just slipping in that direction along the curve in which the ψ of any point of the string increases. Also the pressure R of the curve on the string, when positive, acts outwards. If either of these assumptions is reversed, the sign of μ must be changed. In order that the string may not leave the curve, the sign of C should be such that R acts from the curve towards that side on which the string lies.

To prove these results, we refer to equations (1) and (2) Art. 454. Introducing the pressure R into these equations, we have

$$dT + Fds - \mu Rds = 0, \qquad \frac{Tds}{\rho} + Gds - Rds = 0 \dots\dots\dots\dots\dots(2).$$

Eliminating R, as in Art. 467 $Te^{-\mu\psi} = -\int(F - \mu G)\,\rho e^{-\mu\psi}\,d\psi + C\dots\dots\dots(3).$
When the string is hanging freely, $R = 0$; by eliminating T between the equations (2) we find that $F\rho = \dfrac{d}{d\psi}(G\rho)$ is true along the curve. When the string is constrained to lie on a curve which possesses this property, we can substitute this value of $F\rho$ in the equation (3). We then find $Te^{-\mu\psi} = -e^{-\mu\psi}G\rho + C$. The first result to be proved follows immediately, the second is obtained by substituting this value of T in the second of equations (2).

471. Ex. 1. A uniform heavy string AB is placed on the upper side of a rough curve whose form is a catenary with its directrix horizontal. If the lower extremity is at the vertex, find the least force F which, acting at the upper extremity, will just move the string.

At the upper end of the string we have $T = F$, $G = -g\cos\psi$, at the lower $T = 0$, $G = -g$, $\psi = 0$. Hence by Art. 470 $(F - g\rho\cos\psi)\,e^{\pm\mu\psi} = -gc$, $\therefore F = g\,(y - ce^{\mp\mu\psi})$. The upper sign of μ gives the larger value of F, i.e. the force which will just move the string upwards, the lower sign gives the force which will just sustain the string. Instead of quoting equation (1), the reader should deduce this result from the equations of equilibrium.

Ex. 2. A uniform string AB rests on the circumference of a rough circle under the action of a central force tending to a point O situated at the opposite extremity of the diameter through A. If the force of attraction vary as the inverse cube of the distance, prove that the force F acting at A necessary to prevent the string from slipping is $F = k\,(\sec^2\beta e^{-2\mu\beta} - 1)$, where β is the angle AOB, $\dfrac{2k}{a}$ the force at A, and a is the diameter.

472. Endless and other strings. When a heavy inextensible string rests in equilibrium in contact with a smooth curve without singularities in a vertical plane, the pressure and tension can be found as in Art. 459, with one undetermined constant. This constant is usually found by equating to zero the tension at the free extremity. If, however, the string is either endless or has both its extremities attached to the curve and is tightened at pleasure, there is nothing to determine the constant.

Let us suppose the string to be in contact along the under side of the curve. Let the string be gradually loosed until its length exceeds the length of the arc in contact by an infinitely small quantity. The string is then just on the point of leaving the curve at some unknown point Q, and is then said to *just fit* the curve. If the length of the string were still further increased a finite portion of the string would be off the curve and hang in the form of a catenary. In the same way if the portion of the string under consideration rest with its weight supported on the upper and concave side of the curve, we may conceive the string to be gradually tightened until it separates from the curve at some point Q. If still further tightened or shortened a finite part of the string would hang in the form of a catenary, while the remainder would still rest on the curve.

To determine the position of the point Q we notice that the pressure of the curve on the string measured towards that side on which the string lies must be positive at every point of the curve and zero at Q. The pressure thus measured is therefore a minimum at Q.

Referring to Art. 460, the *outward pressure* R is given by

$$R\rho = T_0 + w\,(y - \rho \cos \psi) \dots\dots\dots\dots\dots\dots\dots\dots\dots(1).$$

Differentiating, and remembering that both R and dR/ds are zero at Q, we find

$$0 = \frac{dy}{ds} - \cos \psi \frac{d\rho}{ds} + \rho \sin \psi \frac{d\psi}{ds},$$

except when ρ is infinite at the point thus determined. Since $dy/ds = \sin \psi$ and $\rho = ds/d\psi$, this gives at once

$$2 \tan \psi = \frac{d\rho}{ds} \dots\dots\dots\dots\dots\dots\dots(2).$$

This equation determines the points at which $R\rho$ is a maximum, a minimum, or stationary. When both R and dR/ds are zero, we have

$$\rho \frac{d^2 R}{ds^2} = \frac{d^2 R\rho}{ds^2} = \cos \psi \left(\frac{2}{\rho} - \frac{d^2\rho}{ds^2} \right) + \sin \psi \frac{1}{\rho} \frac{d\rho}{ds}.$$

The sign of this expression determines whether R is a maximum or a minimum. When the length of the string is finite, some of these maxima or minima may be excluded as being beyond the given limits. But we must then also take into consideration the extremities of the string, for it is manifest that the pressure at either end may be less than that at any point between the limits of the string. *The required point Q is that one of all these points at which the pressure measured towards the string is least.* The undetermined constant T_0 is then found by making the pressure zero at this point.

If the string leave the curve at the lowest point we have $d\rho/ds = 0$, i.e. the radius of curvature ρ must be either a maximum, a minimum, or stationary at that point. Since $R\rho$ must be a minimum or a maximum according as the string is outside or inside, it is also necessary that $d^2 R\rho/ds^2$ should be positive in the first case and negative in the second.

We may express these conditions in a geometrical form. Consider a portion of the string on the under and convex side of a curve, and let it be gradually loosened until it leaves the curve. Let Q be the point whose anti-centre is lowest, and let the constant T_0 be determined by making the statical directrix pass through that anti-centre, Art. 460. If R represent the outward pressure on the string, $R\rho$ is then positive at every point of the string and equal to zero at Q. The string therefore leaves the curve at Q.

Next, let the string rest on the upper and concave side of a curve. If gradually tightened it will leave the curve at the point Q whose anti-centre is highest. For, choosing the constant T_0 so that the statical directrix passes through the anti-centre, and assuming that the whole string is still above the directrix (Art. 460), the value of $R\rho$ is negative at every point of the string and equal to zero at Q.

473. Ex. 1. A heavy string *just fits* round a vertical circle: show that the tension at the highest point is three times that at the lowest.

Let T_0, T_1 be the tensions at the lowest and highest points, and let a be the radius. Then $T_1 - T_0 = 2wa$. Since ρ is constant the only solution of (2) is $\psi = 0$, and this makes the outward pressure R a minimum. The pressure is therefore zero at the lowest point. The weight, viz. wds, of the lowest element is therefore supported by the tensions at each end, i.e. $wds = T_0 ds/a$. These equations give $T_0 = wa$, and $\therefore T_1 = 3wa$.

We may obtain the result more simply by using the geometrical rule given in the last article. The locus of the anti-centre is obviously another circle of radius $2a$ and concentric with the given circle. Taking the tangent at its lowest point for the statical directrix, the altitudes of the highest and lowest points of the given circle are as $3 : 1$, Art. 460. The tensions at these points are therefore also in the same ratio. We see also that if the string be slightly loosened, it will begin to leave the curve *at the lowest point*.

Ex. 2. A heavy string (length $2l$) rests on the inner or concave side of a segment of a smooth sphere (radius a, angle 2β) and hangs down symmetrically over the smooth rim which is in a horizontal plane. Find the conditions of equilibrium.

Since every point of the string must be above the statical directrix, it will be seen on drawing a figure that $l > a(\beta + 1 - \cos\beta)$. Since the string rests on the concave side, the outward pressure R must be negative and therefore every point of the anti-centric curve must be below the statical directrix, hence $l < a(\beta + \cos\beta)$. These two conditions require that β should be less than $\frac{1}{3}\pi$. If the second inequality be reversed the string will leave the spherical segment *at the highest point*.

Ex. 3. A heavy string is attached to two points of the arc of a catenary with its axis vertical, and rests against its under surface. If the string is gradually loosed, show that it will leave the curve at every point at the same instant.

Ex. 4. A heavy string has one end fastened to the lowest point of the arc of a cycloid with the axis vertical and the vertex at the lowest point. The string envelopes the arc outside up to the cusp, and passing over a small smooth pulley has the other end hanging freely. Prove that the least length of the string hanging down which is consistent with equilibrium is equal to six times the radius of the generating circle. Find also in this case the resultant pressure on the cycloid.

[Queens' Coll.]

Ex. 5. A heavy string just fits the under surface of a cycloidal arc, the extremities of the string being attached to the cusps. Show that the pressure is zero at the point Q given by the negative root of the equation $3\sin(2\phi + a) = -\sin a$, where ϕ is the inclination of the normal at Q to the axis of the cycloid, and a is the inclination of the axis to the vertical. Find also the tension at the vertex.

Ex. 6. A heavy string surrounds an oval curve, and is so much longer than the perimeter that a finite portion hangs in the form of a catenary. If the string is gradually shortened until the arc of the catenary is evanescent, show (1) that the curve and the catenary have four consecutive points coincident, and (2) that the evanescent arc is situated at a point of the curve determined by $2 \tan \psi = d\rho/ds$.

Ex. 7. A string is bound tightly round a smooth ellipse, and is acted on by a central repulsive force in the focus varying directly as the square of the distance. Find the law of variation of the tension, and prove that, if the string be slightly loosened, it will leave the curve at the points at a distance from the focus equal to 7/4 times the semi-major axis, provided the eccentricity be greater than 3/4. If the eccentricity be less than 3/4, where will it leave the curve? [Coll. Ex., 1887.]

474. Central forces. *A string of given length is attached to two fixed points, and is under the action of a central force. Find the relation between the form of the curve and the law of force.*

Let the arc be measured from any fixed point A on the string in the direction AB, and let $s = AP$.

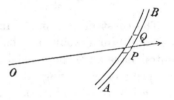

Let O be the centre of force, and let Fds be the force on the element ds estimated positive when acting in the positive direction of the radius vector, i.e. when the force is repulsive.

The element PQ is in equilibrium under the action of the tensions T and $T + dT$ and the central force Fds. Resolving along the tangent at P, we have

$$dT + Fds \cos \phi = 0,$$

where ϕ is the radial angle, i.e. the angle OPA. Since $\cos \phi = dr/ds$,

this reduces to

$$\frac{dT}{dr} + F = 0 \dots\dots\dots\dots\dots\dots (1).$$

We might obtain a second equation by resolving the same forces along the normal at P, but the result is more easily found by taking the moment of the forces which act on the finite portion of string AP. This portion is in equilibrium under the action of the tensions T_0, T and the central force tending from O on each element. Taking moments about O, these latter disappear; we have therefore

$$Tp = A \dots\dots\dots\dots\dots\dots\dots (2),$$

where p is the perpendicular from O on the tangent at P, and A is the moment about O of the tension T_0.

Let the tangents at any two points A, B of the curve meet in C. Then the arc AB is in equilibrium under the action of the tensions at A and B and the resultant

R of the central forces on all the elements. *This resultant force must therefore act along the straight line joining the centre of force O to the intersection C of the tangents at A and B.* Also if OY, OZ are the perpendiculars from O on the tangents at A and B, we see by compounding the tensions that $R = A \cdot \dfrac{YZ}{OY \cdot OZ}$.

As the point P moves from A to B, the foot of the perpendicular on the tangent at P traces out the pedal curve. This curve, when sketched, exhibits to the eye the magnitude of the tension at all points of the catenary.

475. Two cases have now to be considered.

First. Suppose the form of the string to be given, and let the force be required. By known theorems in the differential calculus we can express the equation to the curve in the form $p = \psi(r)$. The equations (1) and (2) then give

$$T = \frac{A}{\psi(r)}, \qquad F = \frac{A\psi'(r)}{\psi(r)^2} \dots\dots\dots\dots (3).$$

The constant A remains indeterminate, for it is evident that the equilibrium would not be affected if the magnitude of the central force were increased in any given ratio. The tension at any point of the string and the pressures on the fixed points of suspension would be increased in the same ratio.

Secondly. Suppose that the force is given, and that the form of the curve is required. Eliminating T between (1) and (2), we find

$$\frac{A}{p} = B - \int F dr \dots\dots\dots\dots\dots\dots\dots (4).$$

This differential equation has now to be solved. Put $u = 1/r$ and $\int F dr = f(u)$; we find by a theorem in the differential calculus

$$A^2 \left\{ u^2 + \left(\frac{du}{d\theta} \right)^2 \right\} = (B - fu)^2 \dots\dots\dots\dots (5).$$

Separating the variables, we have

$$\int \frac{\pm A \, du}{\{(B - fu)^2 - A^2 u^2\}^{\frac{1}{2}}} = \theta + C \dots\dots\dots\dots (6).$$

When this integration has been effected the polar equation to the curve has been found.

There are three undetermined constants, viz. A, B, C, in this equation. To discover their values we have given the polar coordinates $(u_0 \theta_0)$, $(u_1 \theta_1)$ of the points of suspension. After integrating (6) we substitute in turn for $(u\theta)$ these two terminal values, and thus obtain two equations connecting the three constants. We have also given the length of the string. To use this datum we must find the length of the arc. We easily find

$$(ds)^2 = (dr)^2 + (rd\theta)^2 = \frac{1}{u^4}\{(du)^2 + (ud\theta)^2\}.$$

Substituting from (5), we have

$$s = \int \frac{(B - fu)\,du}{u^2\{(B - fu)^2 - A^2u^2\}^{\frac{1}{2}}} \quad \dots \dots \dots \dots (7).$$

Taking this between the given limits of u, and equating the result to the given length of the string, we have a third equation to find the three constants.

The equation (6) agrees with that given by John Bernoulli, *Opera Omnia, Tomus Quartus*, p. 238. He applies the equation to the case in which the force varies inversely as the nth power of the distance, and briefly discusses the curves when $n=0$ and $n=2$.

476. Ex. 1. A string is in equilibrium under the action of a central force. If F be the force at any point per unit of length, prove that the tension at that point $= F\chi$, where χ is the semi-chord of curvature through the centre of force.

Show also that $F = A \dfrac{r}{p^2\rho}$, where A is a constant.

Ex. 2. *A uniform string is in equilibrium in the form of an arc of a circle under the influence of a centre of force situated at any point O. Find the law of force.* Let C be the centre, $OC = c$, $CP = a$. Then $2ap = r^2 + a^2 - c^2$,

$$\therefore F = -A \frac{d}{dr}\frac{1}{p} = 4aA \frac{r}{(r^2 + a^2 - c^2)^2}.$$

If the centre of force is situated at any point of the arc not occupied by the string the law of force is the inverse cube of the distance.

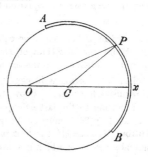

Since $Tp = A$, A is positive, hence F is positive, i.e., the force must be repulsive. If the centre of force is outside the circle, p is negative for that part of the arc nearest O which is cut off by the polar line of O. If the string occupy this part of the arc, A is negative and the force F must be attractive.

We have taken r or u as the independent variable. If the centre of force be at the centre of the circle, this would be an impossible supposition. This case therefore requires a separate investigation. It is however clear that the string will be in equilibrium whatever the law of force may be, provided it is repulsive.

Ex. 3. A uniform string is in equilibrium in the form of the curve $r^n = a^n \cos n\theta$ under a central force F in the origin: prove that $F = \mu u^{n+2}$.

Ex. 4. A string of infinite length has one extremity attached to a fixed point A, and passing through a small smooth fixed ring at B stretches to infinity in a straight line, the whole being under the influence of a central repulsive force $= \mu u^n$, where

$n > 1$. Show that the form of the string between A and B is $r^{n-2} = b^{n-2} \cos(n-2)\theta$. If $n = 2$ the curve is an equiangular spiral.

Ex. 5. A closed string surrounds a centre of force $= \mu u^n$, where $n > 1$ and < 2. Show that, as the length of the string is indefinitely increased so that one apse becomes infinitely distant from the centre of force, the equilibrium form of the string tends to become $r^{n-2} = b^{n-2} \cos(n-2)\theta$. If $n = \frac{3}{2}$ the form of the curve is a parabola.

Ex. 6. *A uniform string of length $2l$ is attached to two fixed points A, B at equal distances from a centre O of repulsive force $= \mu u^2$. If $OA = OB = b$ and the angle $AOB = 2\beta$, prove that the equation to the string is* $\dfrac{M}{r} = 1 + \dfrac{\cos(\theta \sin \alpha)}{\cos \alpha}$,

where the real and imaginary values of M and α are determined from the equations

$$\frac{M}{b} = 1 + \frac{\cos(\beta \sin \alpha)}{\cos \alpha} \qquad \sin \alpha = \pm \frac{b}{l} \sin(\beta \sin \alpha).$$

The equations (1) and (2) of Art. 474 become here $\qquad dT = \mu du, \qquad Tp = A$.

Proceeding as explained in Art. 475, we find $\pm \displaystyle\int \dfrac{A\,du}{\{(B + \mu u)^2 - A^2 u^2\}^{\frac{1}{2}}} = \theta + C$.

This integral is one of the standards in the integral calculus, and assumes different forms according as $A^2 - \mu^2$ is positive, negative or zero. Taking the first assumption, we have after a slight reduction

$$\frac{A^2 - \mu^2}{B} u = \mu \pm A \cos\left(1 - \frac{\mu^2}{A^2}\right)^{\frac{1}{2}} (\theta + C).$$

This formula really includes all cases, for when $A^2 - \mu^2$ is negative we may write for the sine of the imaginary angle on the right-hand side its exponential value.

Proceeding to find the arc in the manner already explained, we easily arrive at

$$Bs = \pm \{(Br + \mu)^2 - A^2\}^{\frac{1}{2}} + D,$$

where the radical must have opposite signs on opposite sides of an apse.

The conditions of the question require that the string should be symmetrical about the straight line determined by $\theta = 0$. We have therefore $C = 0$ and $D = 0$. Putting $A = \mu \sec \alpha$, the equation to the curve reduces to $\dfrac{\mu \tan^2 \alpha}{B} \dfrac{1}{r} = 1 \pm \dfrac{\cos(\theta \sin \alpha)}{\cos \alpha}$.

We also have $\qquad\qquad B^2 l^2 = (Bb + \mu)^2 - \mu^2 \sec^2 \alpha$.

Eliminating B between these equations, we find $l \sin \alpha = \pm b \sin(\beta \sin \alpha)$. We now put M for the coefficient of $1/r$ and include the double sign in the value of α. Since $r = b$ when $\theta = \pm \beta$ the three results given above have been obtained.

Ex. 7. A string is in equilibrium in the form of a closed curve about a centre of repulsive force $= \mu u^2$. Show that the form of the curve is a circle.

Referring to the last example, we notice that, since r is unaltered when θ is increased by 2π, r must be a trigonometrical function of θ. Hence $\sin \alpha = 1$ or 0. Putting $M \cos \alpha = M'$, the first makes $M'/r = \cos \theta$, which is not a closed curve, the second gives $M = r$, which is a circle.

Ex. 8. If the curve be a parabola, and the centre of force at the focus, and if the equilibrium be maintained by fixing two points of the string, find the law of force, and prove that the tension at any point P is $2fr$, where $r = SP$ and f is the force at P per unit of length. [St John's Coll., 1883.]

Ex. 9. An infinite string passes through two small smooth rings, and is acted on by a force tending from a given fixed point and varying inversely as the cube of the distance from that point. Show that the part of the string between the rings assumes the form of an arc of a circle. [Coll. Ex., 1884.]

Ex. 10. If a string, the particles of which repel each other with a force varying as the distance, be in equilibrium when fastened to two fixed points, prove that the tension at any point varies as the square root of the radius of curvature.

[Math. Tripos, 1860.]

Ex. 11. Show that the *catenary of equal strength* for a central force which varies as the inverse distance is $r^n \cos n\theta = a^n$, where $1 - n$ is the ratio of the line density to the tension. Show also that this system of curves includes the circle, the rectangular hyperbola, the lemniscate, and when n is zero the equiangular spiral.

[O. Bonnet, *Liouville's J.*, 1844.]

Ex. 12. A string is placed on a smooth plane curve under the action of a central force F, tending to a point in the same plane; prove that, if the curve be such that a particle could freely describe it under the action of that force, the pressure of the string on the curve referred to a unit of length will be equal to $\dfrac{F \sin \phi}{2} + \dfrac{c}{\rho}$, where ϕ is the angle which the radius vector from the centre of force makes with the tangent, ρ is the radius of curvature, and c is an arbitrary constant.

If the curve be an equiangular spiral with the centre of force in the pole, and if one end of the string rest freely on the spiral at a distance a from the pole, then the pressure is equal to $\dfrac{\mu \sin \phi}{2r} \left(\dfrac{1}{r^2} + \dfrac{1}{a^2} \right)$. [Math. Tripos, 1860.]

Ex. 13. A free uniform string, in equilibrium under the action of a repulsive central force F, has a form such that a particle could freely describe it under a central force F' tending to the same centre. Show that $F = kpF'$, where k is a constant. If v be the velocity of the particle and T the tension of the string, show also that $T = kpv^2$. See Art. 476, Ex. 1.

Ex. 14. It is known that a particle can describe a rectangular hyperbola about a repulsive central force which varies as the distance and tends from the centre of the curve. Thence show that a string can be in equilibrium in the form of a rectangular hyperbola under an attractive central force which is constant in magnitude and tends to the centre of the curve. Show also that the tension varies as the distance from the centre.

For a comparison of the free equilibrium of a uniform string with the free motion of a particle under the action of a central force, see a paper by Prof. Townsend in the *Quarterly Journal of Mathematics*, vol. XIII., 1873.

477. When there are two centres of force the equations of equilibrium are best found by resolving along the tangent and normal. Let r, r' be the distances of any point P of the string from the centres of force; F, F' the central forces, which are to be regarded as functions of r, r' respectively. Let p, p' be the perpendiculars from the centres of force on the tangent at P. We then have

$$dT + F dr + F' dr' = 0 \ldots (1), \qquad \frac{T}{\rho} - F \frac{p}{r} - F' \frac{p'}{r'} = 0 \ldots (2).$$

The first equation gives $\qquad T = B - \int F dr - \int F' dr' \ldots\ldots\ldots\ldots\ldots\ldots (3).$

We may suppose the lower limits of these integrals to correspond to any given point P_0 on the string. If this be done B will be the tension at P_0. Substituting the value of T thus obtained from (1) in (2) and remembering that $\rho = r dr/dp$,

$$\frac{d}{dp} \left(p \int F dr \right) + \frac{d}{dp'} \left(p' \int F' dr' \right) = B \ldots\ldots\ldots\ldots\ldots\ldots (4);$$

on the other hand, if we find T from (2) and substitute in (1), we find after reduction

$$\frac{1}{p}\, d \left(\frac{Fp^2 \rho}{r} \right) + \frac{1}{p'}\, d \left(\frac{F'p'^2 \rho}{r'} \right) = 0 \dots\dots\dots\dots\dots\dots (5).$$

Thus of the four elements, viz. (1) the force F, (2) the force F', (3) the tension T, (4) the equation to the curve, if any two are given, sufficient equations have now been found to discover the other two.

Ex. 1.   A string can be in equilibrium in the form of a given curve under the action of each of two different centres of force.   Show that it is in equilibrium under the joint action of both centres of force, and that the tension at any point is equal to the sum of the tensions due to the forces acting separately.

Ex. 2.   Prove that a uniform string will be in equilibrium in the form of the curve $r^2 = 2a^2 \cos 2\theta$ under the action of equal centres of repulsive force situated at the points, $(a, 0)$, $(-a, 0)$, the force of each per unit of length at a distance R being μ/R.   Prove also that the tension at all points will be the same and equal to $\frac{1}{3}\mu$.

[Coll. Ex., 1891.]

478. String on a surface. *A string rests on a smooth surface under the action of any forces. To find the position of equilibrium.*

Let the equation to the surface be $f(x, y, z) = 0$. Let Rds be the outward pressure of the surface on the string. Let (l, m, n) be the direction cosines of the inward direction of the normal. By known theorems in solid geometry, l, m, n are proportional to the partial differential coefficients of $f(x, y, z)$ with regard to x, y, z respectively.

If the equations are required to be in Cartesian coordinates, we deduce them at once from those given in Art. 455 by including R among the impressed forces.   We thus have

$$\left. \begin{array}{l} \dfrac{d}{ds} \left(T \dfrac{dx}{ds} \right) + X - Rl = 0 \\[2mm] \dfrac{d}{ds} \left(T \dfrac{dy}{ds} \right) + Y - Rm = 0 \\[2mm] \dfrac{d}{ds} \left(T \dfrac{dz}{ds} \right) + Z - Rn = 0 \end{array} \right\}.$$

We have here one more unknown quantity, viz. R, than we had in Art. 455, but we have also one more equation, viz. the given equation to the surface.

479. *Let us next find the intrinsic equations to the string.* Let PQ be any element of the string, PA a tangent at P. Let APB be a tangent plane to the surface, PB being at right angles to PA. Let PN be the normal to the surface. Let PC be the radius of curvature of the string, then PC lies in the plane BPN. Let χ be the angle CPN, then χ is also the angle the osculating plane CPA of the string makes with the normal PN to the surface.

The element PQ is in equilibrium under the action of (1) the forces Xds, Yds, Zds acting parallel to the axes of coordinates, which are not drawn in the figure, (2) the reaction Rds along NP, (3) the tensions at P and Q, which have been proved in Art. 454 to be equivalent to dT along PQ and Tds/ρ along PC.

Resolving these forces along the tangent PA, we have

$$dT + Xds\frac{dx}{ds} + Yds\frac{dy}{ds} + Zds\frac{dz}{ds} = 0,$$

$$\therefore \quad T + \int(Xdx + Ydy + Zdz) = A \quad \dots\dots\dots (1).$$

The forces are said to be *conservative*, when their components X, Y, Z are respectively partial differential coefficients with regard to x, y, z, of some function W which may be called the work function, Art. 209. Assuming this to be the case, the integral in (1) is equal to the work of the forces. It follows from this equation that *the tension of the string plus the work of the forces is the same at all points of the string.* Taking the integral between limits for any two points P, P' of the string, we see that *the difference of the tensions at two points P, P' is in-*

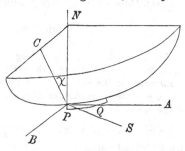

dependent of the length or form of the string joining those points and is equal to the difference of the works at the points P', P taken in reverse order.

We shall suppose that, while ρ is measured inwards along PC, the pressure R of the surface on the string is measured outwards along NP, Art. 457. We shall also suppose that (l, m, n) are the direction cosines of the normal PN measured inwards. With this understanding we now resolve the forces along the normal PN to the surface; we find

$$\frac{Tds}{\rho}\cos\chi + Xds\,l + Yds\,m + Zds\,n - Rds = 0.$$

By a theorem in solid geometry, if ρ' be the radius of curvature of the section of the surface made by the plane NPA, i.e. by a plane containing the normal to the surface and the tangent to the string, then $\rho' \cos \chi = \rho$. We therefore have

$$\frac{T}{\rho'} + Xl + Ym + Zn = R \quad \dots\dots\dots\dots\dots \text{(2)}.$$

It follows from this equation that *the resultant pressure on the surface is equal to the normal pressure due to the tension plus the pressure due to the resolved part of the forces.* The tension at any point P having been found by (1), the pressure on the surface follows by (2), provided we know the direction of the tangent PA to the string. This last is necessary in order to find the value of ρ'.

Lastly, let us resolve the forces along the tangent PB to the surface. Let λ, μ, ν be the direction cosines of PB. Since PB is at right angles to both PN and PA, these direction cosines may be found from the two equations

$$\lambda f_x + \mu f_y + \nu f_z = 0, \qquad \lambda \frac{dx}{ds} + \mu \frac{dy}{ds} + \nu \frac{dz}{ds} = 0.$$

We then have by the resolution

$$\frac{T}{\rho} \sin \chi + X\lambda + Y\mu + Z\nu = 0 \quad \dots\dots\dots\dots \text{(3)}.$$

Ex. An endless string lies along a central circular section of a smooth ellipsoid, prove that $b^4 F^2 = T^2 (b^2 - p^2)$, where F is the force per unit of length which acting transversely to the string in the tangent plane is required to keep the string in its place, p is the perpendicular from the centre on the tangent plane and b is the mean semi-axis.   [Trin. Coll., 1890.]

480. Geodesics. If any portion of the string is not acted on by external forces, we have for that portion $X = 0$, $Y = 0$, $Z = 0$. The equation (1) then shows that *the tension of the string is constant.* The equation (2) shows that *the pressure at any point is proportional to the curvature of the surface along the string.* The equation (3) (assuming the string not to be a straight line) shows that $\chi = 0$, i.e. at every point *the osculating plane of the curve contains the normal to the surface.* Such a curve is called a geodesic in solid geometry.

Conversely, if the string rest on the surface in the form of a geodesic under the action of forces, we see by (3) that they must be such that *at every point of the string their resolved part perpendicular to the osculating plane of the string is zero.*

Returning to the general case in which the string is under the action of forces, we notice that $\sin \chi / \rho$ is the resolved curvature of the string in the tangent plane at P to the surface. When the resolved curvature vanishes and changes sign as P

moves along the string the concavity changes from one side of the string to the other. Such a point may be regarded as a point of geodesic inflexion. It follows from the equation (3) that *a string stretched on a surface can have a point of geodesic inflexion only when the force transverse to the string and tangential to the surface is zero.*

481. A string on a surface of revolution. When the surface on which the string rests is one of revolution, we can replace the rather complicated equation (3) of Art. 479 by a much simpler one obtained by taking moments about the axis of figure. If also the resultant force on each element is either parallel to or intersects the axis of figure, there is a further simplification. This includes the useful case in which the only force on the string is its weight, and the axis of figure of the surface is vertical.

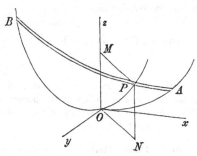

Let the axis of figure be the axis of z, and let (r, θ, ϕ) be the polar coordinates and (r', ϕ, z) the cylindrical coordinates of any point on the string, so that in the figure $r' = ON$, $z = PN$, and $\phi =$ the angle NOx. Then from the equation to the surface we have $z = f(r')$. Let the forces on the element ds be Pds, Qds, Zds when resolved respectively parallel to r', $r'd\phi$, and z.

We shall now take moments about the axis of figure. The moment of R is clearly zero. To find the moment of T, we resolve it perpendicular to the axis and multiply the result by the arm r'. In this way we find that the moment is $Tr' \sin \psi$, where ψ is the angle the tangent to the string makes with the tangent to the generating curve of the surface, i.e. ψ is the curvilinear angle OPA. The equation of moments is therefore

$$d\left(Tr' \sin \psi\right) + Qr'ds = 0 \quad \dots\dots\dots\dots\dots (4).$$

We also have by resolving along the tangent as in Art. 479,

$$dT + Pdr' + Qr'd\phi + Zdz = 0 \quad \dots\dots\dots\dots (5).$$

We have also the geometrical equation expressing $\sin \psi$ in terms of the differentials of the coordinates of P. Let the generating curve OP turn round Oz through an angle $d\phi$ and then intersect the string in P' and a plane drawn through MP parallel to xy in Q. Then $PQ = PP' \sin \psi$, i.e. $r'd\phi = ds . \sin \psi$. We therefore have

$$(r'd\phi)^2 = \{(dr')^2 + (r'd\phi)^2 + (dz)^2\} \sin^2 \psi \;\;\;\ldots\ldots\ldots (6).$$

Eliminating T and $\sin \psi$ between (4), (5) and (6) we have an equation from which the form of the string can be deduced.

If the only force acting on the string is gravity, and if the axis is vertical, the equations take the simple forms

$$Tr' \sin \psi = wB, \qquad T = w(z+A)\ldots\ldots\ldots\ldots\ldots\ldots\ldots(7).$$

Eliminating T and $\sin \psi$, by help of (6), we have

$$(z+A)^2 r'^2 = B^2 \left\{ 1 + \left(\frac{dr'}{r'd\phi} \right)^2 + \left(\frac{dz}{r'd\phi} \right)^2 \right\} \;\;\;\ldots\ldots\ldots\ldots (8).$$

Substituting for z from the equation of the surface, viz. $z = f(r')$, this becomes the polar differential equation of the projection of the string on a horizontal plane. *The outward normal pressure* of the surface on the string may be deduced from equation (2) of Art. 479.

482. Heavy string on a sphere. Using polar coordinates referred to the centre O as origin, the fundamental equations take the simple forms

$$T \sin \theta \sin \psi = wB', \qquad\qquad T = w(a \cos \theta + A),$$

$$(\sin \theta d\phi)^2 = \{(\sin \theta d\phi)^2 + (d\theta)^2\} \sin^2 \psi, \qquad Ra = w(2a \cos \theta + A),$$

where ψ is the angle the string makes with the meridian arc drawn through the summit and $B = aB'$. These give as the differential equation * of the string

$$\left(\frac{d\theta}{d\phi} \right)^2 + \sin^2 \theta = \sin^4 \theta \left(\frac{a \cos \theta + A}{B'} \right)^2 .$$

The tension at any point $P = wz$ where z is the altitude of P above a fixed horizontal plane called the directrix plane, and every point of the string must be above this plane. The plane is situated at a depth A below the centre of the sphere. At each point P let the normal OP be produced to cut in some point S a concentric sphere whose radius is twice that of the given sphere. The point S is *the anti-centre of* P, and *the outward pressure on the string is* wz'/a where z' is the altitude of S above the directrix plane. As already explained every anti-centre must lie above or below the directrix plane according as the string lies on the convex or concave side of the sphere, Art. 460.

The values of the constants A, B depend on the conditions at the ends of the string. We see that $B' = 0$, (1) if either end is free, for then T vanishes at that end, (2) if the string pass through the summit of the sphere, for then $\sin \theta$ vanishes, (3) if a meridian can be drawn from the summit to touch the sphere, for $\sin \psi = 0$ at the point of contact. In all these cases, $\sin \psi$ vanishes throughout the string, i.e. *the string lies in a vertical plane.*

If the string form a closed curve, the three quantities T, $\sin \theta$, $\sin \psi$ cannot

---

* The reduction of the integral giving ϕ in terms of θ to elliptic functions is given by Clebsch in *Crelle's J.*, vol. 57. A model was exhibited at the Royal Society, June 1895, by Greenhill and Dewar of an algebraical spherical catenary. By a proper choice of the constants the projection of the chain on a horizontal plane became a closed algebraical curve of the tenth degree; see also *Nature*, Jan. 10, 1895.

vanish or change sign at any point of the string. The highest and lowest points of the string are therefore given by $\psi = \frac{1}{2}\pi$, hence at these points

$$T \sin \theta = wB', \qquad T = w\,(a\cos\theta + A), \qquad \therefore \; \sin\theta\,(a\cos\theta + A) = B'.$$

These equations yield only two available values of $\cos\theta$; for tracing the two curves whose common abscissa is $\xi = \cos\theta$ and whose ordinates are the *reciprocals* of the two values of T, we have an ellipse and a rectangular hyperbola, which, since T must be positive, give only two intersections. Let $\theta = a$, $\theta = \beta$ be the meridian distances of the highest and lowest points of the string, both being positive. Then

$$-\frac{2A}{a} = \frac{\sin 2a - \sin 2\beta}{\sin a - \sin \beta}, \qquad -\frac{B'}{a} = \sin a \sin \beta \, \frac{\cos a - \cos \beta}{\sin a - \sin \beta}.$$

It follows that the directrix plane passes through the centre of the sphere when a and β are complementary. In general the tensions, and therefore the depths of the directrix plane below the highest and lowest points, are inversely as the distances of those points of the string from the vertical diameter.

It has been proved in Art. 480, that the string can have a point of geodesic inflexion when the transverse tangential force is zero. This requires that the meridian drawn from the summit should touch the string, and this, we have already seen, cannot occur. It follows that *the string must be concave throughout its length on the same side.*

If the form of the string is a circle its plane must be either horizontal or vertical, and in the latter case it must pass through the centre of the sphere. To prove this we give the string a virtual displacement without changing its form, it is easy to see that the altitude of the centre of gravity can be a max-min only in the cases mentioned. In both cases the altitude is a maximum and the equilibrium is therefore unstable. Art. 218. In the same way it may be shown that *any position of equilibrium of a heavy free string on a smooth sphere is unstable.*

Ex. 1. A heavy uniform chain, attached to two fixed points on a smooth sphere, is drawn up just so tight that the lowest point just touches the sphere. Prove that the pressure at any point is proportional to the vertical height of the point above the lowest point of the string.   [Coll. Ex., 1892.]

Ex. 2. A string rests on a smooth sphere, cutting all the sections through a fixed diameter at a constant angle. Show that it would so rest if acted on by a force varying inversely as the square of the distance from the given diameter, and that the tension varies inversely as that distance.   [Coll. Exam., 1884.]

Ex. 3. A string can rest under gravity on a sphere in a smooth undulating groove lying between two small circles whose angular distances from the highest point of the sphere are complementary, without pressing on the sides of the groove. If ψ is the acute angle at which the string cuts the vertical meridian prove that the points at which ψ is a minimum occur at angular distances $\frac{1}{4}\pi$ from the highest point and find the value of ψ at these points.   [Math. T., 1889.]

483. String on a Cylindrical Surface. Ex. 1. A heavy string is in equilibrium on a cylindrical surface whose generators are vertical, the extremities of the string being attached to two fixed points on the surface. Find the circumstances of the equilibrium.

Let $PQ = ds$ be any element, wds its weight. Let the axis of z be parallel to the generators, and let z be measured in the direction opposite to gravity. Resolving along a tangent to the string, we have as in (1) Art. 479, $T - wz = A$. Resolving vertically, we have by Art. 478, $\dfrac{d}{ds}\left(T\dfrac{dz}{ds}\right) - w = 0$. These are the same as the

equations to determine the equilibrium of a heavy string in a vertical plane. The constants, also, of integration are determined by the same conditions in each case. We see therefore that *if the cylinder is developed on a vertical plane, the equilibrium of the string is not disturbed.* The circumstances of the equilibrium may therefore be deduced from the ordinary properties of a catenary.

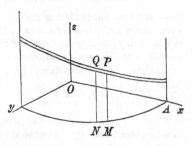

To find the pressure on the cylinder, we either resolve along the normal at P to the surface, or quote the general result found in Art. 479. We thus find $R = T/\rho'$, also $\dfrac{1}{\rho'} = \dfrac{\cos^2\psi}{\rho_1} + \dfrac{\sin^2\psi}{\infty} = \dfrac{\cos^2\psi}{\rho_1}$, by Euler's theorem on curvature, where ρ_1 is the radius of curvature at M of the section AMN of the cylinder made by a horizontal plane, and ψ is the angle the tangent at P to the string makes with the horizontal plane.

Ex. 2. If a string be suspended symmetrically by two tacks upon a vertical cylinder, and if $z_1, z_2, z_3\ldots$ be the distances above the lowest point of the catenary at which the string crosses itself, then $z_1 z_{2n+1} = (z_{n+1} - z_n)^2$.   [Math. Tripos, 1859.]

Ex. 3. If an endless chain be placed round a rough circular cylinder, and pulled at a point in it parallel to the axis, prove that, if the chain be on the point of slipping, the curve formed by it on the cylinder when developed will be a parabola; and find the length of the chain when this takes place.   [Math. Tripos.]

Ex. 4. A heavy uniform string rests on the surface of a smooth right circular cylinder, whose radius is a and whose axis is horizontal. If (a, θ, z) be the cylindrical coordinates of a point on the string, θ being measured from the vertical, prove that

$$T = w\,(b + a\cos\theta),\quad z = \int \frac{ac\,d\theta}{\{(b + a\cos\theta)^2 - c^2\}^{\frac{1}{2}}},\text{ where }b\text{ and }c\text{ are two constants.}$$

It is clear that the tension resolved parallel to z is constant, i.e. $T dz/ds = wc$. Combining this result with the value of T found in Art. 483, Ex. 1, we obtain the second result in the question.

Ex. 5. The extremities of a heavy string are attached to two small rings which can slide freely on a rod which is placed along the highest generator of a right circular horizontal cylinder, and are held apart by two forces each equal to wa. The lowest point of the string just reaches to a level with the axis of the cylinder. If D be the distance between the rings and L the length of the string, prove that

$$\frac{D}{4a} = \int \frac{d\psi}{\sqrt{(3 + \sin^2\psi)}},\qquad \frac{L}{8a} = \int \frac{d\psi}{\sqrt{(3 + \sin^2\psi)}}\,\frac{1}{1 + \sin^2\psi},$$

the limits of integration being 0 to $\frac{1}{2}\pi$.

These follow from the results in the last question. The conditions of the question give $a = b = c$. The integrals are reduced by putting $\tan\frac{1}{2}\theta = \sin\psi$.

Ex. 6. A uniform string rests on a horizontal circular cylinder of radius a with its ends fastened to the highest generator and its lowest point at a depth a below it; prove that the curvature at the lowest point is $1/a$, and that the inclination of the string at any point to the axis is $\sec^{-1}(1+z/a)$, where z is the height of the point above the axis, supposing the string cuts the highest generator at an angle of 60°.

[June Exam.]

Ex. 7. A heavy uniform string has its two ends fastened to points in the highest generator of a smooth horizontal cylinder of radius a, and is of such a length that its lowest point just touches the cylinder. Prove that, if the cylinder be developed, the origin being at one of the fixed points, the curve on which the string lay is given by $c^2 \left(\dfrac{dy}{dx}\right)^2 = a^2 \cos^2 \dfrac{y}{a} + 2ac \cos \dfrac{y}{a}$.   [Math. T., 1883.]

484. String on a right cone. Ex. 1. A string has its extremities attached to two fixed points on the surface of a right cone, and is in equilibrium under the action of a centre of repulsive force F at the vertex. Show that the equilibrium is not disturbed by developing the cone and string on a plane passing through the centre of force.

Let the vertex O be the origin, (r', θ', z) the cylindrical coordinates of any point P on the string. Let $OP = r$. Taking moments about the axis and resolving along the tangent, we have as in Art. 481,

$$Tr' \sin \psi = B, \qquad T + \int F dr = C \quad \dotfill \quad (1).$$

We may imagine the cone divided along a generator and together with the string on its surface unwrapped on a plane. Let (r, θ) be the polar coordinates of the position of P in this plane. Let p be the perpendicular from O on the tangent to the unwrapped string, then $p = r \sin \psi$. The equations (1) become

$$Tp = B', \qquad T + \int F dr = C \quad \dotfill \quad (2).$$

These are the equations of equilibrium of a string in one plane under the action of a central force, and the constants of integration are determined by the same conditions in each case. We may therefore transfer the results obtained in Art. 474 to the string on the cone. In transferring these results we notice that the point (r, θ) on the plane corresponds to $(r'\theta'z)$ on the cone, where $r' = r \sin \alpha$, $\theta' \sin \alpha = \theta$, $z = r \cos \alpha$.

The pressure R is given by $R = \dfrac{T}{\rho'} = \dfrac{\sin \phi}{r^2} \cdot \dfrac{B \cos \alpha}{\sin^2 \alpha}$, since $\dfrac{1}{\rho'} = \dfrac{\cos^2 \phi}{\infty} + \dfrac{\sin^2 \phi}{r' \sec \alpha}$ by Euler's theorem on curvature. Art. 479.

Ex. 2. The two extremities of a string, whose length is $2l$, are attached to the same point A on the surface of a right cone. The equation to the projection of the string on a plane perpendicular to the axis is $\pi r' = l \cos (\theta' \sin \alpha)$, the point A being given by $\theta' = \pi$. Show that the string will rest in equilibrium under the influence of a centre of force in the vertex varying inversely as the cube of the distance.

Ex. 3. A heavy uniform string has its ends fastened to two points on the surface of a right circular cone whose axis is vertical and vertex upwards, the string lying on the surface of the cone. Prove that, if the cone be developed into a plane, the curve on which the string lay is given by $p(a + br) = 1$, the origin being the vertex, p the perpendicular on the tangent, and a, b constants.

[Coll. Ex., 1890.]

485. String on a rough surface. *A string rests on a rough surface under the action of any forces, and every element borders on motion; to find the conditions of equilibrium.*

The required conditions may be deduced from the equations for a smooth surface by introducing the limiting friction. The pressure of the surface on the element ds being Rds, the limiting friction will be μRds. This friction acts in some direction PS lying in the tangent plane to the surface. See figure of Art. 479. Let ψ be the angle SPA. Resolving along the principal axes at any point of the string exactly as in Art. 479, we have

$$dT + Xdx + Ydy + Zdz + \mu Rds\cos\psi = 0$$
$$\frac{T}{\rho'} + Xl + Ym + Zn - R = 0$$
$$\frac{T}{\rho'}\tan\chi + X\lambda + Y\mu + Z\nu + \mu R\sin\psi = 0$$

These three equations express the conditions of equilibrium.

486. The simplest case is that in which the applied forces can be neglected compared with the tension. We then have, putting zero for X, Y, Z,

$$\frac{dT}{ds} + \mu R\cos\psi = 0$$
$$\frac{T}{\rho'} = R$$
$$\frac{T}{\rho'}\tan\chi + \mu R\sin\psi = 0$$

It easily follows from these equations that $\tan\chi + \mu\sin\psi = 0$. This requires that $\tan\chi$ should be less than μ; thus equilibrium is impossible if the string be placed on the surface so that its osculating plane at any point makes an angle with the normal greater than $\tan^{-1}\mu$. Eliminating ψ and R from these equations,

$$\frac{dT}{ds} + \frac{T}{\rho'}(\mu^2 - \tan^2\chi)^{\frac{1}{2}} = 0,$$
$$\therefore \log T = C - \int\frac{ds}{\rho'}(\mu^2 - \tan^2\chi)^{\frac{1}{2}}.$$

Thus, when the string is laid on the surface in a given form and is bordering on motion, the tension at any point can be found.

It also follows from the equations of Art. 486 that, if $\chi = 0$, then $\psi = 0$. If therefore the string is placed along a geodesic line on the surface, the friction must act along a tangent to the string. Putting $\psi = 0$, we have from the two first equations

$$\log T = C - \mu\int\frac{ds}{\rho'}.$$

Since along a geodesic $\rho' = \rho$, we may deduce from this equation the following extension of the theorem in Art. 463. *If a light string rest on a rough surface in a state bordering on motion, and the form of the string be a geodesic, then* (1) *the friction at any point acts along the tangent to the string, and* (2) *the ratio of the tensions at any two points is equal to e to the power of* $\pm \mu$ *times the sum of the infinitesimal angles turned through by a tangent which moves from one point to the other.*

The conditions of equilibrium of a string on a rough surface are given in Jellett's *Theory of Friction.* He deduces from these the equations obtained in Art. 486.

487. Ex. 1. A fine string of inconsiderable weight is wound round a right circular cylinder in the form of a helix, and is acted on by two forces F, F' at its extremities. Show that, when the string borders on motion, $\log \dfrac{F'}{F} = \pm \mu \dfrac{\cos^2 a}{a} s$, where s is the length of the string in contact with the cylinder, a the angle of the helix and a the radius of the cylinder.

Since the helix is a geodesic, this result follows from the equations of Art. 486 by writing for $1/\rho'$ its value $\cos^2 a/a$ given by Euler's theorem on curvature.

Ex. 2. A heavy string AB, initially without tension, rests on a rough horizontal plane in the form of a circular arc. Find the least force F which, applied along a tangent at one extremity B, will just move the string.

Let O be the centre of the arc, let the angle $AOP = \theta$, the arc $AP = s$. Let the element PQ of the string begin to move in some direction PP', where $P'PQ = \psi$; then by the nature of friction the angle ψ must be less than a right angle. The friction at P therefore acts in the opposite direction, viz. $P'P$, and is equal to $\mu w ds$. The equations of equilibrium are

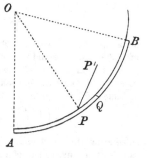

$$dT - \mu w ds \cos \psi = 0 \left.\right\}$$
$$T d\theta - \mu w ds \sin \psi = 0 \left.\right\} \quad \dots\dots\dots (1).$$

Substituting in the first equation the value of T given by the second, we have, since $ds = a d\theta$, $d\psi = d\theta$, and therefore $\psi = \theta + C \dots\dots\dots (2)$.
We have by substituting in (1) $T = \mu w a \sin (\theta + C)$.

If every element of the string border on motion, the equations (1) hold throughout the length. Since T must be zero when $\theta = 0$, we find that $C = 0$. Hence, if aa be the given length of the string AB, the force required to just move it is given by $F = \mu w a \sin a$. It is evident that this result does not hold if the length of the string exceed a quadrant, for then ψ at the elements near B would be greater than a right angle.

Supposing the arc AB to be greater than a quadrant, let the force F acting at B increase gradually from zero. When $F = \mu w a \sin a$, where $a < \frac{1}{2}\pi$, it follows from what precedes that a finite arc EB, terminating at B and subtending at O an angle EOB equal to a, is bordering on motion, and that the tension at E is zero. When $F = \mu w a$ the resolved part of the tension at B along the normal is $\mu w a d\theta$, and is just balanced by the friction. When F increases beyond the value $\mu w a$, the whole friction is insufficient to balance the normal force.

Summing up, the force required to move the string is $F = \mu a w \sin \alpha$ if the length is less than a quadrant. If the length exceed a quadrant, the force is $\mu a w$, and the string begins to move at the extremity at which the force is applied. See Art. 190.

Ex. 3. If a weightless string stretched by two weights lie in one plane across a rough sphere of radius a, show that the distance of the plane from the centre cannot exceed $a \sin \epsilon$, where ϵ is the angle of friction.   [St John's Coll., 1889.]

488. Virtual Work. The equations of equilibrium of a string may be deduced from the principle of virtual work by taking each element separately, and following the general method indicated in Art. 203. In fact the left-hand side of the x equation given in Art. 455, after multiplication by $ds \cdot dx$, is the virtual moment resulting from a displacement dx. This method requires that the tensions at the ends of the element should be included as part of the impressed forces. The principle may also be expressed as a max-min condition (Art. 212) in a form which includes only the given external forces. As an example of this let us consider the following problem.

A heterogeneous string of given length l, fixed at its extremities A, B, is in equilibrium in one plane in a field of force whose potential is V. It is required to find the form of the string.

Supposing $m = f(s)$ to be the line density at a point whose arc distance from A is s, the work function for the whole string is $\int V m \, ds$, the limits being 0 to l. We shall take the arc s as the independent variable and regard x, y as two functions of s connected by the equation

$$\left(\frac{dx}{ds}\right)^2 + \left(\frac{dy}{ds}\right)^2 = 1 \quad\ldots\ldots\ldots\ldots\ldots\ldots\ldots\ldots(1).$$

Following Lagrange's rule we remove the restriction (1) and make

$$u = \int \left\{ V m + \lambda \left(\left(\frac{dx}{ds}\right)^2 + \left(\frac{dy}{ds}\right)^2 - 1 \right) \right\} \, ds \quad\ldots\ldots\ldots\ldots (2),$$

a max-min for all variations of x and y, the quantity λ being an arbitrary function of s, afterwards chosen to make the resulting values of x, y satisfy the condition (1)*.

As the limits are fixed, there is no obvious advantage in varying all the coordinates. We shall therefore take the variation of u on the supposition that x, y are variable and s constant. We have

$$\delta u = \int \left\{ m \left(\frac{dV}{dx} \delta x + \frac{dV}{dy} \delta y \right) + 2\lambda \left(\frac{dx}{ds} \frac{d\delta x}{ds} + \frac{dy}{ds} \frac{d\delta y}{ds} \right) \right\} \, ds.$$

Integrating the third and fourth terms by parts and remembering that δx, δy vanish at the fixed ends of the string, we find

$$\delta u = \int \left\{ \left(m \frac{dV}{dx} - 2 \frac{d}{ds} \left(\lambda \frac{dx}{ds} \right) \right) \delta x + \left(m \frac{dV}{dy} - 2 \frac{d}{ds} \left(\lambda \frac{dy}{ds} \right) \right) \delta y \right\} \, ds.$$

At a max-min, this must be zero for all values of δx, δy, hence

$$m \frac{dV}{dx} - 2 \frac{d}{ds} \left(\lambda \frac{dx}{ds} \right) = 0, \qquad m \frac{dV}{dy} - 2 \frac{d}{ds} \left(\lambda \frac{dy}{ds} \right) = 0 \quad\ldots\ldots\ldots\ldots (3).$$

Restoring the condition (1) we have now three equations from which x, y, and λ

---

* We regard s as the abscissa, x, y as the two ordinates of an unknown curve, which is to be found by making u a max-min for *all* variations of x, y. The rules of the calculus of variations then enable us to write down the equations to find the curve. The equation of this curve contains λ and is made to satisfy (1) by a proper choice of this quantity. Then since (2) is a max-min for all variations of x, y, it follows that $\int V m \, dx$ is a max-min for those variations of x, y which satisfy the condition (1).

may be determined as functions of s. It is evident that these agree with the equations already found in Art. 455, with -2λ written for T.

We may also deduce the value of λ by multiplying the equations (3) respectively by dx/ds and dy/ds and adding. We then find

$$m\frac{dV}{ds} = \frac{1}{\lambda}\frac{d}{ds}\lambda^2\left\{\left(\frac{dx}{ds}\right)^2 + \left(\frac{dy}{ds}\right)^2\right\} = 2\frac{d\lambda}{ds},$$

which agrees with the equation to determine the tension in Art. 479.

If the string is in three dimensions and constrained to rest on a smooth surface, we make $\int Vm\,ds$ a max-min subject to the two conditions

$$x'^2 + y'^2 + z'^2 - 1 = 0, \qquad F(x, y, z) = 0 \quad \ldots\ldots\ldots\ldots\ldots \text{(I)},$$

where accents denote differentiations with regard to s. Following the same method as before we make

$$u = \int\{Vm + \lambda\,(x'^2 + y'^2 + z'^2 - 1) + \mu F(x, y, z)\}\,ds$$

a max-min. Varying only x, y, z and integrating by parts exactly as before, we find on equating the coefficients of δx, δy, δz to zero

$$m\frac{dV}{dx} - 2\frac{d}{ds}\left(\lambda\frac{dx}{ds}\right) + \mu\frac{dF}{dx} = 0, \qquad \&\text{c.} = 0, \qquad \&\text{c.} = 0 \ldots\ldots\ldots\text{(II)},$$

the two latter equations being obtained from the first by writing y and z respectively for x. These three equations joined to the conditions (I) determine $x, y, z, \lambda,$ μ in terms of s. These agree with the equations obtained in Art. 478, when -2λ and $-\mu\,(F_x^2 + F_y^2 + F_z^2)^{\frac{1}{2}}$ are written for T and R.

489. Elastic Strings. The theory of elastic strings depends on a theorem which is usually called *Hooke's law*. This may be briefly enunciated in the following manner. Let an extensible string uniform in the direction of its length have a natural length l_1. Let this string be stretched by the application of two forces at its extremities, and let these forces be each equal to T. Let the stretched length of the string be l. Then it is found by experiment that the extension $l - l_1$ bears to the force T a ratio which is constant for the same string.

If the natural or unstretched length of the string were doubled so as to be $2l_1$, the force T being the same as before, it is clear that each of the lengths l_1 would be stretched exactly as before to a length l. The extension of this string of double length will therefore be twice that of the single string. More generally, we infer that the extension must be proportional to the natural length when the stretching force is the same.

Joining these two results together, we see that

$$l - l_1 = l_1\frac{T}{E},$$

where E is some constant, which is independent of the natural length of the string and of the force by which it is stretched.

It is clear that, if two similar and equal strings are placed side by side, they will together require twice the force to produce the same extension that each string alone would require. It follows that the force required to produce a given extension is proportional to the area of the section of the unstretched string. The coefficient E is therefore proportional to the area of the section of the string when unstretched. The value of E when referred to a sectional area equal to the unit of area is called *Young's modulus*.

To find the meaning of the constant E, let us suppose that the string can be stretched to twice its natural length without violating Hooke's law. We then have $l = 2l_1$, and therefore $E = T$. Thus E is a force, it is the force which would theoretically stretch the string to twice its natural length.

490. This law governs the extension of other substances besides elastic strings. It applies also to the compression and elongation of elastic rods. It is the basis of the mathematical theory of elastic solids. But at present we are not concerned with its application except to strings, wires, and such like bodies.

The law is true only when the extension does not exceed certain limits, called the limits of elasticity. When the stretching is too great the body either breaks or receives such a permanent change of structure that it does not return to its original length when the stretching force is removed. In all that follows, we shall suppose this limit not to be passed.

The reader will find tables of the values of Young's modulus and the limits of elasticity for various substances given in the article *Elasticity*, written by Sir W. Thomson, now Lord Kelvin, for the *Encyclopædia Britannica*.

491. Ex. 1. A uniform rod AB, suspended by two equal vertical elastic strings, rests in a horizontal line; a fly alights on the rod at C, its middle point, and the rod is thereupon depressed a distance h; if the fly walk along the rod, then when he arrives at P, the depression of P below its original level is $2h\,(AP^2 + BP^2)/AB^2$, and the depression of Q, any other point of the rod, is $2h\,(AP \cdot AQ + BP \cdot BQ)/AB^2$

[St John's Coll., 1887.]

Ex. 2. A heavy lamina is supported by three slightly extensible threads, whose unstretched lengths are equal, tied to three points forming a triangle ABC. Show that when it assumes its position of equilibrium the plane of the lamina will meet what would be its position in case the threads were inelastic in the line whose areal equation is $xx_0/E + yy_0/F + zz_0/G = 0$, where E, F, G are the moduli, and x_0, y_0, z_0 the areal coordinates of the centre of gravity of the lamina referred to the triangle ABC. [St John's Coll., 1885.]

492. *A uniform heavy elastic string is suspended by one extremity and has a weight W attached to the other extremity. Find the position of equilibrium and the tension at any point.*

Let OA_1 be the unstretched string, P_1Q_1 any element of its length. Let OA be the stretched string, PQ the corresponding position of P_1Q_1. Let w be the weight of a unit of length of unstretched string, $l_1 = OA_1$, $x_1 = OP_1$; $l = OA$, $x = OP$. The tension T at P clearly supports the weight of PA and W. Hence

$$T = w\,(l_1 - x_1) + W \quad\ldots\ldots\ldots\ldots(1).$$

If PA were equally stretched throughout we could apply Hooke's law to the finite length PA. But as this is not the case we must apply the law to an elementary length PQ. We have therefore

$$dx - dx_1 = dx_1\,\epsilon T \quad\ldots\ldots\ldots\ldots\ldots\ldots(2),$$

where ϵ has been written for the reciprocal of E.

Eliminating T, $\qquad \dfrac{dx}{dx_1} = 1 + \epsilon\,\{w\,(l_1 - x_1) + W\}$.

Integrating, $\qquad x = x_1 + \epsilon\,\{w\,(l_1 x_1 - \tfrac{1}{2}x_1^2) + W x_1\} + C$.

The constant C introduced in the integration is clearly zero, since x_1 and x must vanish together. Putting $x_1 = l_1$, we find

$$l - l_1 = \tfrac{1}{2}\epsilon\,.\,w l_1^2 + \epsilon W l_1.$$

If the string had no weight, the extension due to W would be $\epsilon W l_1$. If there were no weight W at the lower end, the extension would be $\tfrac{1}{2}\epsilon w l_1^2$. *Hence the extension due to the weight of the string is equal to that due to half its weight attached to the lowest point.* We also see that the extension due to the weight of the string and the attached weight is *the sum of the extensions due to each of these treated separately.*

Ex. 1. A heavy elastic string OA placed on a rough inclined plane along the line of greatest slope is attached by one extremity O to a fixed point, and has a weight W fastened to the other extremity A. Find the greatest length of the stretched string consistent with equilibrium.

When the string is as much stretched as possible, the friction on every element acts down the plane and has its limiting value. Let α be the inclination of the plane to the horizon. Let μ, μ' be the coefficients of friction between the plane and the string and between the plane and the weight respectively. If $f = \sin\alpha + \mu\cos\alpha$, then fw replaces w in Art. 492. We therefore find for the whole elongation $l' - l = \tfrac{1}{2}\epsilon f w l_1^2 + \epsilon f' W l$, where f' is what f becomes when μ' is written for μ.

Ex. 2. A heavy elastic string AA' is placed on a rough inclined plane along the line of greatest slope. Supposing the inclination of the plane to be less than $\tan^{-1}\mu$, find the greatest length to which the string could be stretched consistent with equilibrium. Compare also the stretching of the different elements of the string.

The frictions near the lower end A of the string will act down the plane, while those near the upper end A' will act up the plane. There is some point O separating the string into two portions OA, OA' in which the frictions act in opposite directions. Each of these portions may be treated separately by the method used in the last example. An additional equation, necessary to find the unstretched length z of OA, is obtained by equating the tensions at O due to the two portions. The results are

$$z = \frac{l_1}{2}\left(1 - \frac{\tan\alpha}{\mu}\right), \qquad l - l_1 = \tfrac{1}{4}\epsilon\mu w \cos a\, l_1{}^2\left(1 - \frac{\tan^2\alpha}{\mu^2}\right).$$

Ex. 3. A series of elastic strings of unstretched lengths l_1, l_2, l_3... are fastened together in order, and suspended from a point, l_1 being the lowest. Show that the total extension is

$$\tfrac{1}{2}\left(\epsilon_1 w_1 l_1{}^2 + \epsilon_2 w_2 l_2{}^2 + \ldots\right) + w_1 l_1\left(\epsilon_2 l_2 + \epsilon_3 l_3 + \ldots\right) + w_2 l_2\left(\epsilon_3 l_3 + \ldots\right) + \&c.,$$

where w_1, w_2, &c. are the weights per unit of length of unstretched string, ϵ_1, ϵ_2, &c. the reciprocals of the moduli of elasticity. [Coll. Exam., 1888.]

493. Work of an elastic string.
If the length of a light elastic string be altered by the action of an external force, *the work done by the tension is the product of the compression of the string and the arithmetic mean of the initial and final tensions.*

In the standard case let the length be increased from a to a', then $a - a'$ is the shortening or compression of the string. As before, let l_1 be the unstretched or natural length.

By referring to Art. 197, we see that the work required is

$$-\int T dl = -\int E \frac{l - l_1}{l_1}\, dl = -E\frac{(a' - l_1)^2 - (a - l_1)^2}{2 l_1},$$

the limits of the integral being from $l = a$ to $l = a'$. This result may be put into the form $\tfrac{1}{2}(T_1 + T_2)(a - a')$, where T_1 and T_2 represent the values of T when a and a' are written for l. The rule follows immediately. *See the author's Rigid Dynamics* 1877.

This rule is of considerable use in dynamics where the length of the string may undergo many changes in the course of the motion. It is important to notice that the rule holds even if the string becomes slack in the interval, provided it is tight in the initial and final states. If the string is slack in either terminal state, we may still use the same rule provided we suppose the string to have its natural or unstretched length in that terminal state.

Ex. 1. Show that the depth below the point of suspension O of the centre of gravity of the elastic string considered in Art. 492 is $\tfrac{1}{2}l_1 + \epsilon l_1\left(\tfrac{1}{3}S + \tfrac{1}{2}W\right)$, where S is the weight of the string. Show also that the work done by gravity as the string and weight are moved from the unstretched position OA_1 to the stretched position OA, is $\epsilon l_1\left(\tfrac{1}{3}S^2 + SW + W^2\right)$ where $\epsilon = 1/E$.

Ex. 2. Let one end of an elastic string be fixed to the rim of a wheel sufficiently rough to prevent sliding, and let the other be attached to a mass resting on the ground, so that when the string (of length a) is just taut it shall be vertical. Show that the work which must be spent in turning the wheel so as just to lift the mass off the ground is $Mga + Ea \log E/(E + Mg)$, where E is the tension which would double the length of the string, neglecting the weight of the string. [Math. Tripos.]

Ex. 3. A disc of radius r is connected by n parallel equal elastic strings, of natural length l_1, to an equal fixed disc; the wrench necessary to maintain the discs at a distance x apart with the moveable one turned through an angle θ about the common axis, consists of a force X and a couple L given by

$$X = nEx\left(\frac{1}{l_1} - \frac{1}{\xi}\right), \qquad L = 2nEr^2 \sin\theta\left(\frac{1}{l_1} - \frac{1}{\xi}\right),$$

where $\xi^2 = x^2 + 4r^2 \sin^2 \frac{1}{2}\theta$.   [Coll. Exam., 1885.]

One disc being moved to a distance x from the other and turned round through an angle θ, we first show that the length of each string is changed from l_1 to ξ. Using the rule above, the work function is $W = n \cdot \frac{1}{2}T(\xi - l_1) = nE(\xi - l_1)^2/2l_1$.

By Art. 208 we have $Xdx + Ld\theta = \dfrac{dW}{dx}dx + \dfrac{dW}{d\theta}d\theta$.

Effecting the differentiations $X = dW/dx$, $L = dW/d\theta$, we obtain the results given.

494. Heavy elastic string on a smooth curve. Ex. 1. A heavy elastic string is stretched over a smooth curve in a vertical plane: show that the difference between the values of $T + T^2/2E$ at any two points of the string is equal to the weight of a portion of the string whose unstretched length is the vertical distance between the points. It follows from this theorem that any two points at which the tensions are equal are on the same level.

If ds_1 is the unstretched length of any element ds of the string, we have by Hooke's law $ds_1 = dsE/(T + E)$. If then w is the weight per unit of *unstretched length*, the weight of any element ds of the stretched string is equal to $w'ds$, where $w' = wE/(T + E)$. Let us now form the equations of equilibrium, using the same figure and reasoning as in Art. 459, where a similar problem is discussed for an inextensible string. We evidently arrive at the same equations (1) and (2) with w' written for w. Substituting for w' and integrating, we find that (1) leads to the result given above.

Ex. 2. A heavy elastic string is stretched on a smooth curve in a vertical plane: show that $T + \dfrac{T^2}{2E} = wy, \qquad R\rho - \dfrac{T^2}{2E} = wy'$,

where T is the tension at any point P, R the outward pressure of the curve on the string per unit of length of unstretched string, w the weight of a unit of length of unstretched string, and y, y' the altitudes of P and its anti-centre above a fixed horizontal line called *the statical directrix* of the string, Art. 460. Show also that no part of the string can be below the directrix, and that the free ends, if there are any, must lie on it.

Ex. 3. A heavy elastic string rests in equilibrium on a smooth cycloid with its cusps upwards. If one extremity is attached to a point on the curve while the free extremity is at the vertex, prove that the stretched length of any unstretched arc s_1 measured from the vertex is given by $\gamma s = \sinh \gamma s_1$, where $4aE\gamma^2 = w$, and a is the radius of the generating circle.

Ex. 4. An elastic string rests on a smooth curve whose plane is vertical with its ends hanging freely. Show that the natural length σ may be found from the equation $\left(\dfrac{d\sigma}{ds}\right)^2 = \dfrac{b}{2y+b}$, where y is the vertical height above the free extremities, and b the natural length of a portion of the string whose weight is the coefficient of elasticity. If the natural length of each vertical portion be l, and if $(l+b)^2 = 2ab$, and if the curve be a circle of radius a, prove that the natural length of the portion in contact with the curve is $2\sqrt{(ab)}\log(\sqrt{2}+1)$. [June Exam., 1877.]

Ex. 5. An elastic string, uniform when unstretched, lies at rest in a smooth circular tube under the action of an attracting force (μr) tending to a centre on the circumference of the tube diametrically opposite to the middle point of the string. If the string when in equilibrium just occupies a semicircle, prove that the greatest tension is $\{\lambda(\lambda+2\mu\rho a^2)\}^{\frac{1}{2}} - \lambda$, where λ is the modulus of elasticity, a the radius of the tube, ρ the mass of a unit of length of the unstretched string.

[Trinity Coll., 1878.]

Ex. 6. An infinite elastic string, whose weight per unit of length when unstretched is m, and which requires a tension ma to stretch any part of it to double its length (when on a smooth table), is placed on a rough table (coefficient μ) in a straight line perpendicular to its edge. The string just reaches the edge, which is smooth. A weight $\frac{1}{2}ma\mu$ is attached to the end and let hang over the edge. If the weight takes up its position of rest quietly, so that no part of the string re-contracts after having been once stretched, show that the distance of the weight below the edge of the table is $\frac{1}{8}a\mu(3\mu+4)$, and that beyond a distance $\frac{1}{2}a(\mu+2)$ from the edge of the table the string is unstretched. [Trinity Coll.]

495. Light elastic string on a rough curve. Ex. 1. An elastic string is stretched over a rough curve so that all the elements border on motion. If no external forces act on the string except the tensions F, F' at its extremities, then

$\dfrac{F'}{F} = e^{\pm\mu\psi}$, where ψ is the angle between the normals to the curve at its extremities.

This follows by the same reasoning as in Art. 463.

Ex. 2. An elastic string (modulus λ) is stretched round a rough circular arc so that every element of it is just on the point of slipping; if T, T' are the tensions at its extremities, the ratio of the stretched to the unstretched length is

$$\log\frac{T'}{T} : \log\frac{T'(T+\lambda)}{T(T'+\lambda)}.$$ [St John's Coll., 1884.]

Ex. 3. An endless cord, such as a cord of a window blind, is just long enough to pass over two very small fixed pulleys, the parts of the cord between the pulleys being parallel. The cord is twisted, the amount of twisting or torsion being different in the two parts, and the portions in contact with the pulleys being unable to untwist. If the pulleys be made to turn slowly through a complete revolution of the string, show that the quotient of the difference by the sum of the torsions is decreased in the ratio $e^4 : 1$. [Math. Tripos, 1853.]

Ex. 4. An elastic band, whose unstretched length $= 2a$, is placed round four rough pegs A, B, C, D, which constitute the angular points of a square of side $= a$. If it be taken hold of at a point P between A and B, and pulled in the direction AB, show that it will begin to slip round both A and B at the same time if $AP = a/(e^{\frac{1}{2}\mu\pi}+1)$. [May Exam.]

Ex. 5. An endless slightly extensible strap is stretched over two equal pulleys : prove that the maximum couple which the strap can exert on either pulley is $\dfrac{2a(c+\pi a)}{c\coth\frac{1}{2}\mu\pi + 2a/\mu}T$, where a is the radius of either pulley, c the distance of their centres, μ the coefficient of friction, and T the tension with which the strap is put on.   [Math. Tripos, 1879.]

Ex. 6. A rough circular cylinder (radius a) is placed with its axis horizontal, and a string, whose natural length is l, is fastened to a point Q on the highest generator of the cylinder and to an external point P at a distance l from Q, PQ being horizontal and perpendicular to the axis of the cylinder; the cylinder is then slowly turned upon its fixed axis in the direction away from P; show that the string will slip continually along the whole of the length in contact with the cylinder until S (the natural length of the part wound up) $= a/\mu$, when all slipping will cease, and that up to this stage the relation between S and θ (the angle turned through by the cylinder) is $le^{\mu\phi} = (l - a\phi)\, e^{\mu\theta} + a\phi$, where $S = a\phi$.   [Coll. Exam., 1880.]

496. Elastic string, any forces. *To form the equations of equilibrium of an elastic string under the action of any forces.*

Let ds_1 be the unstretched length of any element ds of the string. Then by Hooke's law $ds = ds_1 (T + E)/E$. The forces on the element, due to the attraction of other bodies, will be proportional to the unstretched length. Let then the resolved parts of these forces along the principal axes of the string be Fds_1, Gds_1, Hds_1, as in Art. 454. The equations of equilibrium (1), (2), and (3) of that article are obtained by equating to zero the resolved parts of the forces along the principal axes of the curve; these equations will therefore apply to the elastic string if we replace Fds, Gds, Hds, by Fds_1, Gds_1, Hds_1. The equations of equilibrium for the elastic string may therefore be derived from those for an inelastic string by treating the forces as

$$Fds\,\frac{E}{T+E}, \quad Gds\,\frac{E}{T+E}, \quad Hds\,\frac{E}{T+E},$$

i.e. reducing all the impressed forces in the ratio $E : T + E$.

497. Suppose, for example, that the string rests on any smooth surface. The resolution along the tangent to the string (as in Art. 479) gives

$$\left(1 + \frac{T}{E}\right)dT + Xdx + Ydy + Zdz = 0. \qquad \therefore\ T + \frac{T^2}{2E} + \int(Xdx + Ydy + Zdz) = C.$$

It follows that $T + T^2/2E +$ the work function of the forces is constant along the whole length of the string, Art. 479.

Ex. *When gravity is the only force acting,* show that the equations of equilibrium of an elastic string corresponding to (1), (2), (3) of Art. 479 may be written in the simple forms

$$T + \frac{T^2}{2E} = wz, \qquad R\rho' - \frac{T^2}{2E} = wz', \qquad \left(wz + \frac{T^2}{2E}\right)\tan\chi = w\rho'\sin\theta,$$

where T is the tension at any point P, R the outward pressure of the surface on the string per unit of unstretched length, χ the angle the radius of curvature of the string makes with the normal to the surface, z and z' the altitudes of P and the anti-centre S above a certain horizontal plane, θ the angle the vertical makes with the plane containing the normal to the surface and the tangent to the string, and w the weight of a unit of unstretched length. If PS be a length measured outwards along the normal to the surface equal to the radius of curvature of a normal section of the surface drawn through the tangent at P to the string, S is the anti-centre of P.

If the surface is one of revolution with its axis vertical, we replace the third equation by $Tr' \sin\psi = B$, where r' is the distance of P from the axis of the surface, ψ the angle the tangent to the string makes with the meridian and B is a constant. See Art. 481.

498. To take another example, suppose that *the elastic string is under the action of a central force.* Taking moments about the centre of force, and resolving along the tangent to the string, we find, after integration,

$$Tp = A, \qquad T + \frac{T'^2}{2E} + \int F dr = C.$$

These equations may be treated in a manner somewhat similar to that adopted for inelastic strings.

499. Ex. 1. An elastic string rests in equilibrium in the form of an arc of a circle under the influence of a centre of force at any unoccupied point of the circle. Show that the law of force is $F = \dfrac{\mu}{r^3}\left(1 + \dfrac{\mu}{2E}\dfrac{1}{r^2}\right)$.

Ex. 2. An elastic string, whose elements repel each other with a force proportional to the product of their masses into the square of their distance, rests in equilibrium on a smooth horizontal plane. If T be the tension at a point whose distance from one extremity is y, show that $\dfrac{d^4}{dy^4}(T+E)^2 + \dfrac{c^2}{T+E} = 0$, where c is a constant depending on the nature of the string. Explain also how the constants of integration are to be determined.

Ex. 3. An elastic string, whose elements repel each other with a force which varies as the distance, rests on a smooth horizontal plane. If $2l_1$ and $2l$ be the unstretched and stretched lengths of the string, show that $cl = \tan cl_1$, where $Ec^2 dx$ is the force due to the whole string on an element whose unstretched length is dx when placed at a unit of distance from the middle point of the string.

Ex. 4. A uniform elastic string lying on a rough horizontal plane is fixed to two points, and forms a curve every part of which is on the point of motion. Show that the tension is given by the equation $\left(1+\dfrac{t}{\lambda}\right)^2\left\{\left(\dfrac{dt}{d\psi}\right)^2 + t^2\right\} = \mu^2 w^2 \rho^2$, where w is the weight per unit of length of the unstretched string, μ the coefficient of friction and ρ the radius of curvature. [Math. Tripos, 1881.]

Ex. 5. An elastic string has its two ends fastened to points on the surface of a smooth circular cylinder of which the axis is vertical; show that in the position of equilibrium of the string on the surface the density of the string at any point varies as the tangent of the angle which the osculating plane at that point makes with a normal section of the cylinder through the direction of the string. [Math. T., 1886.]

500. *A heavy elastic string is suspended from two fixed points and is in equilibrium in a vertical plane. To find its equation.*

We may here use the same method as that employed in Art. 443 to determine the form of equilibrium of an inelastic string. Referring to the figure of that article, let the unstretched length of CP (i.e. the arc measured from the lowest point up to any point P) be s_1, and let the rest of the notation be the same as before. Consider the equilibrium of the finite portion CP;

$$T \cos \psi = T_0 \ldots\ldots(1) \qquad T \sin \psi = w s_1 \ldots\ldots(2),$$

$$\therefore \frac{dy}{dx} = \tan \psi = \frac{w s_1}{T_0} = \frac{s_1}{c} \ldots\ldots\ldots\ldots\ldots (3).$$

From these equations we may deduce expressions for x and y in terms of some subsidiary variable. Since $s_1 = c \tan \psi$ by (3), it will be convenient to choose either s_1 or ψ as this new variable.

Adding the squares of (1) and (2), we have

$$T^2 = w^2 (c^2 + s_1^2) \ldots\ldots\ldots\ldots\ldots\ldots (4).$$

Since $dx/ds = \cos \psi$ and $dy/ds = \sin \psi$, we have by (1) and (2)

$$x = \int \frac{T_0}{T} \, ds = \int \frac{wc}{T} \left(1 + \frac{T}{E} \right) ds_1 = \frac{wc}{E} s_1 + c \log \frac{s_1 + \sqrt{(c^2 + s_1^2)}}{c},$$

$$y = \int \frac{w s_1}{T} \, ds = w \int \frac{s_1}{T} \left(1 + \frac{T}{E} \right) ds_1 = \frac{w}{2E} (c^2 + s_1^2) + \sqrt{(c^2 + s_1^2)},$$

where the constants of integration have been chosen to make $x = 0$ and $y = c + c^2 w/2E$ at the lowest point of the elastic catenary. The axis of x is then the statical directrix, Art. 494, Ex. 2.

501. Ex. 1. Prove the following geometrical properties of the elastic catenary

(1) $wy = T + \dfrac{T^2}{2E}$,   (2) $\rho = \dfrac{c^2 + s_1^2}{c} \left\{ 1 + \dfrac{w}{E} \sqrt{(c^2 + s_1^2)} \right\}$,

(3) $s = s_1 + \dfrac{w}{2E} \left\{ s_1 \sqrt{(c^2 + s_1^2)} + c^2 \log \dfrac{s_1 + \sqrt{(c^2 + s_1^2)}}{c} \right\}$,

all of which reduce to known properties of the common catenary when E is made infinite.

Ex. 2. Let M, M' be two points taken on the ordinate PN so that MM' is bisected in N by the statical directrix and let each half be equal to $T^2/2Ew$. If M be above the directrix draw ML perpendicular to the tangent at P. Show that $T = w . PM$, $s_1 = PL$, $c = ML$, $w . MN = T^2/2E$ and that M' is the projection of the anti-centre on the ordinate.

Ex. 3. An elastic string, uniform when unstretched, is hung up by two points. Prove that the intrinsic equation of the catenary in which it will hang under gravity is

$$s = c \tan \psi + \frac{c^2}{2\lambda} \left\{ \tan \psi \sec \psi + \log \tan \left(\frac{\pi}{4} + \frac{\psi}{2} \right) \right\},$$

where c is the natural length of the string whose weight is equal to the tension at the lowest point, from which s is measured, and λ is the natural length of the string whose weight is equal to the modulus of elasticity.   [Coll. Exam., 1880.]

CHAPTER XI.

THE MACHINES.

502. It is usual to regard the complex machines as constructed of certain simple combinations of cords, rods and planes. These combinations are called the *mechanical powers*. Though given variously by different authors, they are generally said to be six in number, viz. the lever, the pulley, the wheel and axle, the inclined plane, the wedge and the screw*.

Mechanical advantage. In the simplest cases they are usually considered as acted on by two forces. One of these, viz. the force applied to work the machine, is usually called *the power*. The other, viz. the force to be overcome, or the weight to be raised, is called *the weight*. The ratio of the weight to the power is called the *mechanical advantage* of the machine.

503. As a first approximation, we suppose that the several parts of the machine are smooth, the cords used perfectly flexible, the solid parts of the machine rigid, and so on. In some of the machines these suppositions are nearly true, but in others they are far from correct. It is therefore necessary, as a second approximation, to modify these suppositions. We take such account as we can of the roughness of the surfaces in contact, the rigidity of the cords and the flexibility of the materials. After these corrections have been made, our result is still only an approximation to the truth, for the corrections cannot be accurately made. For example, in making allowance for friction we assume that the bodies in contact are equally rough throughout, and that the coefficient of friction is properly known. The results however thus obtained are much nearer the real state of things than our first approximation.

504. Efficiency. Suppose a machine to be constructed of a combination of levers, pulleys, &c., each acting on the next in order.

---

* In the descriptions of the machines given in this chapter, the author has derived much assistance from Capt. Kater's *Treatise on Mechanics* in Lardner's Cyclopædia, 1830, Pratt's *Mechanical Philosophy*, 1842, Willis' *Principles of Mechanism*, 1870, and other books.

Let a force P acting at one extremity of the combination produce a force at the other extremity such that it could be balanced by a force Q acting at the same point. Then, for this machine, P may be regarded as the power and Q as the weight.

Let the machine be made to work, so that its several parts receive small displacements consistent with their geometrical relations. Such a displacement is called an *actual displacement* of the machine. Taking this as a virtual displacement, the work of the force P is equal to that of the force Q together with the work of the resistances of the machine. These resistances are friction &c., in overcoming which some of the work done by the power is said to be wasted or lost. The work done by the force Q is called the *useful work* of the machine. *The efficiency of a machine is the ratio of the useful work to that done by the power when the machine receives any small actual displacement.* It appears that the efficiency of a machine would be unity if all its parts were perfectly smooth, the solid parts perfectly rigid, and so on. In all existing machines however the efficiency is necessarily less than unity.

505. Ex. In any machine for raising a weight show that, if the weight remains suspended by friction when the machine is left free, the efficiency is less than one half. If however a force P be required to raise the weight, and a force P' be required to prevent it from descending, show that the efficiency will be $(P+P')/2P$, supposing the machine to be itself accurately balanced. [St John's Coll., 1884.]

When the force P just raises a weight Q, the friction acts in opposition to the power P; on the contrary it assists P' in supporting Q. The frictions in the two cases are evidently the same in magnitude, being the extreme amounts which can be called into play. Let x, y be the virtual displacements of the points of application of P, Q when the machine is worked, and let the same small displacement be given in each case. Let U be the work of the frictions. Then $Px = Qy + U$, and $P'x = Qy - U$. The efficiency of the machine is measured by the ratio Qy/Px. Eliminating U, we easily obtain the result given. If any of the resistances, other than friction, have no superior limit, but continually increase with the increase of the power, it is easy to see by the same reasoning that the efficiency will be less than the value found above.

506. The lever. A lever is a rigid rod, straight or bent, moveable about a fixed axis. The fixed axis is usually called the *fulcrum*. The portions of the lever between the fulcrum and the points of application of the power and the weight are called the *arms of the lever*. The forces which act on the lever are usually supposed to act in a plane which is perpendicular to the fixed axis.

When the forces act in any directions at any points of the body, the problem is one in three dimensions, the solution of which is given in Art. 268. In what follows we shall also neglect the friction at the axis, as that case has already been considered in Art. 179.

507. *To find the conditions of equilibrium of two forces acting on a lever in a plane perpendicular to its axis.*

The axis of the lever is regarded in the first approximation as a straight line; let C be its intersection with the plane of the forces.

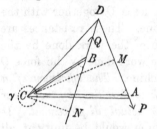
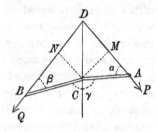

Let the forces be P and Q. Let them act at A and B on the arms CA, CB in the directions DA, DB. When the lever is in its position of equilibrium, the forces P, Q and the reaction at the fulcrum must form a system of forces in equilibrium. Hence the resultant of P and Q must act along DC, and be balanced by the pressure on the fulcrum.

The conditions of equilibrium follow at once from the principles stated in Art. 111. Let CM, CN be perpendiculars drawn from C on the lines of action of the forces. Taking moments about C, we have $P \cdot CM - Q \cdot CN = 0$. It follows that in a lever, *the power and the weight are to each other inversely as the perpendiculars drawn from the fulcrum on their lines of action.*

508. *To find the pressure on the fulcrum,* we find the resultant of the two forces P, Q by any one of the various methods usually employed to compound forces. For example, if the position of D be known, let ϕ be the angle ADB; we then have $R^2 = P^2 + Q^2 + 2PQ \cos \phi$, where R is the required pressure.

Let $CA = a$, $CB = b$, and let α, β be the angles the directions of the forces P, Q make with the arms CA, CB. Let γ be the angle ACB. If these quantities are known, we may find the pressure by another method. Let θ be the angle the line of action of R makes with the arm CA, so that the angle DCA is $\pi - \theta$. Then, resolving the forces along and perpendicular to CA, we have

$$R \cos \theta = P \cos \alpha + Q \cos (\gamma - \beta) \\ R \sin \theta = P \sin \alpha + Q \sin (\gamma - \beta) \Big\}'$$

whence $\tan \theta$ and R can be easily found.

Other relations between P, Q and R may be found by taking moments about A, B or some other point suggested by the data of the question. In the same way other resolutions will sometimes be more convenient than those given above as specimens.

509. When several forces act on the lever, we find *the condition of equilibrium by equating to zero the sum of their moments about the fulcrum*, each moment being taken with its proper sign. The moments are taken about the fulcrum to avoid introducing into the equation the reaction at the axis.

To find the pressure on the fulcrum we transfer each force parallel to itself, in the plane perpendicular to the axis, to act at the fulcrum. We thus obtain a system of forces acting at a single point, viz. the intersection of the axis with the plane of the forces. The resultant of these is the pressure on the axis.

510. In the investigation the weight of the lever itself has been supposed to be inconsiderable compared with the forces P and Q. If this cannot be neglected, let W be the weight of the lever. There are now three forces acting on the body instead of two. These are P, Q acting at A and B, and W acting at the centre of gravity G of the lever. Let the fulcrum be horizontal, and let CL be the perpendicular distance between the fulcrum and the vertical through G. Let us also suppose that in the standard figure the weight W and the force P tend to turn the lever round the fulcrum in the same direction. The equation of moments now becomes $P \cdot CM - Q \cdot CN + W \cdot CL = 0$. The pressure on the fulcrum is found by compounding the forces P, Q, W.

511. Levers are usually divided into three kinds according to the relative positions of the power, the weight, and the fulcrum. In the first kind, the fulcrum is between the power and the weight. In the second kind the weight acts between the fulcrum and the power, and in the third kind the power acts between the fulcrum and the weight. The investigation in Art. 507 applies to all three kinds, the only distinction being in the signs given to the forces and the arms, in resolving and taking moments.

512. The *mechanical advantage* of the lever is measured by the ratio $Q : P$. This ratio has been proved to be equal to $CN : CM$. By applying the power so that its perpendicular distance from the fulcrum is greater than that of the weight, a small power may be made to balance a large weight. Thus a crowbar when used to move a body is a lever of the second kind. The ground is the fulcrum, the weight acts near the fulcrum, and the power is applied at the extreme end of the bar.

513. If the lever be slightly displaced by turning it round its fulcrum through a small angle, the points of application A, B of the forces P, Q are moved through small arcs AA', BB', whose centres are on the fulcrum. Thus the actual displacements of the points of application of the power and the weight are proportional to their distances from the fulcrum. It is however the resolved part of the displacement AA' in the direction of the force P which measures the speed of working. For example, if the force P were applied by pulling a rope attached to the point A, the amount of rope to be pulled in would be measured by the resolved part of AA' in the direction of the length of the rope. The resolved parts of AA', BB' in the direction of the forces P, Q are evidently $AA' \cdot \sin \alpha$, $BB' \cdot \sin \beta$. These are proportional to $CA \sin \alpha$,

$CB \sin \beta$, i.e. to CM, CN. (See fig. of Art. 516.) These resolved displacements are clearly the same as the virtual displacements of the points of application; Art. 64.

If then mechanical advantage is gained by arranging the lever so that the weight is greater than the power, the displacement of the weight is less, in the same ratio, than that of the power, each displacement being resolved in the direction of its own force. It follows that *what is gained in power is lost in speed.*

514. The reader may easily call to mind numerous instances in which levers are used. As examples of levers of the first kind we may mention the common balance, pokers, &c.

Wheelbarrows, nutcrackers, &c. are examples of levers of the second kind. In these the weight is greater than the power. They are used when we wish to multiply the force at our disposal.

In levers of the third kind the weight is less than the power, but the virtual displacement of the weight is greater than that of the power. Such levers therefore are used when economy of force is a consideration subordinate to the speed of working.

515. The most striking example of levers of the third kind is found in the animal economy. The limbs of animals are generally levers of this description. The socket of the bone is the fulcrum; a strong muscle attached to the bone near the socket is the power; and the weight of the limb, together with whatever resistance is opposed to its motion, is the weight. A slight contraction of the muscle in this case gives a considerable motion to the limb: this effect is particularly conspicuous in the motion of the arms and legs in the human body; a very inconsiderable contraction of the muscles at the shoulders and hips giving the sweep to the limbs from which the body derives so much activity.

The treddle of the turning lathe is a lever of the third kind. The hinge which attaches it to the floor is the fulcrum, the foot applied to it near the hinge is the power, and the crank upon the axis of the fly-wheel, with which its extremity is connected, is the weight.

Tongs are levers of this kind, as also the shears used in shearing sheep. In these cases the power is the hand placed immediately below the fulcrum or point where the two levers are connected. *Capt. Kater's Mechanics.*

516. The principle of virtual work may be conveniently used to investigate the conditions of equilibrium in the lever. Let P, Q be two forces acting at A and B, and let C be the fulcrum. If the lever be displaced round C through a small angle $\delta\theta$, so 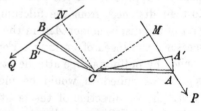 that A, B come into the positions A', B', we have

$$P . AA' \sin \alpha - Q . BB' \sin \beta = 0,$$

where α, β have the same meanings as in Art. 507. This immediately leads to the result $P . CM = Q . CN$.

517. Roberval's Balance. This machine supplies an excellent example of the principle of virtual work. In this balance the four rods AA', $A'B'$, $B'B$, BA are hinged at their extremities and form a parallelogram. The sides AB, $A'B'$ are also hinged at the points C, C' to a fixed

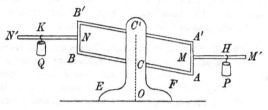

vertical rod OCC'. The line CC' must be parallel to AA' and BB', but need not necessarily be equidistant from them. Two more rods MM', NN' are rigidly attached to AA', BB' so as to be at right angles to them. These support the weights P and Q suspended in scale-pans from any two points H and K. As the combination turns smoothly round the supports C, C', the rods AA', BB' remain always vertical, and MM', NN' are always horizontal.

The peculiarity of the machine is that, if the weights P, Q balance in any one position, the equilibrium is not disturbed by moving either of the weights along the supporting rods MM', NN'. It may also be remarked that, if the machine be turned round its two supports C, C' so that one of the rods MM', NN' descends and the other ascends, the two weights continue to balance each other.

To show this, let the equal lengths CA, $C'A'$ be denoted by a, and the equal lengths CB, $C'B'$ by b. Let the inclination to the horizon of the parallel rods AB, $A'B'$ be θ. If the machine is displaced so that the angle θ is increased by $d\theta$, the rod AA' descends a vertical space $a \cos \theta d\theta$, and the rod BB' ascends a space $b \cos \theta d\theta$. When the weights of all the parts of the machine are neglected in comparison with P and Q, we have by the principle of virtual work $Pa \cos \theta d\theta = Qb \cos \theta d\theta$. This gives $Pa = Qb$; thus the condition of equilibrium is independent of the positions H, K at which P and Q act on the supporting rods, and is also independent of the inclination θ of the rods AB, $A'B'$ to the horizon.

If the balance is so constructed that the weights P, Q are equal, when in equilibrium, we can detect whether any difference in weight exists between two given bodies by simply attaching them to any points of the supporting rods. The advantage of the balance is that no special care is necessary to place them at equal distances from the fulcrum.

Ex. 1. If the weights of the rods AB, $A'B'$ are w, w' and the weights of the bodies $AA'M'$, $BB'N'$ are W, W', prove that the condition of equilibrium is

$$(P+W) a - (Q+W') b + \tfrac{1}{2} (w + w') (a - b) = 0.$$

Thence show that, if the weights P, Q balance in one position, they will as before balance in all positions. Find also the point of application of the resultant pressure of the stand EF on the supporting table.

Ex. 2. If the balance be at rest and horizontal, prove that the horizontal pressure on either support bears to either weight the ratio of the difference of the horizontal distances of the centres of gravity of the weights from the central plane of the balance to the distance between the supports.　　　[Math. Tripos, 1874.]

Let X, Y; X', Y', be the horizontal and vertical components of the reactions at

A, A'. By taking moments about A' for the system $AM'A'$ we have $Xa=Ph$, where $AA'=a$, $MH=h$. We have also $X+X'=0$, $Y+Y'=P$. Thus X, X' are known while the separate values of Y and Y' are indeterminate, Arts. 268, 148. Similarly if X_1, Y_1; X_1', Y_1', are the corresponding components at the points B, B', we have $X_1a=Pk$ where $NK=k$. Since the rod AB is acted on by X, Y; X_1, Y_1 (reversed) at the extremities, the horizontal component of pressure at the pin C is $X-X_1$, which at once leads to the given result.

518. The Common Balance. In the common balance two equal scale-pans E, F are suspended by equal fine strings from the extremities A, B of a straight rod or beam. The rod AB can turn freely about a fulcrum O, with which it is connected by a short rod OC which bisects AB at right angles. The centre of gravity G of the beam AOB lies in the rod OC, and therefore, when the beam and the empty scales are in equilibrium, the straight line AB is horizontal.

The bodies to be weighed are placed in the scale-pans, and if their weights are unequal, the horizontality of the beam AB is disturbed. The centre of gravity G of the beam is now no longer under the point of support, and in the new position of equilibrium the inclination θ of the rod AB to the horizon is such that the moment of the weight of the beam about the fulcrum O is 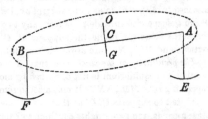 equal to that of the weight of the bodies and the scale-pans. It is therefore evident that the fulcrum should not coincide with the centre of gravity of the beam.

Let P, Q be the weights in the scales E and F, w the weight of either scale, let W be the weight of the beam AOB. Let $OG=h$, $OC=c$, $AB=2a$. Let θ be the inclination of AB to the horizon when the system is in equilibrium. Taking moments about O, we have

$$(P+w)(a\cos\theta+c\sin\theta)-(Q+w)(a\cos\theta-c\sin\theta)+Wh\sin\theta=0.$$

The coefficient of $P+w$ in this equation is the length of the perpendicular from O on the vertical AE, and is easily found by projecting the broken line OC, CA on the horizontal. The other coefficients are found in the same way. We therefore have $\tan\theta=\dfrac{(Q-P)a}{(P+Q+2w)c+Wh}$.

For a minute account of a balance with illustrative diagrams the reader is referred to the tract, "The theory and use of a physical balance," by J. Walker, 1887.

519. *A good balance has three requisites.* The first is that when loaded with equal weights in the pans the rod AB should be horizontal. This is secured by making the arms AC, CB equal. To determine when the beam is horizontal, a small rod called *the tongue* is attached to it at right angles at its middle point. The beam is usually suspended from a point above O, and when the beam is horizontal the direction of the tongue should pass through the point of suspension.

The second requisite is *sensibility*. When the weights P, Q differ by a small quantity, the angle θ should be so large that it can be easily observed. For a given difference $Q-P$ the sensibility increases as $\tan\theta$ increases. We may therefore measure the sensibility by the ratio $\dfrac{\tan\theta}{Q-P}=\dfrac{a}{(P+Q+2w)c+Wh}$. The sensibility is therefore secured by so constructing the balance that the expression on the right-hand side of this equation is as large as possible.

The sensibility is therefore increased (1) by increasing the length of the rod AB, (2) by diminishing the length of the rod OC, (3) by diminishing the weight of the beam. If the balance is so constructed that h and c have opposite signs, the sensibility can be greatly increased. This requires that the fulcrum O should lie between G and C.

The third requisite of a balance is usually called *stability*. When the balance is disturbed, it should return readily to its horizontal position. The beam oscillates about its position of equilibrium, and the quicker the oscillation the sooner can it be determined by the eye whether the mean position of the beam is or is not horizontal. The balance should be so constructed that the times of oscillation are as short as possible. The discovery of the nature of the oscillations is a problem in dynamics, and cannot properly be discussed from a statical point of view.

520. Ex. 1. If one arm of a common balance, whose weight can be neglected, is longer than the other, prove that the true weight of a body is the geometrical mean of the apparent weights when weighed first in one scale and then in the other. [Coll. Exam.]

Ex. 2. A balance has its arms unequal in length and weight. A certain article appears to weigh Q_1 or Q_2 according as it is put in the one scale or the other. Similarly another article appears to weigh R_1 or R_2. Find the true weights of these articles; and show that if an article appears to weigh the same in whichever scale it is put, its weight is $\dfrac{Q_1 R_2 - Q_2 R_1}{Q_1 - Q_2 - R_1 + R_2}$.

[Coll. Exam., 1886.]

Ex. 3. In a false balance a weight P appears to weigh Q, and a weight P' to weigh Q': prove that the real weight X of what appears to weigh Y is given by $X(Q - Q') = Y(P - P') + P'Q - PQ'$. [Math. Tripos, 1870.]

Ex. 4. A true balance is in equilibrium with unequal weights P, Q in its scales. If a small weight be added to P, the consequent vertical displacement of Q is equal to that which would be the vertical displacement of P were the same small weight to be added to Q instead of to P. [Math. Tripos, 1878.]

Looking at the expression for $\tan\theta$ in Art. 518, we notice that the changes produced in θ by altering either P or Q by the same small quantity are equal with opposite signs. The effect of increasing P or Q is therefore to turn the balance the one way or the other through the same small angle. The vertical displacements of the weights are therefore equal in the two cases.

Ex. 5. If the tongue of the balance be very slightly out of adjustment, prove that the true weight of a body is nearly the arithmetic mean of its apparent weights, when weighed in the opposite scales. [Coll. Exam.]

Ex. 6. A delicate balance, whose beam was originally suspended by a knife-edged portion of itself (higher than its centre of gravity) resting upon a horizontal agate plate, has its knife-edge worn down a distance ϵ so that it becomes curved (curvature $=1/r$), and has a corresponding hollow made in the agate plate (curvature $=1/\rho$). If slightly different weights P and Q be placed in the scales (whose weights may be neglected), show that the reciprocal of the sensibility is increased by $(P + Q + W)\left(\epsilon + \dfrac{r\rho}{\rho - r}\right)\dfrac{1}{a}$. [Coll. Exam., 1890.]

521. The Steelyards. The common steelyard is a lever ACB with unequal arms AC, CB, the fulcrum being situated at a point a little above C. The body Q to be weighed is suspended from the extremity B of the shorter arm, and a given weight P is moved along the

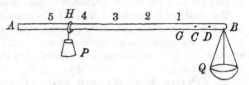

longer arm CA to some point H such that the system balances. Let G be the centre of gravity of the beam, w its weight. The three weights, P acting at H, w at G, and Q at B are in equilibrium. Taking moments about C, we have

$$P \cdot HC + w \cdot GC = Q \cdot CB \dots\dots\dots\dots\dots\dots\dots\dots(1).$$

Let D be a point on the shorter arm CB, such that $w \cdot GC = P \cdot CD$; the equation (1) then becomes $\qquad P \cdot HD = Q \cdot CB \dots\dots\dots\dots\dots\dots(2).$

Thus the weight of Q is determined by measuring the distance HD. To effect this easily, we measure from D towards A a series of lengths DE_1, E_1E_2, E_2E_3, &c. each equal to CB. The weight of the body Q is therefore equal to P, $2P$, $3P$, &c. according as the weight P is placed at the points E_1, E_2, E_3, &c. when the system is in equilibrium. The intervals E_1E_2, E_2E_3, &c. are usually graduated into smaller divisions, so that the length HD can be easily read. The points E_1, E_2, &c. are marked 1, 2, &c. in the figure.

An instrument of this form was used by the Romans and is therefore often called the Roman steelyard.

522. In the Danish steelyard the weights P and Q act at fixed points of the lever, but the fulcrum or point of support C is made to slide along the rod AB until the system balances. The weight P, being fixed, can be conveniently joined to that of the lever. Let,

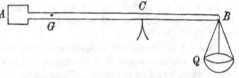

then, P' be the weight of the instrument, so that $P' = P + w$, and let G be the centre of gravity. Taking moments about C, we evidently have $P' \cdot GC = Q \cdot CB$, and

$\therefore BC = \dfrac{P' \cdot BG}{P' + Q}$. This expression enables us to calculate the values of BC when $Q = P'$, $2P'$, $3P'$, &c. Marking these points of the rod AB with the figures 1, 2, 3, &c., the weight of any body placed at B can be read off when the place of the fulcrum C has been found by trial.

If C, C' be two successive marks of graduation when the weights suspended at B are Q and $Q + S$, we easily find that $\dfrac{1}{BC'} - \dfrac{1}{BC} = \dfrac{S}{P' \cdot BG}$; since the right-hand side is constant when S is given, we infer that the marks of graduation on the bar are such that their distances from B form a harmonical progression when the weights form an arithmetical progression. Thus *in the common steelyard the distances of the graduations from a certain point are in arithmetical progression, and in the Danish steelyard in harmonical progression.*

523. *The advantages of a steelyard over the balance* are, (1) the exact adjustment of the instrument is made by moving a single weight P along the rod, (2) when the body to be weighed is heavier than the fixed weight the pressure on the point of support is less than in the balance. The steelyard is therefore better adapted to measure large weights. There is on the other hand this advantage in the balance, that by using numerous small weights the reading can be effected with greater precision than by subdividing the arm of the steelyard.

524. Ex. 1. The weight of a common steelyard is w, and the distance of its fulcrum from the point from which the weight hangs is a when the instrument is in perfect adjustment; the fulcrum is displaced to a distance $a + \alpha$ from this end; show that the correction to be applied to give the true weight of a body which in the imperfect instrument appears to weigh W is $(W + P + w)\,a/(a + \alpha)$, P being the moveable weight. [Math. Tripos, 1881.]

Ex. 2. In a weighing machine constructed on the principle of the common steelyard the pounds are read off by graduations reaching from 0 to 14, and the stones by weights hung at the end of the arm; if the weight corresponding to one stone be 7 oz., the moveable weight $\frac{1}{2}$ lb., and the length of the arm one foot, prove that the distances between the graduations are $\frac{3}{4}$ in. [Math. Tripos.]

Ex. 3. In graduating a steelyard to weigh pounds, marks are made with a file, a weight x being removed for each notch. With the moveable weight P at the end of the beam, n lbs. can be weighed after the graduation is completed, $(n+1)$ before it is begun. Show that $n(n+1)x = 2P$, and find the error made in weighing m pounds. The centre of gravity of the steelyard is originally under the point of suspension. [Coll. Exam., 1885.]

Ex. 4. Show that, if a steelyard be constructed with a given rod whose weight is inconsiderable compared with that of the sliding weight, the sensibility varies inversely as the sum of the sliding weight and the greatest weight which can be weighed. [Math. Tripos, 1854.]

Ex. 5. A common steelyard is graduated on the assumptions that its weight is Q, and that the moveable weight is W, both which assumptions are incorrect. If two masses whose real weights are P and R appear to weigh $P + X$ and $R + Y$, then the weight of the steelyard and the moveable weight are less than their assumed values by $\dfrac{W}{D}(X - Y)$ and $\dfrac{Q}{D}(X - Y) + \dfrac{a}{bD}(PY - RX)$, where b, a are the distances from the fulcrum to the centre of gravity of the bar and to the point of attachment of the substance to be weighed, and $D = P - R + X - Y$. [Math. Tripos, 1887.]

Ex. 6. The sum of the weight of a certain Roman steelyard and of its moveable weight is S, the fulcrum is at the point C and the body to be weighed is hung at the end B. The steelyard is graduated and after graduation the fulcrum is shifted towards B to another point C'. A body is then weighed, the old graduation being used, and the apparent weight is W. Prove that the true weight is greater than the apparent weight by $(S + W)\,CC'/BC'$. [Trin. Coll., 1889.]

Ex. 7. If, on a common steelyard, the moveable weight P, which forms the power, be increased in the ratio $1 + k : 1$, prove that the consequent error in Q, the weight to be found, is kY, where Y is the weight which must be removed from Q in order to preserve equilibrium when P is moved close to the fulcrum. [Coll. Exam., 1885.]

Ex. 8. In the Danish steelyard, if a_n be the distance of the fulcrum from that end of the steelyard at which the weight is suspended, the weight being n lbs., prove that $\dfrac{1}{a_{n+2}} - \dfrac{2}{a_{n+1}} + \dfrac{1}{a_n} = 0$. [Math. Tripos, 1859.]

Ex. 9. An old Danish steelyard, originally of weight W lbs., and accurately graduated, is found coated with rust. In consequence of the rust, the apparent weights of two known weights of X lbs. and Y lbs. are found when weighed by the steelyard to be $(X-x)$ lbs., $(Y-y)$ lbs. respectively. Prove that the centre of gravity of the rust divides the graduated arm in the ratio $W(x-y) : Yx - Xy$; and that its weight is, to a first approximation, $\dfrac{W+Y}{X-Y}x + \dfrac{W+X}{Y-X}y$. [Math. Tripos, 1885.]

Ex. 10. A brass figure $ABDC$, of uniform thickness, bounded by a circular arc BDC (greater than a semicircle) and two tangents AB, AC inclined at an angle $2a$, is used as a letter-weigher as follows. The centre of the circle, O, is a fixed point about which the machine can turn freely, and a weight P is attached to the point A, the weight of the machine itself being w. The letter to be weighed is suspended from a clasp (whose weight may be neglected) at D on the rim of the circle, OD being perpendicular to OA. The circle is graduated, and is read by a pointer which hangs vertically from O: when there is no letter attached, the point A is vertically below O and the pointer indicates zero. Obtain a formula for the graduation of the circle, and show that, if $P = \frac{1}{2}w\sin^2 a$, the reading of the machine will be $\frac{1}{4}w$ when OA makes with the vertical an angle equal to $\tan^{-1}\left\{\dfrac{(\pi+2a)\sin^2 a + 2\sin a\cos a}{(\pi+2a)\sin^3 a + 2\cos a}\right\}$.

[Math. Tripos, 1878.]

525. The Pulley. The common pulley consists of a wheel which can turn freely on its axis. A rope or cord runs in a groove formed on the edge of the wheel, and is acted on by two forces P and P' one at each end. If the pulley is smooth and the weight of the string infinitesimal, the tension is necessarily the same throughout the arc of contact. It follows that the forces P, P' acting at the extremities of the string are equal to each other and to the tension. See fig. 1 of Art. 527. The same thing is true if the pulley is rough and circular, but can turn freely about a smooth axis; Art. 197.

526. When the axis of the pulley is fixed one of the forces P, Q is the power and the other is the weight. Thus a fixed pulley has no mechanical advantage in the technical sense. A machine, however, which enables us to give the most advantageous direction to the moving power is as useful as one which enables a small power to support a large weight.

527. A moveable pulley can however be used to obtain mechanical advantage. Suppose a perfectly flexible string to be fixed at A, pass under a pulley C of weight Q, and to be acted on at B by a force P; see fig. 2. In the position of equilibrium the strings on each side of the pulley meet in the line of action of the force Q (Art. 34), and must therefore make equal angles with the vertical (Art. 27).   Let α be the inclination of either string to the vertical, then $2P \cos \alpha = Q.$

Fig. 1.   Fig. 2.

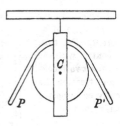 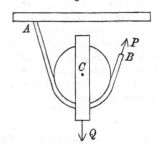

The mechanical advantage is therefore $2 \cos \alpha$.   Unless α is less than 60° the mechanical advantage is less than unity.   When the strings are parallel, we have $2P = Q$.

528. Ex. 1. In the single moveable pulley with parallel strings a weight W is supported by another weight P attached to the free end of the string and hanging over a fixed pulley. Show that, in whatever position the weights hang, the position of their centre of gravity is the same.   [Math. Tripos, 1854.]

Ex. 2. A string is attached to the centre of a heavy circular pulley of radius r and is then passed over a fixed peg, then under the pulley, and afterwards passes over a second fixed peg vertically over the point where the string leaves the pulley and has a weight W attached to its extremity. The second peg is in the same horizontal line as the first peg and at a distance $\frac{8}{5}r$ from it. If there is equilibrium, prove that the weight of the pulley is $\frac{8}{5}W$, and find the distance between the first peg and the centre of the pulley.   [Coll. Exam., 1886.]

Ex. 3. An endless string without weight hangs at rest over two pegs in the same horizontal plane, with a heavy pulley in each festoon of the string; if the weight of one pulley be double that of the other, show that the angle between the portions of the upper festoon must be greater than 120°.   [Math. Tripos, 1857.]

529. Systems of pulleys may be divided into two classes, (1) those in which a single rope is used; and (2) those in which there are several distinct ropes. We begin with the first of these systems.

Two blocks are placed opposite each other, containing the same number of pulleys in each.   Three are represented in each block in the figure.   The string passes over the pulleys in the order $ADBECF$, and has one extremity attached to one of the blocks.   The power P acts at the other extremity of the string, while the weight Q acts on a block.

Let n be the number of pulleys in either block, W the

weight of the lower block; we then have $Q + W$ supported by $2n$ tensions. Since the tension of the string is the same throughout, and equal to P, we have by resolving vertically $2nP = Q + W$.

If the pulleys were all of the same size, and exactly under each other, some difficulty might arise in their arrangement so that the cords should not interfere with each other. For this, and other reasons, the parts of the string not in contact with the pulleys cannot be strictly parallel. Except when the two blocks are very close to each other the error arising from treating the strings as parallel is very slight, and may evidently be neglected when we take no account of the other imperfections of the machine; Art. 503.

We may also deduce the relation between the power and the weight from the principle of virtual work. If the lower block, together with the weight Q, receive a virtual displacement upwards equal to q, it is clear that each string is slackened by the same space q. To tighten the string, P must descend a space q for each separate portion of string, i.e. P must descend a space $2nq$. We have therefore by the principle of work

$$P \cdot 2nq = (Q + W)\, q.$$

The result follows immediately.

530. In some arrangements of this system the pulleys on each block have a common axis, but each pulley turns on the axis independently of the others. This change however does not affect the truth of the relation just established between the power and the weight.

When the system works, it is clear that all the pulleys, if of equal size, do not move with equal angular velocities. To give greater steadiness to the several parts of the machine, it has been suggested that the pulleys in each block should not only have a common axis, but be of such radii that each turns with the same angular velocity. When this has been effected, the pulleys in each block may be welded into one and the string made to run in grooves cut out of the same wheel.

To understand how this may be done, we notice that if the lower block rises one foot, each string would be slackened one foot. To tighten the string between C and F on the right hand the pulley F must be turned round so that one foot of rope may pass over it. The string on the left hand between C and F is now slackened by two feet, hence the pulley C must be turned round so that two feet of rope may pass over it. In the same way the pulley E must be turned round so that three feet of rope may pass over it, and so on. If then the wheels in the upper block are constructed so that their radii are in the proportion $2:4:6:\&c.$, and those in the lower block so that the radii are in the proportion $1:3:5:\&c.$, the wheels in each block will turn with the same angular velocity.

When very accurately constructed this arrangement works well. It is found

however that a very slight deviation from the true proportion of the radii will
cause the rope to be unequally stretched, even the thickness of the rope must be
allowed for. Some parts of the rope are therefore unduly tight, and others become
nearly slack. This mode of arranging the pulleys is due to White. It is not
now much used.

531. Ex. In that system of pulleys in which the same cord passes round all
the pulleys it is found that on account of the rigidity of the cord and the friction
of the axle a weight of P lbs. requires $aP+p$ lbs. to lift it by a cord passing over
one pulley. Prove that when there are n parallel cords in the above system a
power P can support a weight $Q = a\dfrac{a^n-1}{a-1}P + \dfrac{a(a^n-1)-n(a-1)}{(a-1)^2}p$, and find the
additional weight required to be added to P to raise Q.   [Math. Tripos, 1884.]

The rigidity of cordage was made the subject of many experiments by Coulomb,
Art. 170. The discussion of these would require too much space, but the general
result may be shortly stated. Suppose a cord $ABCD$ to pass over a pulley of
radius r, touching it at B and C, and moving in the direction $ABCD$. Then
the rigidity of the portion AB of the cord which is about to be rolled on the
pulley may be allowed for, by regarding the cord as perfectly flexible and applying
a retarding couple to the pulley whose moment is $a+bT$, where a and b are constants
which depend on the nature and size of the cord, but are sensibly independent
of the velocity. If T' be the tension of the portion CD of the cord which is
being unwound from the pulley, its rigidity may be represented in the same way by
the application of a couple equal to $a'+b'T'$. The values of a', b' are so much less
than those of a, b, that this last correction is generally omitted. Taking moments
about the centre this gives $T' - T = \dfrac{a+bT}{r}$, where r is the radius.

532. When several cords are used pulleys may be combined in
various ways to produce mechanical advantage. Two systems are
usually described in elementary books, both of which are represented in the figure.

In fig. (1) each pulley is supported by a separate string, one end

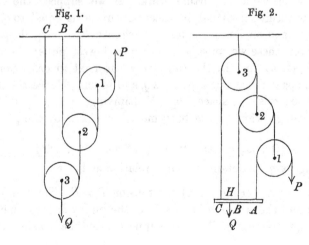

Fig. 1.   Fig. 2.

of which is attached to a fixed point of support, and the other to the pulley next in order. In fig. (2) the string resting on each pulley has one end attached to the weight and the other to the pulley next in order. The two systems resemble each other in the arrangement of the pulleys, but to a certain extent each is the inversion of the other.

Let w_1, w_2, &c. be the weights of the pulleys M_1, M_2, &c., T_1, T_2, &c. the tensions of the strings which pass over them. In the figures only the suffixes of M_1, M_2, &c. are marked on the pulleys to save space.

Considering fig. (1), the tension $T_1 = P$. The tensions of the parts of the string on each side of the pulley M_1 support the weight of that pulley and the tension T_2, we have therefore

$$T_2 = 2T_1 - w_1 = 2P - w_1.$$

Considering the pulleys M_2, M_3, we have in the same way

$$T_3 = 2T_2 - w_2 = 2^2P - 2w_1 - w_2,$$
$$T_4 = 2T_3 - w_3 = 2^3P - 2^2w_1 - 2w_2 - w_3,$$

and so on through all the pulleys. It is evident that the right-hand side of each equation is twice that of the one above with a w subtracted. We therefore have finally

$$Q = 2T_n - w_n = 2^nP - 2^{n-1}w_1 - 2^{n-2}w_2 - \&c. - 2w_{n-1} - w_n.$$

If all the pulleys are of equal weight this gives

$$Q = 2^nP - (2^n - 1)w.$$

The relation between the power and the weight follows easily from the principle of virtual work. If we suppose the lowest pulley to receive a virtual displacement upwards equal to q, each of the strings on its two sides is slackened by an equal space q. To tighten these we must raise the next lowest pulley through a space equal to $2q$. In the same way, the next in order must be raised a space twice this last, i.e. 2^2q, and so on. Hence the power P must be raised a space 2^nq. Multiplying each weight by the space through which it has been moved, we have, by the principle of work

$$(Q + w_n)q + w_{n-1}2q + w_{n-2}2^2q + \ldots = P \cdot 2^nq.$$

Dividing by q we obtain the same relation as before.

533. Considering fig. (2), the tension $T_1 = P$. The tensions of the parts of the string on each side of the pulley M_1, together with the weight of that pulley, are supported by the tension T_2, we therefore have $T_2 = 2T_1 + w_1 = 2P + w_1$. Taking the other pulleys in order, we see that we have the same results as before except that the w's have opposite signs. We thus have

$$T_3 = 2T_2 + w_2 = 2^2P + 2w_1 + w_2,$$
$$T_4 = 2T_3 + w_3 = 2^3P + 2^2w_1 + 2w_2 + w_3,$$

and so on. Since the pulleys are all attached to the weight we have $T_1 + T_2 + \ldots + T_n = Q + W$, where W is the weight of the bar.

Substituting the values of T_1, T_2, &c. in this last equation, we find $Q + W = (2^n - 1)P + (2^{n-1} - 1)w_1 + (2^{n-2} - 1)w_2 + \ldots + w_{n-1}$.

If all the pulleys are of equal weight this reduces to

$$Q + W = (2^n - 1)(P + w) - nw.$$

When the pulleys are arranged as in fig. (1), the mechanical advantage is decreased by increasing the weights of the pulleys. In fig. (2) the reverse is the case, for the weights of the pulleys assist the power in sustaining the weight.

To deduce the relation between the power and the weight from the principle of virtual work, let us first imagine the bar to be held at rest and the highest pulley to be moved downwards through a space q. Each of the strings on the two sides of that pulley is equally slackened by the space q. To tighten the string, the second highest pulley must be moved downwards through a space $2q$, and so on. The power must descend a space $2^n q$. To restore the upper pulley to its original position let us now suppose the whole system to be moved upwards through a space equal to q, Art. 65. On the whole, the weight Q, together with the bar ABC, has ascended a space q; the downward displacements of the several pulleys in order, counting from the highest, are respectively 0, $(2-1)q$, $(2^2-1)q$, ...; while the downward displacement of the power P is $(2^n - 1)q$. The principle of work at once yields the equation

$$(Q + W)q = w_{n-1}(2-1)q + w_{n-2}(2^2-1)q + \ldots$$
$$+ w_1(2^{n-1} - 1)q + P(2^n - 1)q.$$

Dividing by q we have the same relation as before.

534. We notice that the bar ABC will not remain horizontal unless the weight Q is fastened to it at the proper point. The bar is acted on at the points A, B, &c. by the tensions T_1, T_2, &c., and these are to be in equilibrium with the weight Q acting at some point H and the weight W of the bar at its middle point G. The intervals AB, BC, &c. depend on the radii of the pulleys. If the radii be a_1, a_2, &c.

we have $AB = 2a_2 - a_1$, $BC = 2a_3 - a_2$, and so on. Taking moments about A we have

$$T_2 . AB + T_3 . AC + \&c. = Q . AH + W . AG.$$

This equation determines the position of H.

If the weights of the strings or ropes cannot be neglected, we may suppose the weight of the portion of string between the pulleys M_1, M_2 included in the weight w_1, that of the portion between the pulleys M_2, M_3 included in w_2, and so on. The portions of string which join the points A, B, C, &c. to the pulleys are supported by the fixed beam ABC, &c. in fig. (1), and may be included in the weight of the bar in fig. (2). The weight of the string wound on any pulley may be included in the weight of that pulley.

The system of pulleys represented in fig. (1) of Art. 532 is sometimes called the *first system*. That represented in Art. 529 is the *second system*; while the one drawn in fig. (2) of Art. 532 is the *third system*.

535. When the weights of the pulleys are neglected and each hangs by a separate string, we can easily find the relation between the power and the weight when the strings are not parallel.

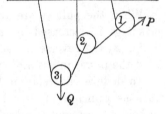

Let $2a_1$, $2a_2$, $2a_3$, &c. be the angles between free parts of the strings which pass over the pulleys M_1, M_2, M_3, &c. respectively. Let also T_1, T_2, T_3, &c. be the tensions. Then by the same reasoning as before

$$T_1 = P, \quad T_2 = 2T_1 \cos a_1, \quad T_3 = 2T_2 \cos a_2, \quad \&c.$$

If there are n pulleys we easily obtain $Q = 2^n P . \cos a_1 . \cos a_2 . \&c. \cos a_n$.

536. Ex. 1. In that system of pulleys in which all the strings are attached to the weight, if the weight of the lowest pulley be equal to the power P, of the second $3P$, and so on...that of the highest moveable pulley being $3^{n-2}P$, the ratio of $P : W$ will be $2 : 3^n - 1$.   [Math. Tripos, 1856.]

Ex. 2. In that system of pulleys in which each hangs by a separate string from a horizontal beam the weights of the pulleys, beginning with the highest, are in arithmetical progression, and a power P supports a weight Q; the pulleys are then reversed, the highest being placed lowest, and the second highest placed lowest but one, and so on, and now Q and P when interchanged are in equilibrium; show that $n(Q + P) = 2W$, where W is the total weight of the pulleys, and n the number of pulleys.   [Coll. Exam., 1882.]

Ex. 3. In a system of n pulleys where a separate string goes round each pulley and is attached to the weight, if the string which goes over the lowest have the end, at which the power is usually hung, passed under another moveable pulley and then over a fixed pulley, and attached to the weight Q; and if the weight of each pulley be w and no other power be used, prove that $Q = (3 . 2^{n-1} - n - 1) w$, and find the point of the beam at which Q must be hung.   [Math. Tripos, 1876.]

Ex. 4. In that system of pulleys in which each of the strings, supposed parallel, is attached to the weight, if the power be equal to the weight of the lowest pulley, and if each pulley weigh three times as much as the one immediately below it, prove that the weight of each pulley is equal to the tension of the string passing over it.   [Coll. Exam.]

Ex. 5. In the system of pulleys in which each hangs by a separate string, all the strings being vertical, if W be the weight supported, and w_1, $w_2 \ldots \ldots w_n$ the weights of the moveable pulleys, there will be no mechanical advantage unless

$$W - w_n + 2\,(W - w_{n-1}) + 2^2\,(W - w_{n-2}) + \ldots \ldots + 2^{n-1}\,(W - w_1)$$

be positive. 　　　　　　　　　　　　　　　　　　　　　[Math. Tripos, 1869.]

Ex. 6. In the system of n heavy pulleys in which each hangs by a separate string, P is the power (acting upwards), Q the weight, and R the stress on the beam from which the pulleys hang : show that R is greater than $Q\,(1 - 2^{-n})$ and less than $(2^n - 1)\,P$. 　　　　　　　　　　　　　　　　[Math. Tripos, 1880.]

Ex. 7. If there be two pulleys, without weight, which hang by separate strings, the fixed ends only of the string being parallel, and the power horizontal, prove that the mechanical advantage is $\sqrt{3}$. 　　　　　　　[St John's Coll., 1883.]

Ex. 8. In that system of pulleys, in which all the strings are attached to the weight, if the power be made to descend through one inch, through what distance will the weight rise? Illustrate by reference to this system of pulleys the principle which is expressed by the words, "In machines, what is gained in power is lost in time." 　　　　　　　　　　　　　　　　　　　　[Math. Tripos, 1859.]

Ex. 9. In the system of pulleys in which all the strings are attached to the weight Q, prove that, if the pulleys be small compared with the lengths of the strings, the necessary correction for the weight of the strings is the addition to Q, w_1, $w_2 \ldots w_{n-1}$ respectively, of the weights of lengths

$$h_1 + h_2 + \ldots + h_{n-1} + h,\ 2\,(h_1 - h),\ 2\,(h_2 - h_1), \ldots 2\,(h_{n-1} - h_{n-2})$$

of string; where h_1, h_2, $h_3 \ldots h_n$ are the heights of the n pulleys (whose weights are w_1, $w_2 \ldots w_n$ respectively) above the line of attachment, supposed horizontal, of the strings to the weight Q, and h the height of the point of attachment of the power above the same line. 　　　　　　　　　　　　　　　　[Math. Tripos, 1877.]

Ex. 10. In that system of pulleys in which the strings are all parallel, and the weights of the pulleys assist the power, show that, if there are n pulleys, each of diameter $2a$ and weight w, the distance of the point of suspension of the weight from the line of action of the power is equal to

$$n \frac{2^{n+1}\,Q + [(n-3)\,2^n + n + 3]\,w}{2\,(2^n - 1)\,Q}\,a,$$

where Q is the weight. 　　　　　　　　　　　　　　[Math. Tripos, 1883.]

Ex. 11. In a system of four pulleys, arranged so that each string is attached to a bar carrying the weight, the string which usually carries the power is attached to one end of the same bar, and the fourth string to the other end. The weight and diameter of each pulley are respectively double of those of the pulley below it, and the strings are all parallel. The weight being 33 times that of the lowest pulley, find at what point of the bar it is hung. 　　　　　　　　　[Trin. Coll., 1885.]

Ex. 12. In the system of pulleys, in which each pulley hangs by a separate string with one end attached to a fixed beam, there are n moveable pulleys of equal weight w. The rth string, counting from the string round the highest pulley, cannot bear a greater tension than T. Prove that the greatest weight which can be sustained by the system is $2^{n-r+1}\,T - (2^{n-r+1} - 1)\,w$. 　　[Trin., 1890.]

Ex. 13. It is found that any force P being applied to the extremity of a string passing over a pulley can just raise a weight $P\,(1 - \theta)$. In the system of pulleys in which each hangs by a separate string a weight Q is just supported, the weight of each pulley being aQ. If a and θ are small quantities, whose squares and products may be neglected, show that an additional power equal to $n\theta Q/2^n$ can be applied without affecting the equilibrium. 　　　　　　　　　[Coll. Exam., 1888.]

537. The Inclined Plane. *To find the relation between the power and the weight in the inclined plane.*

Let AB be the inclined plane, C any particle situated on it. Let CN be a normal to the plane and CV vertical; let α be the inclination of the plane to the horizon, then the angle $NCV = \alpha$. Let Q be the weight of C, P a force acting on C in the direction CK, where the angle $NCK = \phi$. It is supposed that CK lies in the vertical plane VCN.

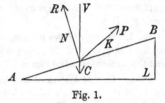

Fig. 1.

If the plane is smooth the reaction R of the plane on the particle acts along the normal CN. We then have by Art. 35

$$\frac{P}{\sin \alpha} = \frac{Q}{\sin \phi} = \frac{R}{\sin (\phi - \alpha)} \dots\dots\dots\dots\dots(1).$$

It is necessary for equilibrium that R should be positive, for otherwise the particle would leave the plane. It follows from these equations that ϕ must be greater than α. This follows also an examination of fig. (1), for Q acting along VC and R along CN cannot be balanced by a force P unless its direction lies within the angle formed by CV and NC produced. If P act up the plane, $\phi = \frac{1}{2}\pi$ and $P = Q \sin \alpha$, $R = Q \cos \alpha$. If P act horizontally, $\phi = \frac{1}{2}\pi + \alpha$, and $P = Q \tan \alpha$, $R = Q \sec \alpha$.

538. If the plane is rough, let $\mu = \tan \epsilon$ be the coefficient of friction. With the normal CN as axis describe a right cone whose semi-angle is ϵ; this is the cone of friction, Art. 173. The resultant action R' of the plane on the particle lies within this cone; let CH be its line of action and let the angle $NCH = i$; then i lies between $\pm \epsilon$. Let the standard case be that in which α is greater than ϵ, and ϕ greater than either; this is represented in fig. (2). We therefore have

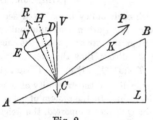

Fig. 2.

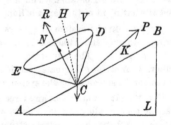

Fig. 3.

$$\frac{P}{\sin (\alpha - i)} = \frac{Q}{\sin (\phi - i)} = \frac{R'}{\sin (\phi - \alpha)} \dots\dots\dots\dots\dots (2).$$

When the force P is so great that the particle is on the point of ascending the plane, the reaction R' acts along CE, and $i = -\epsilon$. Let P_1 be this value of P, then

$$\frac{P_1}{\sin(\alpha+\epsilon)} = \frac{Q}{\sin(\phi+\epsilon)} = \frac{R'}{\sin(\phi-\alpha)} \dots\dots\dots\dots\dots\dots\dots (3).$$

When the force P is so small that the particle is only just sustained, the reaction R' acts along CD, and $i = \epsilon$. Let P_2 be the value of P, then

$$\frac{P_2}{\sin(\alpha-\epsilon)} = \frac{Q}{\sin(\phi-\epsilon)} = \frac{R'}{\sin(\phi-\alpha)} \dots\dots\dots\dots\dots\dots\dots (4).$$

If $\alpha > \epsilon$ as in fig. (2), it is clear that the particle will slide down the plane if not supported by some force P, Art. 166. When the particle is just supported the reaction R' acts along CD and Q along VC; it is clear that these forces could not be balanced by any force P unless its direction lay within the angle made by CV and DC produced. Accordingly we see from (4) that R' is negative unless $\phi > \alpha$. In the same way it is impossible to pull the particle up the plane (without pulling it off) by any force whose direction does not lie between CV and EC produced. Assuming $\phi > \alpha$, the least force required to keep the particle at rest is given by (4), and the greatest by (3).

If $\epsilon > \alpha$ as in fig. (3), the particle will rest on the plane unless disturbed by some force P. To just pull the particle up the plane the force must act within the angle formed by CV and EC produced, and its magnitude is given by (3). In order that the particle may be just descending the plane the force must act within the angle formed by CV and DC produced, and its magnitude is given by (4).

539. Ex. 1. If a power P acting parallel to a smooth inclined plane and supporting a weight Q produce on the plane a pressure R, then the same power acting horizontally and supporting a weight R will produce a pressure Q. [Coll. Ex., 1881.]

Ex. 2. Find the direction and magnitude of the least force which will pull a particle up a rough inclined plane.

By (3) we see that P_1 is least when $\phi + \epsilon = \frac{1}{2}\pi$, i.e. when the force makes an angle with the inclined plane equal to the angle of friction.

Ex. 3. Find the direction and magnitude of the least force which will just support a particle on a rough inclined plane.

Ex. 4. A given particle C rests on a given smooth inclined plane and is supported by a force acting in a given direction. If the inclined plane is without weight and has its side AL moveable on a smooth horizontal table, find the force which when acting horizontally on the vertical face BL will prevent motion. Find also the point of application of the resultant pressure on the table.

Ex. 5. A heavy body is kept at rest on a given inclined plane by a force making a given angle with the plane; show that the reaction of the plane, when it is smooth, is a harmonic mean between the greatest and least reactions, when it is rough. [Math. Tripos, 1858.]

Ex. 6. A heavy particle is attached to a point in a rough inclined plane by a fine rigid wire without weight, and rests on the plane with the wire inclined at an angle θ to a horizontal line in the plane. Determine the limits of θ, the angle of inclination of the plane being $\tan^{-1}(\mu \sec \beta)$. [Coll. Exam.]

Ex. 7. Two equal particles on two inclined planes are connected by a string which lies wholly in a vertical plane perpendicular to the line of junction of the planes, and passes over a smooth peg vertically above this line of junction. If, when the particles are on the point of motion, the portions of the string make equal angles with the vertical, show that the difference between the inclinations of the planes must be twice the angle of friction. [Math. Tripos, 1878.]

540. Wheel and Axle. *To find the relation between the power and the weight in the wheel and axle.*

Let a be the radius of the axle AB, c that of the wheel. The power P acts by means of a string which passes round the wheel several times and is attached to a point on the circumference. The weight Q acts by a string which passes similarly round the axle. Taking moments round the central line of the axle, we have $Pc = Qa$. The mechanical advantage is equal to c/a.

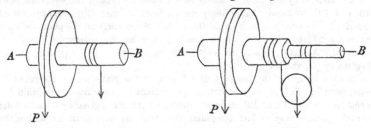

Fig. 1.   Fig. 2.

If p, q be the spaces which the power and weight pass over while the wheel turns through any angle, we have

$$p/q = c/a = Q/P.$$

541. When a great mechanical advantage is required we must either make the radius of the wheel large or that of the axle small. If we adopt the former course the machine becomes unwieldy, if the latter the axle may become too weak to bear the strain put on it. In such a case we may adopt the plan represented in fig. (2). The two parts of the axle are made of different thicknesses, and the rope carried round both. As the power P descends, the rope which supports the weight is coiled on the thicker part of the axle and uncoiled from the thinner. Let a, b be the radii of these two portions of the axis. If Q be the weight attached to the pulley, the tension of the string is $\frac{1}{2}Q$. Taking moments about the central line of the axis, we have $Pc = \frac{1}{2}Q(a - b)$. The mechanical advantage is therefore equal to the radius of the wheel divided by half the difference of the radii of the axle. By making the radii of the two portions of the axis as nearly equal as we please, we can increase the mechanical advantage without decreasing the strength of the machine. This arrangement is called the *differential axle*.

542. Ex. 1. A rope passes round a pulley, and its ends are coiled opposite ways round two drums of different radii on the same horizontal axis. A person pulls vertically upon one part of the rope with a force P. What weight attached to the pulley can he raise, supposing the parts of the rope parallel? [Coll. Exam.]

Ex. 2. In the differential axle if the ends of the chain, instead of being fastened to the axles, are joined together so as to form another loop in which another pulley and weight are suspended, find the least force which must be applied along the chain in order to raise the greater weight, the different parts of the chain being all vertical. [Math. Tripos.]

543. When both the power and the weight act on the circumference of wheels there are various methods of connecting the two wheels besides that of putting them on a common axis. Sometimes, when the wheels are at a distance from each other, they are connected by a strap passing over their circumferences. In some other cases one wheel works on the other by means of teeth placed on their rims.

544. Toothed Wheels. *To obtain the relation between the power and the weight in a pair of toothed wheels.*

Let A, B be the centres of two wheels which act on each other by means of teeth, the teeth on the axis of one wheel working into those on the circumference of the other at the point C. Let a_1, a_2 be the radii of the axles, b_1, b_2 those of the wheels.

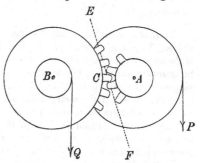

Let p, q be the virtual velocities of the power P and weight Q, then $Pp = Qq$. If the teeth are small the average velocities of the points near C on the two wheels are equal, and the common direction is perpendicular to the straight line AB. If then θ_1, θ_2 are the angles turned through by the wheels when the power P receives a small displacement, we have $a_1\theta_1 = b_2\theta_2$. But $p = b_1\theta_1$, $q = a_2\theta_2$. It follows that $\dfrac{Q}{P} = \dfrac{b_1 b_2}{a_1 a_2}$. We have here omitted the work lost in overcoming the friction at the teeth in contact and at the points of support.

545. Let a tooth on one wheel touch the corresponding tooth on the other in some point D, and let EDF be a common normal to the two surfaces in contact at D. The point D is not marked in the figure because the teeth are not fully drawn, but it is necessarily situated near C. The actual velocities of the points of the teeth in contact at D when resolved in the direction EDF are equal. If, then, h and k are the perpendiculars drawn from A, B on EDF, it is clear that $\theta_1 h = \theta_2 k$. As the wheels turn, the lengths h and k alter, and if the ratio h/k is not constant, there is more or less irregularity in the working of the machine. To correct this defect, the teeth are sometimes cut so that the normal at every point of the boundary of a tooth is a tangent to the circle to which the tooth is attached. When this is done, the line EDF is always a common tangent to the two circles. The ratio h/k is therefore constant throughout the motion and equal to the ratio of the radii of the circles. One cause of irregularity will thus be removed and the motion will be made more uniform. This method is commonly ascribed to Euler.

If the normal at every point of the surface of a tooth is a tangent to a circle, each of the two halves of that tooth is bounded by an arc of an involute of the circle. The two involutes are unwrapped from the circle in opposite directions and portions of each form the sides of the tooth.

When the centres of the toothed wheels are given, and the ratio of the angular velocities at which they are to work, we may determine their radii in the following manner. Let A, B be the given centres; divide AB in C so that $AC \cdot \theta_1 = BC \cdot \theta_2$. Through C draw a straight line ECF, which should not deviate very much from a perpendicular to AB. With A and B as centres describe two circles touching the straight line ECF. The sides of the teeth are to be involutes of these circles. By this construction the common normal to two teeth pressing against each other at D is the straight line ECF. As the wheels turn round, and the teeth move with them, the point of contact D travels along the fixed straight line ECF. The perpendiculars h and k are equal to the radii of these circles and are constant during the motion. Their ratio also is evidently equal to the ratio of AC to BC, i.e. of θ_2 to θ_1.

It has already been shown that $Pp = Qq$, and $p = b_1\theta_1$, $q = a_2\theta_2$. Since $\theta_1 h = \theta_2 k$, we find as before $\dfrac{Q}{P} = \dfrac{b_1 b_2}{a_1 a_2}$.

We may notice that, if the distance between the centres A and B is slightly altered, the pair of wheels will continue to work without irregularity and the ratio of the angular velocities will be the same as before. To prove this, we observe that the common normal to two teeth pressing against each other is still a common tangent to the two circles, though in their displaced positions. Thus, though the inclination to AB of the straight line ECF is altered, the lengths of the perpendiculars h and k are the same as before.

That the teeth should be made of the proper form is a matter of importance to the even working of the machine. Many other considerations enter into the theory besides that mentioned above. Thus defects may arise from the wearing of the teeth if the pressure be very great at the point of contact. There may also be jolts and jars when the teeth meet or separate. But the subject is too large to be treated of in a division of a chapter. The reader who is interested in this matter is referred to books on the principles of mechanism. In Willis' *Principles of Mechanism* (2nd edition, 1870) five different methods of constructing the teeth are described, in three of which epicycloids are used; the advantages and disadvantages of these constructions are also compared.

546. Ex. 1. In a train of n wheels, the teeth on the axle of each wheel work on those on the circumference of the next in order. Show that the power and weight are connected by the relation $\dfrac{Q}{P} = \dfrac{b_1 b_2 \dots b_n}{a_1 a_2 \dots a_n}$, where a_1, a_2 &c. are the radii of the axles and b_1, b_2 &c. those of the wheels.

Ex. 2. In a pair of toothed wheels show that, if the ratio of the power and weight is to be approximately constant, the height and breadth of the teeth must both be small relatively to the radius of each wheel.

Two equal and similar wheels, with straight narrow radial teeth, are started with a tooth of each in contact and in the same straight line; show that they will work together without locking, provided that the distance of their centres be greater than $2a \cos 2\pi/n$ and less than $2a \cos \pi/n$, where a is the radius of either wheel measured to the summit of a tooth, and n the number of teeth. [Math. T., 1872.]

Ex. 3. Investigate the relation $Q/P = b_1 b_2/a_1 a_2$ for a pair of toothed wheels without using the principle of virtual work.

The reaction R between two teeth acts along the straight line EDF. Taking moments in turn about A and B, we have $Pb_1 = Rh$, $Qa_2 = Rk$. As before, we have when the teeth are small $h/k = a_1/b_2$. The result follows at once.

547. The Wedge. *To find the relation between the power and the weight in the wedge.*

Let M, N be two obstacles which it is intended to separate by inserting a wedge ABC between them. For the sake of distinctness these obstacles are represented in the figure by two equal boxes placed on the floor, but it is obvious they may be of any kind.

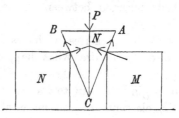

We shall suppose that the wedge used is isosceles, and that it has its median line CN vertical. Let the angle ACB be 2α. Let D, E be the points of contact with the obstacles (not marked in the figure), R, R the normal reactions at these points, F, F the frictions. When the wedge is on the point of motion we have $F = R \tan \epsilon$, where $\tan \epsilon$ is the coefficient of friction.

Let P be a force acting vertically at N urging the wedge downwards. Supposing P to prevail, the frictions on the wedge act along CA, CB; we therefore find by resolving vertically

$$P = 2R (\sin \alpha + \tan \epsilon \cos \alpha) = 2R \sin (\alpha + \epsilon) \sec \epsilon.$$

The resultant reaction R' at D is then found by compounding R and μR.

If the obstacle M can only move horizontally, the whole of the reaction R' is not effective in producing motion. The horizontal component of R' tends to move M, but the vertical component presses the box on the floor and possibly tends to increase the limiting friction between the box and the floor. Let X be the horizontal component of R'; we find

$$X = R \cos \alpha - R \tan \epsilon . \sin \alpha = R \cos (\alpha + \epsilon) \sec \epsilon.$$

The mechanical advantage X/P is therefore equal to $\frac{1}{2} \cot (\alpha + \epsilon)$.

548. It may be noticed that the mechanical advantage of the wedge is increased by making the angle α more and more acute. There is of course a practical limit to the acuteness of this angle, for that degree of sharpness only can be given to the wedge which is consistent with the strength required for the purpose to which it is to be applied.

As examples of wedges we may mention knives, hatchets, chisels, nails, pins, &c. Generally speaking, wedges are used when a large power can be exerted through a small space. This force is usually applied in the form of an impulse.

It has not been considered necessary to consider separately the case in which the wedge is smooth, as the results obtained on so erroneous a supposition have no practical bearing.

549. If the force is applied in the form of a blow so that the wedge is driven forwards between the obstacles, the problem to determine its motion is properly one in dynamics. Our object here is merely to find the conditions of equilibrium of a triangular body inserted between two rough obstacles and acted on by a force P.

When a series of blows is applied to the wedge, we may however enquire what happens in the interval between two impulses. The wedge may either stick fast, held by the friction, or begin to return to its original position, being pressed back by the elasticity of the materials. Assuming that these forces of restitution may be represented by two equal pressures R, R, acting on the sides of the wedge, let P_1 be the force necessary to hold the wedge in position. The friction now acts to assist the power. To determine P_1 we write $-\epsilon$ for ϵ in the equations of equilibrium. We therefore have

$$P_1 = 2R \sin(\alpha - \epsilon) \sec \epsilon.$$

If α is greater than ϵ, P_1 is positive and therefore some force is necessary to hold the wedge in position. If α is less than ϵ, P_1 is negative, thus the friction is more than sufficient to hold the wedge fast. A force equal to this value of P_1 with the sign changed is necessary to pull the wedge out. The result is that the wedge will stick fast or come out according as the angle ACB is less or greater than twice the angle of friction.

Ex. 1. Referring to the figure of Art. 547, show that if either of the equal angles A or B of the wedge is less than the angle of friction, no force P however great could separate the obstacles M, N.

If the angle A is less than ϵ, we find that $a + \epsilon$ is greater than a right angle, and therefore that X is negative. It is easy also to see that, if the angle A is equal to ϵ, the resultant reaction between one side of the wedge and an obstacle is vertical. The wedge therefore merely presses the obstacle against the floor.

Ex. 2. If the obstacles M, N are not of the same altitude and are unequally rough, the position of the wedge when in equilibrium is such that the force P_1 and the *resultant actions* R_1', R_2' across the faces meet in a point. Supposing the force P_1 to act perpendicularly to the face AB of the wedge and to be just sufficient to hold the wedge at rest, show that $\dfrac{P_1}{\sin(2a - \epsilon_1 - \epsilon_2)} = \dfrac{R_2'}{\cos(a - \epsilon_1)} = \dfrac{R_1'}{\cos(a - \epsilon_2)}$, assuming the obstacles to be of such form that the wedge must slip at both simultaneously. Show also that, if the wedge be such that the angle C is less than the sum of the angles $\epsilon_1 + \epsilon_2$, the wedge can be held fast by the frictions without the application of any force.

Ex. 3. Deduce from the principle of virtual work the relation between the force X and the power P in a smooth isosceles wedge as represented in the figure of Art. 547. Discuss the two cases in which (1) one obstacle is immovable and (2) both move equally when the wedge makes an actual displacement.

550. The Screw. *To find the relation between the power and the weight in the screw.*

Let AB be a circular cylinder with a uniform projecting ridge running round its surface, the tangents to the directions of the ridges making a constant angle α with a plane perpendicular to the axis of the cylinder. The screw thus formed fits into a hollow cylinder with a corresponding groove on its internal surface, in which the ridge works. The grooves on the hollow cylinder have not been sketched, but are included in the beam EF.

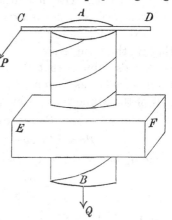

The position of the ridge on the cylinder is easily understood by the following construction. Let a sheet of paper be cut into the form of a right-angled triangle LMN, such that the altitude MN is equal to the altitude of the cylinder AB and the angle the base LM makes with the hypothenuse LN is equal to α. Let this sheet of paper be wrapped round the cylinder AB; if the base LM is long enough to go several times round the base of the cylinder, the hypothenuse will appear to wind gradually round the cylinder. The line thus traced by the hypothenuse is the curve along which the ridge lies.

Let P be the power applied perpendicularly at the end of a lever CD. Let $AC = a$, and let b be the radius of the cylinder. Supposing the body EF in which the screw works to be fixed in space, the end B of the cylinder will be gradually moved as C describes a circle round AB. Let Q be the force acting at B.

Let σ be any small length of the screw which is in contact with an equal length of the groove. Let $R\sigma$ be the normal reaction between these small arcs, $\mu R\sigma$ the friction.

In some screws the ridge is rectangular, so that it may be regarded as generated by the motion of a small rectangle moving round the cylinder with one side in contact with the surface and its plane passing through the axis. When the ridge has this form, the line of action of R lies in the tangent plane to the cylinder and its direction makes with the axis of the cylinder an angle equal to α. In other screws the section of the ridge has some other form, such, for example, as a triangle. In such cases the line of action of R makes some angle θ with the tangent plane to the cylinder. We therefore resolve R into two components, one intersecting at right angles the axis of the cylinder and the other lying in the tangent plane. The magnitude of the latter is $R \cos \theta$, and its direction makes with the axis of the cylinder an angle equal to α. Since the ridge is uniform the angle θ will be the same throughout the length of the screw.

Let us suppose that the power P is about to prevail, then the friction acts so as to oppose the power. Resolving parallel to the axis of the cylinder and taking moments about it, we have

$$Q = \Sigma R\sigma \,.\, \cos\theta \cos\alpha - \Sigma R\sigma \,.\, \mu \sin\alpha,$$
$$Pa = \Sigma R\sigma \,.\, b \cos\theta \sin\alpha + \Sigma R\sigma \,.\, \mu b \cos\alpha.$$

Dividing one of these equations by the other we have

$$\frac{Q}{P} = \frac{\cos\theta \cos\alpha - \mu \sin\alpha}{\cos\theta \sin\alpha + \mu \cos\alpha} \cdot \frac{a}{b}.$$

551. If it be possible to neglect the friction and treat the screw as smooth we put $\mu = 0$. We then find for the mechanical advantage the expression $(a \cot a)/b$. If a point travelling along the ridge or thread of the screw make one complete revolution of the cylinder, it advances parallel to the axis a space equal to the distance h between the ridges. This distance is therefore $h = 2\pi b \tan a$. Substituting for $\tan a$, we find that the mechanical advantage of a smooth screw is c/h, where c is the circumference described by the power and h is the distance between two successive threads of the screw measured parallel to the axis.

552. We may easily deduce the relation between the power and the weight in a smooth screw from the principle of virtual work. When the power has turned the handle AC through a complete circle, the screw and the attached weight have advanced a space h equal to the distance between two threads of the screw measured parallel to the axis. When therefore friction is neglected and no work is otherwise lost in the machine, we have $Pc = Qh$, where c is the circumference of the circle described by P.

When the friction between the ridge and the groove is taken account of we see by Art. 550 that the efficiency of the machine is given by $\dfrac{Qh}{Pc} = \dfrac{\cos\theta - \mu \tan\alpha}{\cos\theta + \mu \cot\alpha}$

When the thread of the screw is rectangular the angle θ is zero. In that case the expression for the efficiency takes the simple form $\dfrac{Qh}{Pc} = \dfrac{\tan\alpha}{\tan(\alpha+\epsilon)}$, where ϵ is the angle of friction.

If the weight Q is about to prevail over the power, we change the signs of μ and ϵ in these formulæ.

553. **Ex. 1.** What force applied at the end of an arm 18 inches long will produce a pressure of 1000 lbs. upon the head of a smooth screw when 11 turns cause the head to advance two-thirds of an inch?   [Trin. Coll., 1884.]

Ex. 2. A screw with a rectangular thread passes into a fixed nut: show that no force applied to the end of the screw in the direction of its length will cause it to turn in the nut, if the pitch of the screw is not greater than ϵ, where ϵ is the angle of friction.   [Coll. Exam., 1878.]

Ex. 3. A rough screw has a rectangular thread: prove that the least amount of work will be lost through friction when the pitch of the screw is $\frac{1}{4}(\pi - 2\epsilon)$, where ϵ is the angle of friction.   [St John's Coll., 1889.]

Ex. 4. The vertical distance between two successive threads of a screw is h, its radius is b, and the power acts perpendicularly to an arm a. If the thread be square and of small section, and the friction of the thread only be taken into account, show that if a and h are given, the efficiency of the machine is a maximum when $2\pi b = h\tan(\frac{1}{4}\pi + \frac{1}{2}\epsilon)$, ϵ being the limiting angle of friction.   [Math. Tripos, 1867.]

Ex. 5. The axis AB of a screw is fixed in space and the beam EF through which the cylinder passes is moveable. The power P, acting at the end of a lever CD, tends to turn the cylinder, while a force Q, acting on EF parallel to the axis AB, tends to prevent motion. Show that the relation between P and Q is the same as that given in Art. 550.

Ex. 6. A weight is supported on a rough vertical screw with a rectangular thread without the application of any power. If l be the length and b the radius of the cylinder on which the thread lies, show that the screw has at least $\dfrac{l\cot\epsilon}{2\pi b}$ turns.

NOTE ON SOME THEOREMS IN CONICS REQUIRED IN ARTS. 126, 127.

THE following analytical proof of the two theorems in conics which are assumed in these articles requires a knowledge only of such elementary equations as those of the normal or of the chord joining two points.

Let ϕ, ϕ' be the eccentric angles of two points P, Q on the conic. Taking the principal axes of the curve as the axes of coordinates, the equations of the normals at these points are

$$\frac{a\xi}{\cos\phi} - \frac{b\eta}{\sin\phi} = a^2 - b^2, \qquad \frac{a\xi}{\cos\phi'} - \frac{b\eta}{\sin\phi'} = a^2 - b^2.$$

The ordinate η of their intersection is therefore given by

$$\frac{b\eta}{a^2 - b^2} = -\frac{\sin\frac{1}{2}(\phi + \phi')}{\cos\frac{1}{2}(\phi - \phi')}\sin\phi\sin\phi' \dots\dots\dots\dots\dots(1).$$

The ordinate of the middle point of the chord PQ is

$$\bar{y} = \tfrac{1}{2}b(\sin\phi + \sin\phi') = b\sin\tfrac{1}{2}(\phi + \phi')\cos\tfrac{1}{2}(\phi - \phi'),$$

$$\therefore\ \frac{b^2}{a^2 - b^2}\frac{\eta}{\bar{y}} = \frac{-\sin\phi\sin\phi'}{\cos^2\frac{1}{2}(\phi - \phi')} = \frac{\cos^2\frac{1}{2}(\phi + \phi')}{\cos^2\frac{1}{2}(\phi - \phi')} - 1 \dots\dots\dots\dots (2).$$

Again, the equation to the chord PQ is

$$\frac{x}{a}\cos\tfrac{1}{2}(\phi + \phi') + \frac{y}{b}\sin\tfrac{1}{2}(\phi + \phi') - \cos\tfrac{1}{2}(\phi - \phi') = 0 \dots\dots\dots\dots (3).$$

If p, p' and q are the perpendiculars on the chord from the foci and the centre, we have the usual formula for the length of a perpendicular

$$\frac{pp'}{q^2} = \frac{\{\cos\frac{1}{2}(\phi - \phi') - e\cos\frac{1}{2}(\phi + \phi')\}\{\cos\frac{1}{2}(\phi - \phi') + e\cos\frac{1}{2}(\phi + \phi')\}}{\cos^2\frac{1}{2}(\phi - \phi')}.$$

It follows by an easy reduction that

$$\left(\frac{\eta}{\bar{y}} - 1\right)\frac{b^2}{a^2} = -\frac{pp'}{q^2} \dots\dots\dots\dots\dots\dots\dots\dots\dots\dots(4).$$

It is explained in the text that the corresponding form for ξ is an inconvenient one because the foci on the minor axis are imaginary. If the chord cut the axes in L and M, we find, from the equation to the chord PQ given above, that

$$\frac{CL}{a} = \frac{\cos\frac{1}{2}(\phi - \phi')}{\cos\frac{1}{2}(\phi + \phi')}, \qquad \frac{CM}{b} = \frac{\cos\frac{1}{2}(\phi - \phi')}{\sin\frac{1}{2}(\phi + \phi')}.$$

We have immediately from (2)

$$\frac{b^2}{a^2}\left(\frac{\eta}{\bar{y}} - 1\right) = -\frac{CL^2 - a^2 + b^2}{CL^2}, \qquad \frac{a^2}{b^2}\left(\frac{\xi}{\bar{x}} - 1\right) = -\frac{CM^2 - b^2 + a^2}{CM^2} \dots\dots(5).$$

The second follows from the first by changing the letters. These are the formulæ used in Art. 126, Ex. 3. By introducing CM into the right-hand side of (1) we find

$$-\frac{CM \cdot \eta}{a^2 - b^2} = \sin\phi\sin\phi', \qquad \frac{CL \cdot \xi}{a^2 - b^2} = \cos\phi\cos\phi' \dots\dots\dots\dots (6).$$

When the points P, Q coincide, ξ, η become the coordinates of the centre of curvature at P. We then deduce from (1) the well-known formulæ

$$-\frac{b\eta}{a^2-b^2}=\sin^3\phi, \qquad \frac{a\xi}{a^2-b^2}=\cos^3\phi \quad \dots\dots\dots\dots\dots\dots (7).$$

The coordinates $\bar{x}$, $\bar{y}$ of the middle point G of the chord being given, the chord itself is determinate. The equation to the chord is

$$\frac{(\xi-\bar{x})\,\bar{x}}{a^2}+\frac{(\eta-\bar{y})\,\bar{y}}{b^2}=0.$$

We then readily find the intercepts CL, CM. We deduce from (2) or (5)

$$\left\{\begin{array}{l}\left\{\dfrac{b^2}{a^2-b^2}\dfrac{\eta}{\bar{y}}+1\right\}\left\{\dfrac{\bar{x}^2}{a^2}+\dfrac{\bar{y}^2}{b^2}\right\}^2=\dfrac{\bar{x}^2}{a^2}\\[2mm]\left\{-\dfrac{a^2}{a^2-b^2}\dfrac{\xi}{\bar{x}}+1\right\}\left\{\dfrac{\bar{x}^2}{a^2}+\dfrac{\bar{y}^2}{b^2}\right\}^2=\dfrac{\bar{y}^2}{b^2}\end{array}\right\} \dots\dots\dots\dots\dots (8).$$

Let X, Y be the coordinates of the intersection T of the tangents at P, Q, then

$$\frac{X}{\bar{x}}=\frac{Y}{\bar{y}}, \qquad \frac{\bar{x}X}{a^2}+\frac{\bar{y}Y}{b^2}=1,$$

because G is the intersection of the straight line joining the origin to T with the polar line of T. We easily find $\bar{x}$, $\bar{y}$ in terms of X, Y, and the equations (7) then become

$$\frac{\eta}{Y}=\frac{(a^2-b^2)(X^2-a^2)}{a^2Y^2+b^2X^2}, \qquad \frac{\xi}{X}=-\frac{(a^2-b^2)(Y^2-b^2)}{a^2Y^2+b^2X^2} \quad \dots\dots\dots (9),$$

which are the equations used in Art. 127.

Ex. 1. A uniform rod, whose ends are constrained to remain on a smooth elliptic wire, is in equilibrium under the action of a centre of force situated in the centre C and varying as the distance, see Art. 51. Show that the centre of gravity G must be either in one of the axes or at a distance from the centre equal to $CR^2/(a^2+b^2)^{\frac{1}{2}}$, where CR is the semidiameter drawn through G. Show that in the latter case half the length of the rod is equal to $CD^2/(a^2+b^2)^{\frac{1}{2}}$, where CD is conjugate to CR. Show also that the tangents at the extremities of the rod are at right angles. Find the lengths of the shortest and longest rods which could be in equilibrium.

Ex. 2. One extremity of a string is tied to the middle point of a rod whose extremities are constrained to lie on a smooth elliptic wire. If the string is pulled in a direction perpendicular to the rod, show that there cannot be equilibrium unless the rod is parallel to an axis of the curve.

Ex. 3. When the conic is a parabola, show that the equations (5), (8), (9) take the simpler forms,

$$\eta=2\bar{y}\cdot\frac{AR}{m} \quad =\frac{2\bar{y}}{m}\left(\bar{x}-\frac{\bar{y}^2}{m}\right)=-\frac{2}{m}XY,$$

$$\xi=2\bar{x}-AR+m=\bar{x}+\frac{\bar{y}^2}{m}+m \quad =-X+\frac{2Y^2}{m}+m,$$

where A is the vertex, R the intersection of the chord with the axis, $2m$ the latus rectum, and the rest of the notation is the same as before.

Ex. 4. Show that the length L of a chord, when expressed in terms of its focal distances p, p', is given by

$$L=\frac{2R^2}{a}\sqrt{1-\frac{pp'}{b^2}}, \qquad \frac{a^2b^2}{R^2}=b^2+\left(\frac{p-p'}{2}\right)^2,$$

where R is the length of the semi-diameter parallel to the chord.

Ex. 5. Two chords of a conic are drawn parallel to any two conjugate diameters and touch a given confocal. Show that the sum of their lengths is constant.

Ex. 6. If the normals at four points P, Q, R, S meet in a point whose coordinates are (ξ, η), prove that the middle points of the six chords which join the points P, Q, R, S two and two lie on the conic

$$(a^2 - b^2)(a^2y^2 - b^2x^2) + a^2b^2(\xi x + \eta y) = 0.$$

This follows at once from (8).

Ex. 7. A heavy uniform rod is in equilibrium with both ends pressing against the interior surface of a smooth ellipsoidal bowl. If one axis of the bowl is vertical, show that the rod must lie in one of the principal planes.

The ellipsoid being referred to its axes, the normals at the extremities of the rod are $\dfrac{a^2}{x}(\xi - x) = \dfrac{b^2}{y}(\eta - y) = \dfrac{c^2}{z}(\zeta - z),$ $\dfrac{a^2}{x'}(\xi - x') = \dfrac{b^2}{y'}(\eta - y') = \dfrac{c^2}{z'}(\zeta - z').$

It is necessary for equilibrium that each of these should be satisfied by $\eta = \tfrac{1}{2}(y + y')$, $\zeta = \tfrac{1}{2}(z + z')$. Substituting, we find that $y'/y = z'/z$, unless either both the y's or both the z's are zero. Putting $y' = \rho y$, $z' = \rho z$, the equations become

$$\frac{2a^2}{x}(\xi - x) = b^2(\rho - 1) = c^2(\rho - 1), \qquad \frac{2a^2}{x'}(\xi - x') = b^2\frac{1-\rho}{\rho} = c^2\frac{1-\rho}{\rho}.$$

Unless $b^2 = c^2$, these give $\rho = 1$. It easily follows that $y' = y$, $z' = z$, $x' = x$ so that the two ends of the rod coincide. As this is impossible, we must have either both the y's or both the z's equal to zero. The rod must therefore be in a principal plane.

END OF VOLUME I.

INDEX.

The numbers refer to the articles.

Cambridge:

PRINTED BY J. AND C. F. CLAY,

AT THE UNIVERSITY PRESS.

Printed in the United States
By Bookmasters